VOLUME ONE HUNDRED AND THIRTY TWO

ADVANCES IN PROTEIN CHEMISTRY AND STRUCTURAL BIOLOGY

Disorders of Protein Synthesis

VOLUME ONE HUNDRED AND THIRTY TWO

ADVANCES IN PROTEIN CHEMISTRY AND STRUCTURAL BIOLOGY

Disorders of Protein Synthesis

Edited by

ROSSEN DONEV

MicroPharm Ltd,
United Kingdom

ELSEVIER

ACADEMIC PRESS

An imprint of Elsevier

Academic Press is an imprint of Elsevier
50 Hampshire Street, 5th Floor, Cambridge, MA 02139, United States
525 B Street, Suite 1650, San Diego, CA 92101, United States
The Boulevard, Langford Lane, Kidlington, Oxford OX5 1GB, United Kingdom
125 London Wall, London, EC2Y 5AS, United Kingdom

First edition 2022

Notices
Knowledge and best practice in this field are constantly changing. As new research and experience broaden our understanding, changes in research methods, professional practices, or medical treatment may become necessary.

Practitioners and researchers must always rely on their own experience and knowledge in evaluating and using any information, methods, compounds, or experiments described herein. In using such information or methods they should be mindful of their own safety and the safety of others, including parties for whom they have a professional responsibility.

To the fullest extent of the law, neither the Publisher nor the authors, contributors, or editors, assume any liability for any injury and/or damage to persons or property as a matter of products liability, negligence or otherwise, or from any use or operation of any methods, products, instructions, or ideas contained in the material herein.

ISBN: 978-0-323-99780-5
ISSN: 1876-1623

For information on all Academic Press publications
visit our website at https://www.elsevier.com/books-and-journals

Publisher: Zoe Kruze
Acquisitions Editor: Leticia Lima
Developmental Editor: Naiza Ermin Mendoza
Production Project Manager: Abdulla Sait
Cover Designer: Victoria Pearson

Typeset by STRAIVE, India

Contents

Contributors

Wardah A. Alhoqail
Department of Biology, College of Education, Majmaah University, Al Majma'ah, Saudi Arabia

Antara Banerjee
Department of Medical Biotechnology, Faculty of Allied Health Sciences, Chettinad Academy of Research and Education (CARE), Chettinad Hospital and Research Institute (CHRI), Chennai, India

Sanjay Basak
Molecular Biology Division, ICMR-National Institute of Nutrition, Indian Council of Medical Research, Hyderabad, India

Alessio Branchini
Department of Life Sciences and Biotechnology, University of Ferrara, Ferrara, Italy

Brenda Chimal-Vega
Departamento de Bioquímica, Facultad de Medicina Mexicali; Laboratorio Multidisciplinario de Estudios Metabólicos y Cáncer, Universidad Autónoma de Baja California, Mexicali, México

Fabio Di Domenico
Department of Biochemical Sciences "A. Rossi Fanelli", Sapienza University of Rome, Rome, Italy

Qihua Duan
Institute for Global Food Security, School of Biological Sciences, Queen's University Belfast, Belfast, United Kingdom

Asim K. Duttaroy
Department of Nutrition, Institute of Basic Medical Sciences, Faculty of Medicine, University of Oslo, Oslo, Norway

Victor García-González
Departamento de Bioquímica, Facultad de Medicina Mexicali; Laboratorio Multidisciplinario de Estudios Metabólicos y Cáncer, Universidad Autónoma de Baja California, Mexicali, México

C. George Priya Doss
Laboratory of Integrative Genomics, Department of Integrative Biology, School of Bio Sciences and Technology, Vellore Institute of Technology, Vellore, Tamil Nadu, India

R. Gnanasambandan
Laboratory of Integrative Genomics, Department of Integrative Biology, School of Bio Sciences and Technology, Vellore Institute of Technology, Vellore, Tamil Nadu, India

Claudio D. Gonzalez
CEMIC University Hospital, Unit of Translational Medical Research; CEMIC University Institute, School of Medicine, Department of Pharmacology, Buenos Aires, Argentina

Chiara Lanzillotta
Department of Biochemical Sciences "A. Rossi Fanelli", Sapienza University of Rome, Rome, Italy

Silvia Lombardi
Department of Biotechnology and Biosciences, University of Milano-Bicocca, Milan, Italy

N. Madhana Priya
Department of Biotechnology, Sri Ramachandra Institute of Higher Education and Research (DU), Chennai, Tamil Nadu, India

R. Magesh
Department of Biotechnology, Sri Ramachandra Institute of Higher Education and Research (DU), Chennai, Tamil Nadu, India

Rahul Mallick
A.I. Virtanen Institute for Molecular Sciences, University of Eastern Finland, Finland

Florencia Mitchell
CEMIC University Institute, School of Medicine, Department of Pharmacology, Buenos Aires, Argentina

Andrea Muñoz-Ayala
Departamento de Bioquímica, Facultad de Medicina Mexicali; Laboratorio Multidisciplinario de Estudios Metabólicos y Cáncer, Universidad Autónoma de Baja California, Mexicali, México

Anuraj Nayarisseri
In Silico Research Laboratory, Eminent Biosciences, Indore, Madhya Pradesh, India

Surajit Pathak
Department of Medical Biotechnology, Faculty of Allied Health Sciences, Chettinad Academy of Research and Education (CARE), Chettinad Hospital and Research Institute (CHRI), Chennai, India

Mirko Pinotti
Department of Life Sciences and Biotechnology, University of Ferrara, Ferrara, Italy

Muthuraja Arun Pravin
Computer Aided Drug Design and Molecular Modeling Lab, Department of Bioinformatics, Science Block, Alagappa University, Karaikudi, Tamil Nadu, India

Francisco Rivera
CEMIC University Institute, School of Medicine, Department of Pharmacology, Buenos Aires, Argentina

Chandrabose Selvaraj
Computer Aided Drug Design and Molecular Modeling Lab, Department of Bioinformatics, Science Block, Alagappa University, Karaikudi, Tamil Nadu, India

Nishaat Shaikh
Laboratory of Integrative Genomics, Department of Integrative Biology, School of Bio Sciences and Technology, Vellore Institute of Technology, Vellore, Tamil Nadu, India

Sanjeev Kumar Singh
Computer Aided Drug Design and Molecular Modeling Lab, Department of Bioinformatics, Science Block, Alagappa University, Karaikudi, Tamil Nadu, India

R. Siva
Laboratory of Integrative Genomics, Department of Integrative Biology, School of Bio Sciences and Technology, Vellore Institute of Technology, Vellore, Tamil Nadu, India

Qiaozhu Su
Institute for Global Food Security, School of Biological Sciences, Queen's University Belfast, Belfast, United Kingdom

Maria Francesca Testa
Department of Life Sciences and Biotechnology, University of Ferrara, Ferrara, Italy

D. Thirumal Kumar
Faculty of Allied Health Sciences, Meenakshi Academy of Higher Education and Research, Chennai, Tamil Nadu; Meenakshi Academy of Higher Education and Research (Deemed to be University), Chennai, India

S. Udhaya Kumar
Laboratory of Integrative Genomics, Department of Integrative Biology, School of Bio Sciences and Technology, Vellore Institute of Technology, Vellore, Tamil Nadu, India

Maria I. Vaccaro
University of Buenos Aires, CONICET, Institute for Biochemistry and Molecular Medicine, Department of Pathophysiology; CEMIC University Hospital, Unit of Translational Medical Research, Buenos Aires, Argentina

Henry Wade
Institute for Global Food Security, School of Biological Sciences, Queen's University Belfast, Belfast, United Kingdom

CHAPTER ONE

Translation termination codons in protein synthesis and disease

Silvia Lombardi[a], Maria Francesca Testa[b], Mirko Pinotti[b], and Alessio Branchini[b,*,†]

[a]Department of Biotechnology and Biosciences, University of Milano-Bicocca, Milan, Italy
[b]Department of Life Sciences and Biotechnology, University of Ferrara, Ferrara, Italy
*Corresponding author: e-mail address: brnlss@unife.it

Contents

Abstract

Fidelity of protein synthesis, a process shaped by several mechanisms involving specialized ribosome regions and external factors, ensures the precise reading of sense as well as stop codons (UGA, UAG, UAA), which are usually localized at the 3′ of mRNA and drive the release of the polypeptide chain.

However, either natural (NTCs) or premature (PTCs) termination codons, the latter arising from nucleotide changes, can undergo a recoding process named ribosome or translational readthrough, which insert specific amino acids (NTCs) or subset(s) depending on

† Senior author.

Advances in Protein Chemistry and Structural Biology, Volume 132
ISSN 1876-1623
https://doi.org/10.1016/bs.apcsb.2022.06.001

the stop codon type (PTCs). This process is particularly relevant for nonsense mutations, a relatively frequent cause of genetic disorders, which impair gene expression at different levels by potentially leading to mRNA degradation and/or synthesis of truncated proteins. As a matter of fact, many efforts have been made to develop efficient and safe readthrough-inducing compounds, which have been challenged in several models of human disease to provide with a therapy. In this view, the dissection of the molecular determinants shaping the outcome of readthrough, namely nucleotide and protein contexts as well as their interplay and impact on protein structure/function, is crucial to identify responsive nonsense mutations resulting in functional full-length proteins. The interpretation of experimental and mechanistic findings is also important to define a possibly clear picture of potential readthrough-favorable features useful to achieve rescue profiles compatible with therapeutic thresholds typical of each targeted disorder, which is of primary importance for the potential translatability of readthrough into a personalized and mutation-specific, and thus patient-oriented, therapeutic strategy.

1. Ribosome fidelity and translation termination

Translation of mRNA is a complex process catalyzed by ribosomes that results in the synthesis of a polypeptide chain driven by correct base pairing between mRNA codons and anticodons of aminoacyl-tRNAs (aa-tRNAs) (Ogle, 2001; Rodnina & Wintermeyer, 2001). This molecular event consists of three main phases, namely, initiation, elongation, and termination.

In the first phases of the translation process, interaction among mRNA, initiator tRNA (placed in ribosomal P site), translation initiation factors and the small ribosome subunit leads to the recruitment of the large subunit, ultimately resulting in the formation of the translation initiation complex. Then, at each mRNA codon, a process of aa-tRNA sampling occurs during the elongation phase. Each aa-tRNA enters the ribosomal A site as a ternary complex with a GTPase elongation factor (EF-Tu in prokaryotes; eEF1A in eukaryotes) and an associated GTP molecule. During this phase, the ribosome can discriminate with high fidelity among cognate (perfect pairing), near-cognate (mismatched pairing) and non-cognate (altered pairing) ternary complexes by initial selection and proofreading strategies. Both steps, separated by the irreversible hydrolysis of GTP associated with elongation factors, rely on the different stability of codon-anticodon matches (Fig. 1) (Hopfield, 1974; Ruusala, Ehrenberg, & Kurland, 1982).

During initial selection, non-cognate aa-tRNAs are efficiently rejected on the basis of two out of three mismatches altering the codon-anticodon pairing, which leads to dissociation of the incorrect aa-tRNA ternary

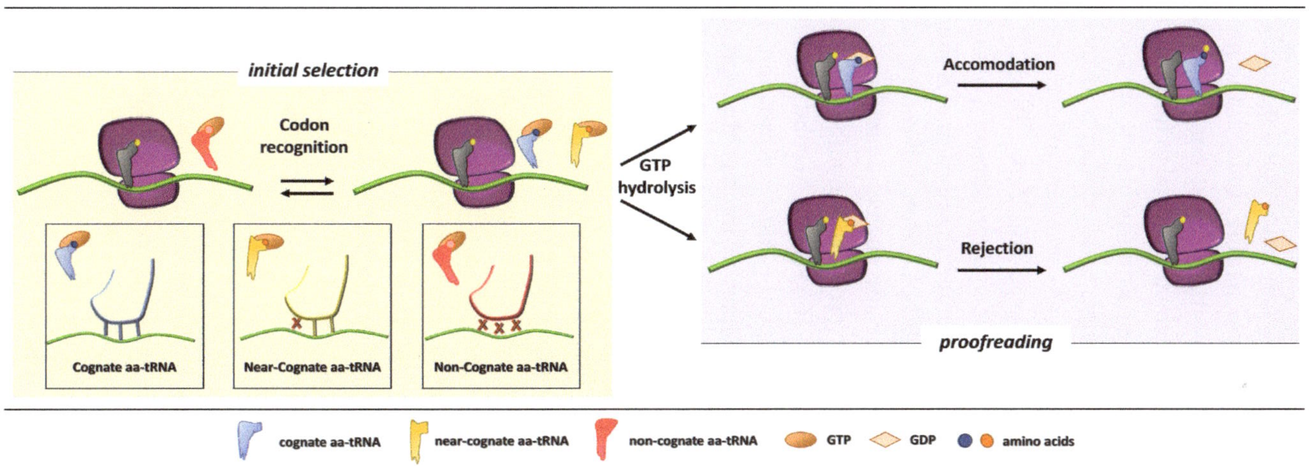

Fig. 1 Initial selection and proofreading activity of the translating ribosome. Insertion of cognate or near-cognate, as well as rejection of non-cognate, aa-tRNAs is exerted by the ribosome during the initial selection step (left panel, with codon-anticodon interactions depicted in the boxes reported in the lower part). Upon GTP hydrolysis driven by proper codon-anticodon interactions, the cognate aa-tRNA is efficiently accommodated, with formation of the peptide bond and progression of protein synthesis (right panel, upper part), followed by a new aa-tRNA selection step. The presence of a near-cognate aa-tRNA, failing to be accommodated, results in rejection of the aa-tRNA through the proofreading step (right panel, lower part). *Adapted from Lombardi, S., Testa, M. F., Pinotti, M., & Branchini, A. (2020). Molecular insights into determinants of translational readthrough and implications for nonsense suppression approaches.* International Journal of Molecular Sciences, 21*(24), E9449. https://doi.org/10.3390/ijms21249449.*

complex with no costs in terms of GTP (Pape, Wintermeyer, & Rodnina, 1999). However, although efficient, this kinetic mechanism alone is not sufficient to discriminate between cognate and near-cognate aa-tRNA complexes, only slightly differing in base pairing. Indeed, the prokaryotic (16S) and eukaryotic (18S) small ribosomal subunit contain a specialized region called decoding site (or center), which is pivotal to finely discriminate between cognate and near-cognate base pairing and to ensure a key quality control phase. Interactions of decoding center elements with both the mRNA codon and the tRNA anticodon are necessary to increase ribosome accuracy (Wohlgemuth, Pohl, & Rodnina, 2010; Yarus, 1992). Ribosome fidelity during translation has been described in prokaryotes, but it is similar in eukaryotes. When a ternary complex enters the ribosomal A site, binding of a cognate aa-tRNA anticodon to a mRNA codon induces crucial changes in the position and/or orientation of specific nucleotide residues. First, two conserved adenines, A1492 and A1493 (prokaryote numbering), are flipped out from an internal loop of helix 44 in 16S rRNA. Second, the universally conserved G530 switches from the *syn* to the *anti*-conformation. This new arrangement enables the two key adenines to interact through hydrogen bonds with the first and second base pairs of the codon-anticodon helix, while the G530 interacts with the second anticodon position and the third codon position. As a result, the induced changes enable the discrimination of correct Watson-Crick geometries, characterized by standard base pairing, at the first and second codon positions, whereas the "wobble" base pair (between the third base of the codon and first base of the anticodon) may accommodate other non-standard geometries (Fischer et al., 2016; Ogle, 2001). Therefore, cognate aa-tRNA recognition results in a local conformational change in the decoding site that elicits a transition from an open to a closed form of the small ribosomal subunit, ensuring the subsequent GTP hydrolysis (Ogle et al., 2001; Ogle, Murphy, Tarry, & Ramakrishnan, 2002). Although it has been defined how recognition of a cognate aa-tRNA occurs, it is still unclear what the entry of near cognate aa-tRNA in ribosomal A site entails. One hypothesis is that the decoding site does not sense a correct Watson-Crick codon-anticodon base pairing, with the lack of stabilization for the near-cognate complex and its preferential rejection. Furthermore, structural studies revealed that binding of a near-cognate aa-tRNA does not induce the closed conformation of the small subunit, which rather remains in an open or partially-closed conformation (Agirrezabala et al., 2011). However, it seems quite clear that a near-cognate aa-tRNA escapes from the initial selection step when a mismatch in the codon-anticodon helix mimics a Watson-Crick geometry (Demeshkina, Jenner, Westhof, Yusupov, & Yusupova, 2012; Rozov, Demeshkina, et al., 2016; Rozov, Westhof, Yusupov, & Yusupova, 2016).

After codon recognition, resulting in the closed conformation of the small ribosomal subunit, a series of rearrangements occur in the aa-tRNA. This leads to the activation of the elongation factor GTPase, whose docking into the sarcin-ricin loop (SRL) located in the large ribosomal subunit triggers GTP hydrolysis (Fischer et al., 2016). It has been demonstrated that, through local conformational changes in the small ribosomal subunit, the binding of a cognate aa-tRNA accelerates GTP hydrolysis in comparison with a near-cognate aa-tRNA (Blanchard, Gonzalez, Kim, Chu, & Puglisi, 2004; Geggier et al., 2010). Following detachment of the GDP-bound elongation factor, an accommodation step of the aa-tRNA takes place, whose mechanism help explaining how the aa-tRNA proofreading step occurs. The aa-tRNA moves into the peptidyl-transferase center, located within the large ribosomal subunit, through reversible fluctuations of its elbow region, acceptor arm and 3′CCA end from the A to the P site (Geggier et al., 2010; Polikanov et al., 2015). It seems that a cognate aa-tRNA accommodates more rapidly than a near-cognate aa-tRNA due to more stable codon-anticodon base pairing during the proofreading step. On the other hand, unfavorable Watson-Crick geometry and weak codon–anticodon interactions seem to facilitate rejection of a near-cognate aa-tRNA, which circumvented the initial selection step (Rozov, Demeshkina, et al., 2016; Zaher & Green, 2010). Overall, ribosome fidelity strategies strictly monitor the aa-tRNA selection during the translation process.

The final translation termination phase is triggered when a stop codon (UGA, UAG, UAA) enters the ribosomal A site (Dever & Green, 2012; Jackson, Hellen, & Pestova, 2012). Extra-ribosomal proteins, named release factors (RFs), are recruited to promote the release of the nascent polypeptide (Schuller & Green, 2018). In eukaryotes, an essential role is played by the two main effectors eRF_1 and eRF_3. The eRF_1 protein consists of three domains, namely, an N-terminal domain directly recognizing all stop codons, a middle domain containing a conserved GGQ (Gly-Gly-Gln) motif involved in the stimulation of polypeptide chain hydrolysis, and a C-terminal eRF_3 binding domain (Hellen, 2018). Among the different domains of eRF_3, the most relevant is the G domain, which binds GTP and assists the termination process through hydrolysis of GTP (Salas-Marco & Bedwell, 2004). Similarly to the elongation phase, sampling of aa-tRNA ternary complexes occurs when a stop codon enters the ribosomal A site. At the same time, the eRF_1-eRF_3-GTP ternary complex competes for stop codon recognition driven by the eRF_1 N-terminal domain that establishes multiple contacts with the 40S subunit (Keeling, Xue, Gunn, & Bedwell, 2014). The key step of GTP hydrolysis, prevented by the middle domain of eRF_1 during formation of the pre-termination complex, is triggered by interactions involving eRF_3 and SRL (des Georges et al., 2014). Upon GTP hydrolysis, the positioning of eRF_1 GGQ motif into the peptidyl transferase center stimulates hydrolysis of the ester bond between tRNA and the polypeptide chain (Shao et al., 2016).

2. Nonsense mutations and the mRNA quality control

A relevant detrimental impact on protein synthesis may be exerted by nonsense mutations, which account for 11% of all gene alterations responsible for inherited human genetic diseases (Mort, Ivanov, Cooper, & Chuzhanova, 2008). Nonsense mutations are defined as single-base pair substitutions affecting gene coding regions, which convert an mRNA sense codon into an in-frame premature termination codon (PTC).

Natural and premature termination codons impact in a different manner on the translational termination process based on their sequence context features and localization (Atkinson & Martin, 1994). Statistical analysis through the Human Genome Mutation Database (HGMD, http://www.hgmd.cf.ac.uk/ac/index.php) (Cooper, Stenson, & Chuzhanova, 2006; Stenson et al., 2003) showed that the three natural nonsense triplets occur with different frequencies, namely, 49.5% (TGA), 28.1% (TAA) and 22.4% (TAG) (Fig. 2A, left panel) (Jacobs, Rackham, Stockwell, Tate, & Brown, 2002). Interestingly, the distribution of the three natural stop codons is significantly different than their pathological counterparts. Indeed, the proportion of TAG PTCs is higher (41.9%) than the corresponding natural TAG. In contrast, a lower proportion has been observed for TGA (36.1%) and TAA (22%) PTCs in comparison with naturally occurring stop codons (Fig. 2A, right panel). Among the 61 sense codons, 18 out of 61 can give rise to nonsense triplets through single-base substitutions. The TAG stop codon mainly stems from CAG (Gln, 44.7%) or TGG (Trp, 14.6%) sense codons, TGA from CGA (Arg, 55.5%) or TGG (Trp, 16.9%), and TAA from GAA

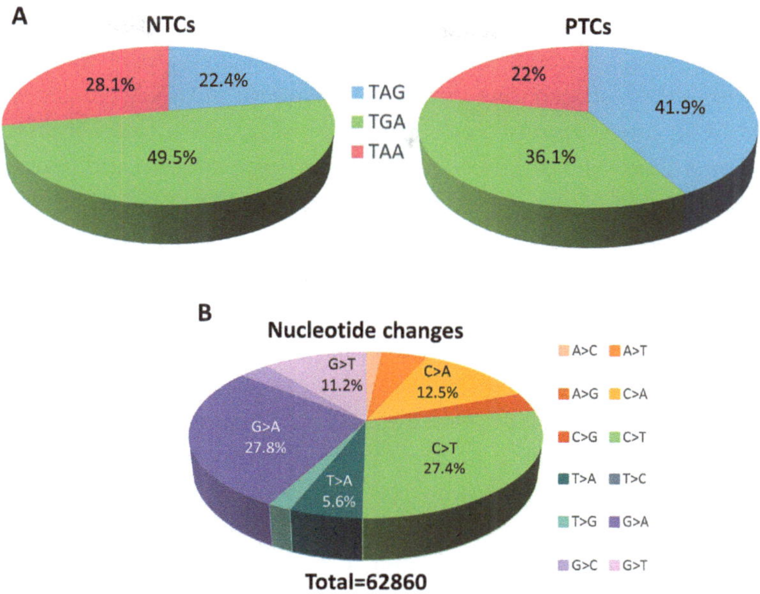

Fig. 2 Distribution of normal or premature termination codons and corresponding nucleotide changes resulting in nonsense mutations. Relative frequency of TAG, TGA and TAA stop codons as normal (A, left) or premature (A, right) termination signals, and corresponding nucleotide changes (B). The reported analysis was obtained by inspection of the ClinVar archive (https://www.ncbi.nlm.nih.gov/clinvar/). The total number of entries reported in the archive is indicated for nucleotide changes resulting in premature termination codons.

(Glu, 26.9%) or CAA (Gln, 26.7%) (HGMD, http://www.hgmd.cf.ac.uk/ac/index.php). In particular, the CGA(Arg) > TGA and CAG(Gln) > TAG changes are those being most associated with nonsense mutations, which can be explained by the highly frequent C > T transitions (27.4% of all single-base pair substitutions) (Fig. 2B). In the human genome, most cytosines at CpG islands are methylated and are the most susceptible to C-to-T transitions, due to a higher rate of spontaneous deamination (Tomkova, McClellan, Kriaucionis, & Schuster-Böckler, 2018), compared to those unmethylated or hydroxymethilated (Tomkova, McClellan, Kriaucionis, & Schuster-Boeckler, 2016).

The impairment of gene expression caused by PTC-containing transcripts is related to (i) degradation of truncated proteins, with decreased stability or loss-of-function features; or (ii) reduction in the steady-state level of the cytoplasmic mRNA pool by the quality control system exerted by nonsense-mediated mRNA decay (NMD) (Behm-Ansmant et al., 2007; Khajavi, Inoue, & Lupski, 2006; Kuzmiak & Maquat, 2006). However, it has been demonstrated that 5–25% of aberrant mRNAs might escape NMD, with production of truncated proteins that, in autosomal disorders, might interfere with the function of the wild-type protein and give rise to potential dominant-negative effects (Isken & Maquat, 2007). In particular, the NMD mechanism degrades PTC-bearing transcripts through the interplay of *cis* and *trans* factors. The first signal is provided by exon junction complexes (EJCs), a multiprotein set deposited 20–24 nucleotides upstream of each exon–exon junction as a marker of the occurrence of splicing events (Ballut et al., 2005; Le Hir, Izaurralde, Maquat, & Moore, 2000; Lejeune, Li, & Maquat, 2003). PTC-bearing mRNAs bound to the cap-binding complex (CBC, consisting of the cap-binding proteins CBP80-CBP20) can be recognized as NMD substrates during the so-called pioneer round of translation, namely, a first cycle in which an mRNA scan is performed by ribosomes. During pioneer round, if the stop codon is located at the 3′ of the coding region, typical of natural termination codons (NTCs), the ribosome displaces all EJCs allowing the subsequent replacement of CBC with eIF4E, which makes the mRNA immune to NMD (Fig. 3, left). Conversely, if a PTC is detected within the mRNA sequence, which causes stalling of the ribosome, the NMD pathway is engaged due to the presence of downstream EJC(s) that cannot be displaced (Popp & Maquat, 2013). Accordingly, mRNAs bound to eIF4E did not co-purify with factors required for NMD, thus supporting that NMD takes place during the pioneer round of translation (Kashima et al., 2006; Kim, Kataoka, & Dreyfuss, 2001).

Among the proteins involved in mRNA decay, the major orchestrator is Upf1, an ATP-dependent RNA helicase that is recruited on the PTC together with eRF_1, eRF_3 and SMG1 (phosphatidylinositol 3-kinase-related protein kinase) in the so-called SURF (SMG1-Upf1-eRF_1-eRF_3) complex (Kashima et al., 2006) (Fig. 3, right). Other components are required for NMD activation, such as Upf3 or 3×, which are generally associated with EJCs in the nucleus, and Upf2, which is assembled to EJCs after mRNA export in the cytoplasm (Kim et al., 2001; Lejeune, Ishigaki, Li, & Maquat, 2002; Lykke-Andersen, Shu, & Steitz, 2001). Once the PTC-bearing transcript is identified by the NMD machinery, several rearrangements and conformational changes occur, including the Upf1-mediated formation of the decay-inducing complex (DECID) (Hwang, Sato, Tang, Matsuda, & Maquat, 2010; Yamashita et al., 2009),

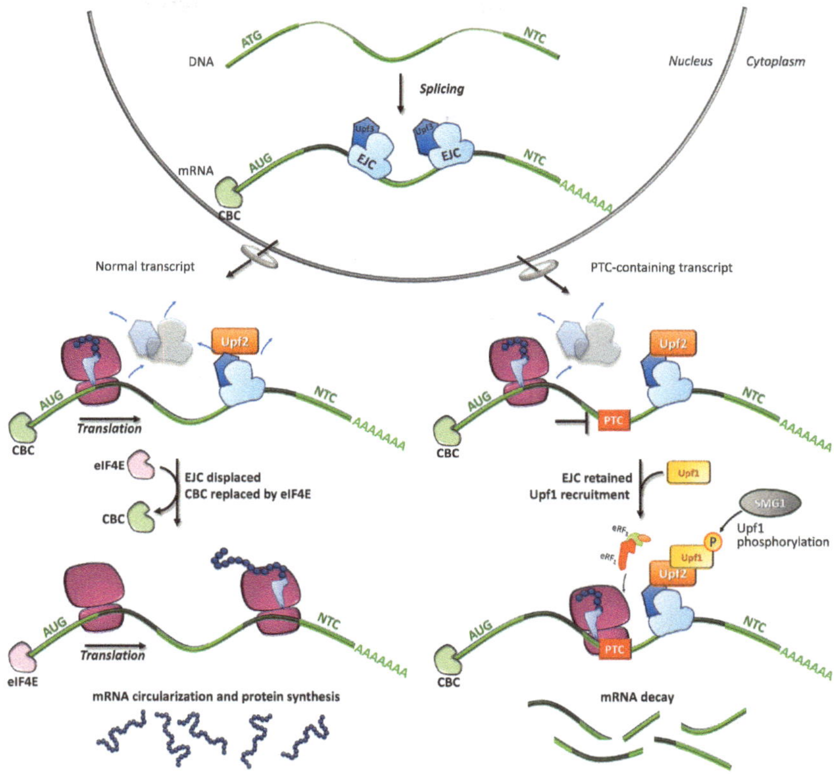

Fig. 3 Fates of normal or PTC-bearing mRNA transcripts during the first round of translation. Splicing of pre-mRNA into the nucleus results in a mature mRNA transcript bound to exon-junction complexes (EJCs) as well as other key components such as the Upf3 (in the nucleus) and Upf2 (in the cytoplasm) proteins. Once transported into the cytoplasm, the mRNA undergoes a first (pioneer) round of translation. In normal conditions, the ribosome displaces all EJCs, resulting in the replacement of CBC with eIF4E, mRNA circularization and efficient protein synthesis, which proceeds until the natural termination codon (NTC) is reached. If a premature termination codon (PTC) is present, EJCs are not efficiently removed by the ribosome, resulting in recruitment of eRF_1, eRF_3, Upf1 and the SMG1 kinase, which leads to Upf1 phosphorylation and degradation of the UPF1-tagged mRNA. *Adapted from Lombardi, S., Testa, M. F., Pinotti, M., & Branchini, A. (2020). Molecular insights into determinants of translational readthrough and implications for nonsense suppression approaches.* International Journal of Molecular Sciences, 21*(24), E9449. https://doi.org/10.3390/ijms21249449.*

which ultimately leads to degradation of the tagged mRNA in a process driven by phosphorylated Upf1 (Kurosaki & Maquat, 2016; Okada-Katsuhata et al., 2012). Finally, NMD factors are disassembled by Upf1, which in turn is converted to its unphosphorylated form (Chiu, Serin, Ohara, & Maquat, 2003).

3. Ribosome readthrough

3.1 Mechanism, programmed readthrough and natural vs premature stop codons

The central role of protein synthesis has forced cells to develop accurate mechanisms to avoid translational errors. However, although effective, the process of translation termination, during which a competition occurs for the recognition of the stop codon (Fig. 4A), is not 100% efficient. At stop codons located at the mRNA 3′ end, release factors highly outcompete aa-tRNAs, thus driving polypeptide chain release. In rare cases, a stop codon can be decoded by an aa-tRNA, with addition of an amino acid to the polypeptide chain, thus allowing elongation to proceed until the next in-frame stop codon is reached (Fig. 4B) (Rodnina et al., 2019). This mechanism is called ribosome or translational readthrough and can be driven by a near-cognate aa-tRNA (i.e., a tRNA whose anticodon is complementary to two out of three positions; see Fig. 1), or by recoding events mediated by specialized tRNAs carrying the 21st selenocysteine or 22nd pyrrolysine amino acids (Cobucci-Ponzano, Rossi, & Moracci, 2012; Namy, Rousset, Napthine, & Brierley, 2004; Zhang, Baranov, Atkins, & Gladyshev, 2005). In particular, pyrrolysine is encoded by the UAG stop codon and is restricted to several microbes, whereas selenocysteine is encoded by the UGA codon and can be found in bacteria, archaea and eukaryotes. Incorporation of selenocysteine requires a specific insertion sequence (selenocysteine inserting sequence, SECIS) and is necessary for the synthesis of selenoproteins (Fig. 4C) (Allmang & Krol, 2006; Cobucci-Ponzano et al., 2012; Zhang et al., 2005). Approximately 100 selenoprotein families have been discovered, among which the human selenoproteome comprises 25 genes, with half of them encoding proteins with known functions (Hatfield, Tsuji, Carlson, & Gladyshev, 2014; Kryukov et al., 2003). These include peroxidase (glutathione-dependent), reductase (NADPH-dependent) and deionidase (metabolism of thyroid hormones) activities, and the specialized selenoprotein P, bearing 10 selenocysteine residues, produced by the liver and involved in the delivery of selenium to other organs (Hatfield et al., 2014; Labunskyy, Hatfield, & Gladyshev, 2014; Reeves & Hoffmann, 2009).

Ribosome readthrough was first identified in viruses as a way to expand their genetic information (Beier & Grimm, 2001). It was observed that, in *Escherichia coli* infected by RNA phage Qβ, the $tRNA^{Trp}$ stimulated readthrough over the UGA stop codon at the end of the coat protein cistron,

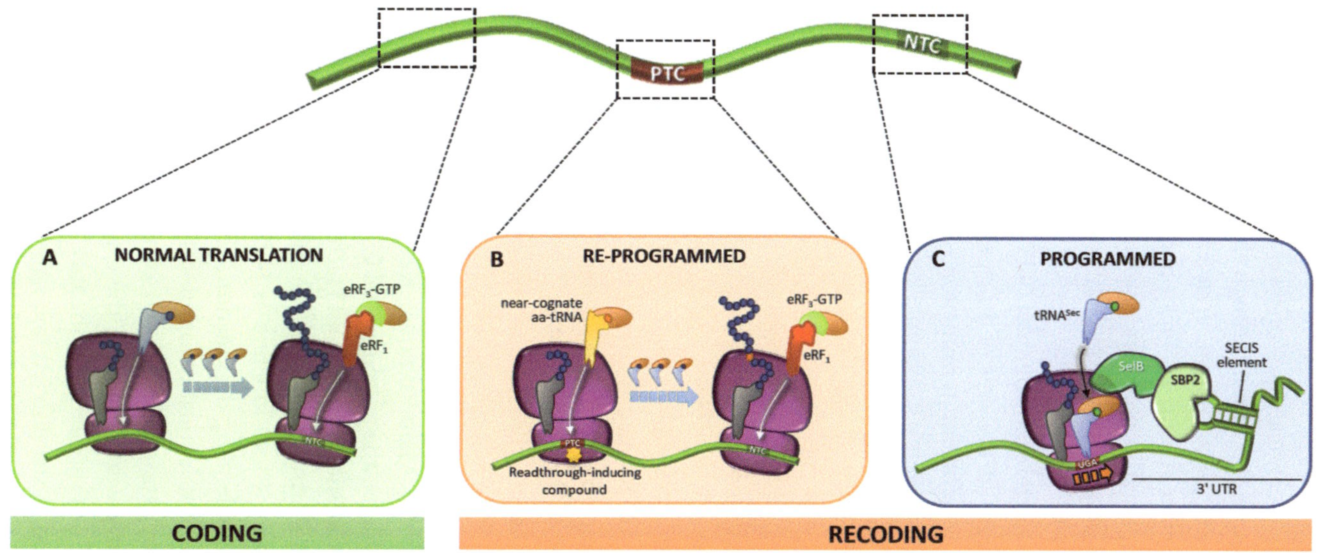

Fig. 4 Translation termination and recoding. Upper part. Schematic representation of mRNA, with indicated a premature termination codon (PTC), potentially arising from a point mutation (see also Fig. 1), and the normal termination codon (NTC) at the 3′ end. (A) Translation events in normal protein synthesis, driven by incorporation of cognate aa-tRNAs, and termination elicited by eRF_1 and eRF_3 at the NTC. (B) Readthrough-mediated suppression of a PTC through incorporation of a near-cognate aa-tRNA, occurring spontaneously or induced by compounds, with translation termination at the NTC upon PTC suppression. (C) Readthrough by terminal UGA recoding through incorporation of a selenocysteine-carrying aa-tRNA ($tRNA^{Sec}$) mediated by a specialized elongation factor (SelB) and a downstream hairpin structure (selenocysteine inserting sequence, SECIS) recognized by the SECIS-binding protein 2 (SBP2).

resulting in a longer coat protein essential for the production of infective Qβ particles (Hirsh, 1971). Since then, stop codon readthrough has been found to play important biological roles in several organisms, providing a mechanism to regulate gene expression through the production of different proteins from the same gene (Beznosková, Wagner, Jansen, von der Haar, & Valášek, 2015; Eswarappa et al., 2014). This phenomenon has been called programmed readthrough and its prevalence varies among organisms, with a few examples found also in humans (Loughran et al., 2018; Rodnina et al., 2019; Stiebler et al., 2014).

Nevertheless, unregulated readthrough of NTCs is generally a rare event and the termination signals at the end of an open reading frame evolved several ways to promote efficient termination, especially among highly expressed transcripts. A paradigmatic example is represented by tandem stop codons, which act as "back-up" termination signals in case of readthrough (Fleming & Cavalcanti, 2019; Liang, Cavalcanti, & Landweber, 2005). Moreover, the nonstop decay surveillance mechanism recognizes and degrades those transcripts with stalling ribosomes bound to the poly(A) tail, thus preventing the generation and accumulation of unwanted C-terminally extended proteins (Klauer & van Hoof, 2012). In addition, recent studies suggested that translation into 3′ UTRs might reduce C-terminally extended protein levels through 3′ UTR-encoded peptides that destabilize the attached protein, either co- or post-translationally (Arribere et al., 2016).

This scenario changes in the presence of a PTC, inserted for instance by a nonsense mutation. In this case, the frequency of spontaneous readthrough is about 10-fold higher (0.01–1%) than at NTCs (0.001–0.1%) (Bonetti, Fu, Moon, & Bedwell, 1995; Cassan & Rousset, 2001; Keeling, Wang, Conard, & Bedwell, 2012; Manuvakhova, Keeling, & Bedwell, 2000). This difference is likely due to the presence (NTCs) or lack (PTCs) of interactions between the termination complex and factors bound to the 3′ UTR of the mRNA. Indeed, during translation the mRNA is maintained in a closed-loop conformation whose function is both to protect the ends of the transcript from exonucleolytic degradation (Chen & Shyu, 2011), and to enhance recycling of translational components, thus increasing the frequency of translation initiation. This closed conformation is maintained by the association of three actors, namely, the cap-binding protein eIF4E, which is bound to the 5′-cap structure of the mRNA, the poly(A)-binding protein (PABP), attached to the poly(A) tail, and eIF4G, which binds both eIF4E and PABP resulting in mRNA circularization. Along with its role in the formation of the closed-loop complex, PABP also promotes translation termination by interacting with eRF_3 and stimulating polypeptide chain release (Cosson et al., 2002). In the presence of a PTC, which is usually distant from the NTC and thus from the poly(A) tail, the interaction between PABP and eRF_3 will likely be less efficient, leading to prolonged ribosomal pausing and increased aa-tRNA sampling, which in turn makes the PTC more susceptible to readthrough (Amrani et al., 2004).

3.2 Readthrough-inducing compounds

Readthrough-mediated suppression of PTCs is promoted certain small molecular weight compounds (Bidou, Allamand, Rousset, & Namy, 2012; Lee & Dougherty, 2012; Linde & Kerem, 2008), thus highlighting their potential application to treat genetic disorders caused by nonsense mutations (Fig. 4B).

The first and most studied readthrough-inducing molecules are aminoglycosides, a class of structurally related antibiotics characterized by a common 2-deoxystreptamine core (ring II) linked to a glucopyranosyl (ring I) at position 4. Aminoglycosides exert their antibiotic activity by binding to the bacterial ribosome and inhibiting protein synthesis (Fan-Minogue & Bedwell, 2007). In particular, they bind to specific nucleotides in the decoding center inducing A1492 and A1493 to flip out from the internal loop of helix 44, inducing a conformational change that partially resembles that caused by cognate aa-tRNA binding (Fig. 5), thus resulting in extensive amino acid misincorporation and translation inhibition (Ogle, 2001). In humans, aminoglycosides can be safely used as antibiotics because their binding to the eukaryotic ribosome is much less efficient. Extensive mutational analyses revealed that the major determinants of the different sensitivity to aminoglycoside between prokaryotes and eukaryotes are the yeast G1645 and A1754 residues, corresponding to *E. coli* A1408 and G1491 residues. These divergent nucleotides cause the aminoglycoside binding pocket within helix 44 to be shallower in eukaryotes, thus preventing stable insertion and binding of the drug (Fan-Minogue & Bedwell, 2007). Nevertheless, in the presence of a PTC, the impact of aminoglycosides on the eukaryotic ribosome is sufficient to reduce discrimination between near-cognate aa-tRNAs and release factors, thus enhancing amino acid (mis)incorporation at PTCs and thus readthrough (Keeling & Bedwell, 2002; Manuvakhova et al., 2000).

The molecular basis of aminoglycosides effects is still unclear, but recent structural analysis of yeast ribosomes in complex with different

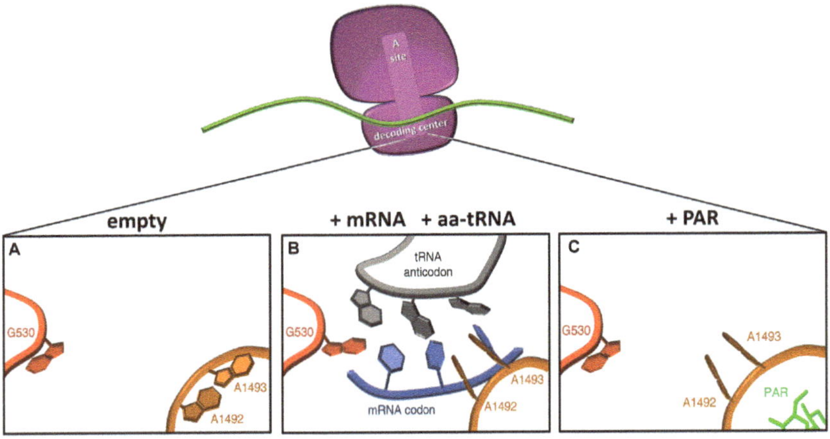

Fig. 5 Conformational changes in the decoding center upon tRNA or aminoglycoside binding. Schematic representation of the ribosomal decoding center as derived from PDB 1IBM (A), 1IBL (B) and 1IBK (C). In the native 30S ribosomal subunit, adenines 1492 and 1493 are stacked inside helix 44, and G530 in the *syn* conformation (A). When the correct codon-anticodon loop, formed upon entering of a cognate aa-tRNA, enters the A site, the A1492 and A1493 residues flip out and G530 switches to the *anti*-conformation (B). In the presence of paromomycin, the A1492 and A1493 residues are similarly flipped out into the A site mimicking the presence of a cognate aa-tRNA. Red, G530 loop; orange, helix 44, gray, tRNA anticodon, blue, mRNA codon, green, paromomycin (PAR). *Panel (A–C) adapted from Ogle, J. M., Brodersen, D. E., Clemons, W. M., Tarry, M. J., Carter, A. P., & Ramakrishnan, V. (2001). Recognition of cognate transfer RNA by the 30S ribosomal subunit. Science (New York, N.Y.), 292(5518), 897–902. https://doi.org/10.1126/science.1060612; Lombardi, S., Testa, M. F., Pinotti, M., & Branchini, A. (2020). Molecular insights into determinants of translational readthrough and implications for nonsense suppression approaches.* International Journal of Molecular Sciences, 21*(24), E9449. https://doi.org/10.3390/ijms21249449.*

aminoglycosides revealed that these compounds interact with the eukaryotic ribosome at multiple sites and that the mechanism by which each aminoglycoside induces PTC readthrough is specific (Prokhorova et al., 2017). In particular, aminoglycosides containing 6′-OH substituent in ring I (e.g., G418) bind to the pocket within helix 44, thus facilitating near-cognate aa-tRNA accommodation and competition with the release factor. On the other hand, aminoglycosides containing a 6′-NH_2 substituent (e.g., gentamicin) employ an alternative mechanism based on inter-subunit rotation effects that are likely to hamper the interaction between release factors and the ribosome (Prokhorova et al., 2017). However, in the context of a therapeutic approach based on suppression of nonsense mutations, the lifelong or even prolonged administration of aminoglycosides is hardly feasible due to persistent and serious ototoxicity and nephrotoxicity as severe side effects (Guthrie, 2008; Lopez-Novoa, Quiros, Vicente, Morales, & Lopez-Hernandez, 2011). The mechanism underlying this toxicity is currently unclear, but generation of reactive oxygen species is thought to be the main cause of cell death (Xie, Talaska, & Schacht, 2011).

Several approaches have been tempted to overcome aminoglycosides toxic effects, such as liposome encapsulation (Schiffelers, 2001) and co-administration with antioxidants (Campbell et al., 2007), as well as other compounds able to reduce the interaction with cellular components (Du, Keeling, Fan, Liu, & Bedwell, 2009) or to increase the readthrough-inducing activity of aminoglycosides (Baradaran-Heravi et al., 2016). In addition, a different approach stemmed from the idea that distinct structures of aminoglycosides are responsible for nonsense suppression properties and toxicity. Therefore, aminoglycosides have been chemically modified to reduce toxicity and increase their therapeutic potential. In particular, the paromomycin derivatives NB30 and NB54 showed up to 15-fold reduced toxicity (Nudelman et al., 2009), whereas the G418 derivatives NB74 and NB84 showed reduced toxicity, increased activity and the ability to cross the blood–brain barrier in mice (Gunn et al., 2014). Moreover, NB124 was shown to reduce NMD-mediated degradation of the target transcript (Bidou, Bugaud, Belakhov, Baasov, & Namy, 2017), and the neomycin derivative pyranmycin (TC007) was able to increase the lifespan of spinal muscular atrophy mice (Mattis, Tom Chang, & Lorson, 2012).

A different approach based on high-throughput screening of small molecular weight compound libraries led to the identification of non-aminoglycoside molecules with readthrough induction activity (Borgatti, Altamura, Salvatori, D'Aversa, & Altamura, 2020; Du et al., 2009; Welch et al., 2007). One paradigmatic example is PTC124 (also known as ataluren or Translarna®), an orally bioavailable oxadiazole compound with PTC suppression activity that has been shown to be safe, with minimal off-target side-effects and no antibacterial activity (Welch et al., 2007). Recent studies demonstrated that ataluren acts by inhibiting the release factor complex activity (Huang et al., 2022; Ng, Li, Ghelfi, Goldman, & Cooperman, 2021) and, although its mechanism of action has yet to be completely clarified, encouraging results have been obtained. Different diseases are currently being evaluated in clinical trials (aniridia, NCT02647359; colorectal and endometrium cancers, NCT04014530; epilepsy, NCT0275826), but results are unfortunately conflicting. Ataluren has been approved by the European Medicines Agency (EMA) for Duchenne muscular dystrophy patients with nonsense mutations (Haas et al., 2015; McDonald et al., 2017), whereas the lack of conclusive results obtained in pre-clinical and clinical trials led to discontinuation of the development of ataluren for cystic fibrosis (Kerem et al., 2014). Other promising readthrough-inducing compounds identified include RTC13 and RTC14 (Du, Damoiseaux, et al., 2009), amlexanox (Gonzalez-Hilarion et al., 2012), PTC414 (Moosajee et al., 2016), 2,6-diaminopurine (Trzaska et al., 2020), SRI-41315 (Sharma et al., 2021), TCP-1109 (Otani et al., 2022), NV848 (Bezzerri et al., 2022), and AC1903 (Baradaran-Heravi et al., 2022). In addition, the aminoglycoside analog ELX-02 (previously referred to as NB124) is under clinical investigation (Crawford, Alroy, Sharpe, Goddeeris, & Williams, 2020; Kerem, 2020; Leubitz et al., 2019).

3.3 The determinants of readthrough

The establishment of therapeutic approaches based on nonsense suppression has been hampered to date by their extremely variable outcome. Indeed, the efficacy of readthrough induction depends on several factors, including the amount of the target transcript(s) (related to the occurrence of NMD), the efficiency of PTC suppression, and the features of the full-length proteins arising from readthrough events. Moreover, a recent study by Karki and colleagues explored the tissue-specificity of translational readthrough, which seems to be regulated by the expression levels and alternative splicing isoforms of termination factors (e.g., eRF_1), and by the relative abundance of readthrough-prone suppressor tRNAs (Karki et al., 2021). Here, we will focus on the molecular determinants (nucleotide/protein contexts) of readthrough (Fig. 6).

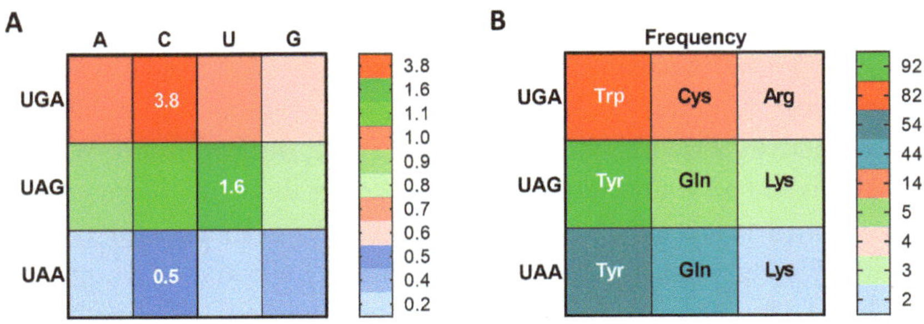

Fig. 6 Codon-dependent (nucleotide and protein) sequence determinants. Readthrough score of the nucleotide sequence contexts (Manuvakhova et al., 2000), referred as the nonsense triplet and the following nucleotide (A), and relative frequency of the inserted amino acid (Blanchet, Cornu, Argentini, & Namy, 2014) as a function of PTC type (B), are represented as heat maps. The legend on the rights of each panel indicates the relative readthrough score (A) or the percentage of insertion of the corresponding amino acids (B).

3.3.1 The sequence context

The PTC sequence context, namely, the stop codon and the surrounding sequences, is a major determinant of readthrough efficiency (Tate, Cridge, & Brown, 2018). It is widely established that the three stop codons are decoded with different accuracy by the ribosome, and are thus differently susceptible (leakiness) to readthrough with the rank order of UGA > UAG > UAA (Fig. 6A) (Bidou et al., 2004; Du et al., 2002; Loughran et al., 2014; Manuvakhova et al., 2000). Early statistical studies of natural termination sequences had suggested that the actual termination signal is a tetranucleotide, thus highlighting the influence of the nucleotide downstream of the stop codon (+4 position, with the first nucleotide of the termination codon marked as +1) (McCaughan, Brown, Dalphin, Berry, & Tate, 1995). Later studies confirmed that the +4 position is strongly involved in termination fidelity and readthrough efficiency, with an up to sixfold difference conferred solely by the 4th nucleotide (Cridge, Crowe-McAuliffe, Mathew, & Tate, 2018; Manuvakhova et al., 2000). Structural studies of the eukaryotic ribosomal complex containing eRF_1 revealed that, when a stop codon is placed in the ribosomal A site, the base at position +4 is pulled into the A site by mRNA compaction, where it is stabilized by stacking against G626 of 18S rRNA (Ben-Shem et al., 2011; Brown, Shao, Murray, Hegde, & Ramakrishnan, 2015; Matheisl, Berninghausen, Becker, & Beckmann, 2015; McCaughan et al., 1995). Stacking of G626 is less stable for pyrimidines, which help explaining the readthrough-favorable role of cytosine and uracil at the +4 position. Extensive studies in yeast showed that the UGAC and UAGU signals are the most efficient in mediating readthrough, followed by UGAN and UAGN, whereas the UAAN signals, and in particular the UAAA sequence, mediate the most accurate termination (Fig. 6A) (Manuvakhova et al., 2000). Besides the tetranucleotide termination signal, both 5′ and 3′ surrounding sequences (from −2 to +9 positions) showed a non-random distribution of nucleotides in natural contexts (Bonetti et al., 1995). Concerning the 5′ sequence, the presence of two adenines in positions −2 and −1 has been associated with the highest levels of readthrough (Harrell, 2002; Tork, 2004), whereas the 3′ sequence context modulates termination fidelity, with a significant but complex role of the +5, +6 and +8 nucleotides (Cridge et al., 2018). The structural basis for these effects is unclear, although the P site codon (including positions −2 and −1) appears to directly contact the top of helix 44 of 18S rRNA (Tork, 2004), and the +4 and +5 nucleotides to stack respectively with G626 and C1698 bases of 18S rRNA (Brown et al., 2015). This highlights the presence of a complex network of interactions between the ribosome and the sequences surrounding the stop codon, although the effects of 5′ and 3′ sequences are subtle and subordinated to the stop codon present. Overall, readthrough efficiency is hardly predictable based strictly on the sequence context, but most authors agree on the high susceptibility of the UGA stop codon followed by UAG and UAA, the latter being the less responsive.

Even though the levels of aminoglycoside-induced readthrough were believed to be simply superimposed on the basal level allowed by each sequence context (Manuvakhova et al., 2000), it has been recently suggested that the sequence context also affects the optimal readthrough-inducing drug type as well as the optimal drug concentration (Schilff, Sargsyan, Hofhuis, & Thoms, 2021). This should be considered to identify susceptible nonsense mutations and optimal treatment in the context of a personalized PTC suppression approach, whose choice should be driven by experimental evidence whenever possible.

3.3.2 The reinserted amino acid

Another critical but largely underestimated factor affecting the readthrough outcome is the amino acid reinserted at the PTC position. Readthrough occurs when a stop codon is misrecognized by the ribosome, which incorporates an aa-tRNA instead of terminating translation. Extensive studies in yeast and mammalian expression systems have shown that defined subsets of aa-tRNAs, and thus amino acids, are actually introduced into the growing polypeptide by the ribosome during readthrough, with UAA and UAG codons that can be suppressed by either glutamine, tyrosine or lysine, whereas UGA codons can be decoded by tryptophan, arginine or cysteine (Fig. 6B) (Blanchet et al., 2014, 2018; Roy, Leszyk, Mangus, & Jacobson, 2015). Comparison of several studies showed that, even though the set of incorporated amino acids is the same, different experimental systems and/or readthrough-inducing compounds can affect the relative frequencies of amino acid insertion at a specific PTC (Blanchet et al., 2014; Pranke et al., 2018; Roy et al., 2015). In particular, a comprehensive study performed in yeast showed that different readthrough-inducing conditions such as loss of Upf factors, defective release factors, or the presence of aminoglycosides, were all able to change the proportion of amino acids reinserted during readthrough (Roy et al., 2015).

The structural basis for aa-tRNA selection has been partly unveiled. Recent studies showed that the ribosomal decoding center tolerates non-standard Watson-Crick base pairs in the A site, with the shape of the base pair being crucial, rather than the number of hydrogen bonds formed (Westhof, Yusupov, & Yusupova, 2014). In particular, aa-tRNA insertion at a PTC occurs by mispairing at either position 1 or 3 of the stop codon. The UAG codon showed a preference for mismatches at position 1, whereas UGA favored mispairings at position 3, and UAA showed equal distribution at both positions (Roy et al., 2015). Moreover, identification of the amino acids reinserted during readthrough suggested that, at the third codon position, the A-C mispairing is favored over A-G and G-G, whereas in the first position the U-G mispairing is predominant, likely due to its geometrical mimicry of a standard base pairing (Demeshkina, Jenner, Westhof, Yusupov, & Yusupova, 2013; Roy et al., 2015; Westhof et al., 2014).

The amino acids reinserted during PTC readthrough dictate the features of the resulting full-length proteins produced, and thus represent a crucial determinant of the outcome in the view of a nonsense suppression approach. Indeed, the result of readthrough is an assortment of full-length proteins with (potential) different missense changes at the site of the PTC, which can have major impact on protein stability and/or function. Reinsertion of the wild-type residue or a tolerant non-conserved position would favor the production of a functional protein. On the other hand, if the PTC affects an essential position (e.g., a catalytic residue), the protein arising from readthrough would be full-length but probably dysfunctional or non-functional. Therefore, careful evaluation of each PTC should be performed to drive the selection of nonsense mutations potentially amenable for suppression.

Overall, the specific features of each PTC (sequence context and role of the residue in the protein) should be carefully considered to help predicting and interpreting the results of a nonsense suppression approach based on readthrough.

4. Readthrough induction as a therapeutic approach for PTC suppression

The ability of some drugs to interact with the ribosomal decoding center and to alter translation fidelity has shed light on the therapeutic potential of the readthrough-inducing approach to suppress nonsense mutations and lead to the synthesis of full-length proteins.

A relevant body of work has focused on the qualitative or final output of readthrough (e.g., through reporter assays), mainly considering the nucleotide sequence context as the major driving force. However, the underlying complexity of this process would benefit from an integrated experimental approach on the molecular determinants dictating the resulting protein output. In this view, prediction of readthrough-favorable features, and models to characterize protein variants bearing predicted amino acid insertions, may shed light on those mutations being eligible for treatment with readthrough-promoting drugs.

The scenario depicted by the several studies on readthrough shows that relevant levels of functional suppression, in terms of correction of protein synthesis and activity, is related to a series of events that may exert complementary effects on the resulting protein output. First, the degree of susceptibility (leakiness) of PTC sequence to the readthrough process itself or by its drug-mediated induction (Fig. 7, left panel), as well as the type of amino acid(s) inserted at the PTC (Fig. 7, middle panel), shape the resulting protein output, whose additional determinants can be further detailed as follows (Fig. 7, right panel).

(i) Degree of re-insertion of the original residue allowing the production of the wild-type polypeptide, representing the most favorable event;

(ii) Tolerated amino acid insertions in terms of protein synthesis, trafficking/secretion and function, or the occurrence of rare gain-of-function effects;

(iii) Favorable localization of the suppressed PTCs in protein regions that are removed during protein processing, and thus outside of the mature protein, which predict counteracting effects on the potentially detrimental impact of amino acid substitutions;

(iv) Tolerated effects of readthrough-derived amino acid insertions on higher-level organization, such as oligomerization, of the target protein.

This complex network of molecular events, resulting in a different degree of functional or productive readthrough, indicates that the nucleotide and protein contexts contribute to restrict the number of responsive PTCs, particularly for proteins with enzymatic activity. On the other hand, the interplay between different molecular determinants might have led to underestimate the potency of the available or under-development readthrough-inducing drugs. Nevertheless, the availability of compounds able to efficaciously induce readthrough with, in parallel, relative low side-effects is a primary goal to achieve for therapeutic purposes.

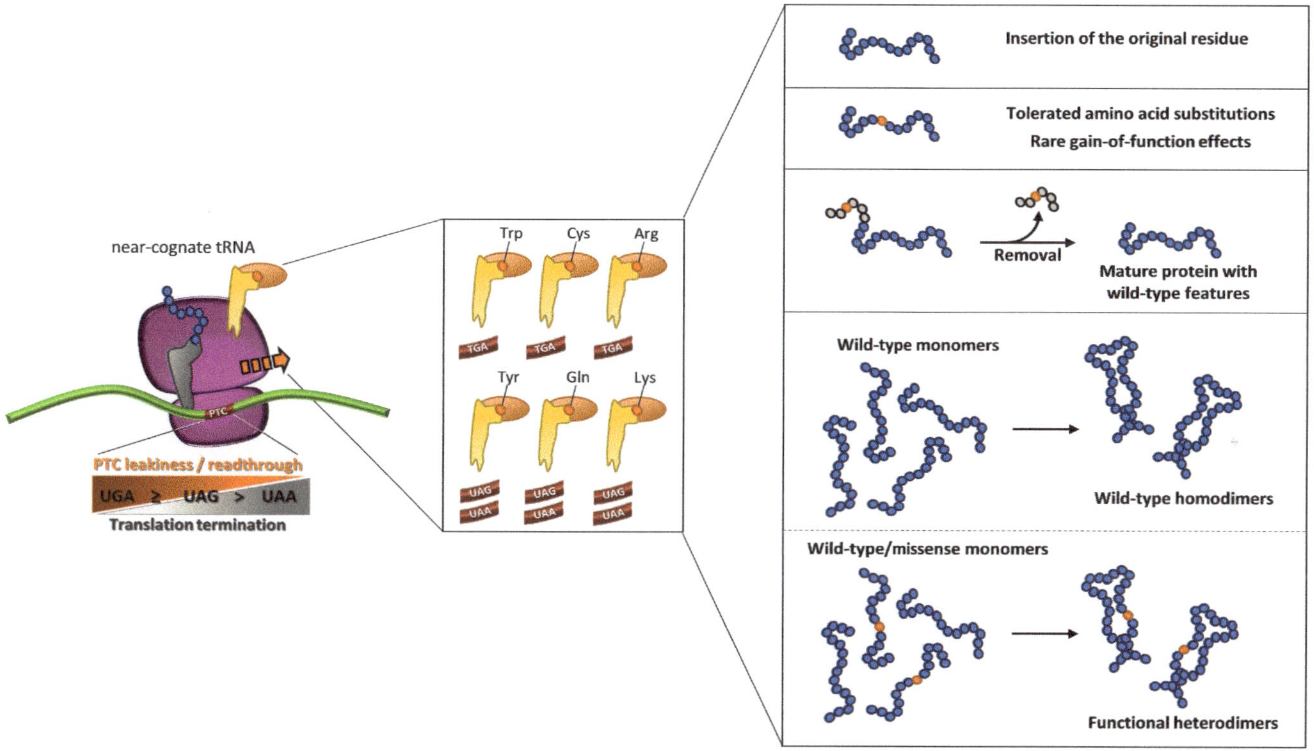

Fig. 7 Outcome of readthrough as an integrated interplay of molecular determinants. Readthrough activity of translating ribosome is driven by PTC type, which leads to the differential insertion of subset(s) of amino acids brought by near-cognate aa-tRNAs. A scenario of the resulting readthrough-mediated protein output is depicted in the right panels (see text for details). *Adapted from Lombardi, S., Testa, M. F., Pinotti, M., & Branchini, A. (2020). Molecular insights into determinants of translational readthrough and implications for nonsense suppression approaches.* International Journal of Molecular Sciences, 21*(24), E9449. https://doi.org/10.3390/ijms21249449.*

These considerations, which are pivotal for potential therapeutic purposes, would also include two additional elements, namely, (i) whether the readthrough approach is compatible with the estimated threshold needed to ameliorate the disease phenotype, and (ii) the possibility that, for proteins involved in the formation of oligomers, missense variants arising from readthrough could affect protein structure/function (i.e., through dominant-negative effects). Interestingly, recently a series of anticodon engineered suppressor tRNAs, driving the insertion of the original amino acid without affecting normal translation termination, have been developed to suppress PTCs and redirect protein synthesis (Lueck et al., 2019; Porter, Heil, & Lueck, 2021).

The literature on the characterization and correction of pathogenic nonsense mutations mediated by readthrough induction is wide and ever-increasing, and encompasses a large number of disease models, with tens of recent examples (Beryozkin et al., 2022; Du, Damoiseaux, et al., 2009; Du et al., 2013; Fazzari, Frasca, Bifari, & Landsberger, 2019; Has et al., 2022, pag.; Heier & DiDonato, 2009; Liu, Zhang, Zhang, Gao, & Qiu, 2020, pag. 6; Martínez-Santamaría et al., 2022, pag. 1; Mattis et al., 2006, 2012; Popescu, Sidorova, Zhang, & Eubanks, 2010; Rebibo-Sabbah, Nudelman, Ahmed, Baasov, & Ben-Yosef, 2007; Richardson, Smart, Tracey-White, Webster, & Moosajee, 2017; Sánchez-Alcudia, Pérez, Ugarte, & Desviat, 2012; Schwarz et al., 2015; Vecsler et al., 2011). Among these, the major experimental evidence at different layers has been produced for cystic fibrosis, Duchenne muscular dystrophy, coagulation factor deficiencies and lysosomal storage disorders.

4.1 Cystic fibrosis

Cystic fibrosis (CF, OMIM # 219700) is an autosomal recessive disorder caused by variants in the *CFTR* gene, which encodes the homonymous epithelial ion channel that normally transports chloride and bicarbonate ions. Altered or reduced CFTR activity causes impaired mucus hydration and clearance, which leads to chronic pulmonary infection and inflammation, pancreatic exocrine insufficiency and male infertility (Shteinberg, Haq, Polineni, & Davies, 2021).

More than 2000 *CFTR* variants have been identified to date (Cystic Fibrosis Mutation Database, http://www.genet.sickkids.on.ca/; HGMD, http://www.hgmd.cf.ac.uk/ac/all.php), grouped in six classes based on their consequences at the protein level. Class I variants cause the complete absence of CFTR protein and are associated with the most severe CF forms. They encompass mainly nonsense variants and affect around 1 out of 10 CF patients, therefore, major efforts have focused on the development of a therapeutic approach for these mutations.

Indeed, CF was the first disease challenged with a nonsense suppression approach when, in 1996, Howard and colleagues demonstrated that treatment with either G418 or gentamicin resulted in CFTR expression and activity rescue. This study, performed in cells expressing the p.G542X and p.R553X variants, demonstrated for the first time that aminoglycoside-mediated suppression of PTCs may be exploited as a therapy (Howard, Frizzell, & Bedwell, 1996). Unfortunately, an effective pharmacological option for CF patients carrying nonsense mutations has not been reached yet.

Early pilot clinical trials showed that both topical or parenteral administration of gentamicin was able to rescue CFTR protein synthesis and function in the nasal epithelium of CF patients (Clancy et al., 2001; Wilschanski et al., 2003). However, these first attempts did not consider the identity of the PTC present, and the levels of restored protein were not sufficient to allow for a long-term clinical benefit. Few years later, a multidisciplinary study evaluated readthrough efficiency both in vitro, through a reporter assay, as well as in CF patients by measuring clinical, functional, and immunological parameters (Sermet-Gaudelus et al., 2018). The authors focused on four nonsense variants, namely, p.Y122X, p.G542X, p.R1162X and W1282X, but only p.Y122X was significantly responsive to gentamicin treatment, both in the reporter assay and in patients. This study demonstrated for the first time the correlation between in vitro and in vivo readthrough efficacy and suggested that PTCs should be screened by in vitro studies before clinical trials to identify patients who are most likely responsive to this therapeutic approach.

The identification of PTC124 (also referred to as ataluren), a safe and orally bioavailable readthrough-inducing compound, held great promise for the treatment of CF patients. Indeed, treatment with ataluren was able to restore CFTR expression and function in humanized transgenic mouse models (Du et al., 2008) and these results led to phase 1 and 2 clinical trials, which resulted in promising safety and efficacy data in CF patients (Kerem et al., 2008; Sermet-Gaudelus et al., 2010; Wilschanski et al., 2011). Unfortunately, the following randomized, double-blind, placebo-controlled phase 3 clinical trial showed that ataluren restored a level of CFTR activity that was not sufficient to improve lung function in CF patients (Kerem et al., 2014), thus leading to the discontinuation of ataluren development for CF.

ELX-02 (previously referred to as NB124) is a novel non-antibiotic aminoglycoside analog that has been developed for the suppression of PTCs (Kandasamy et al., 2012; Nudelman et al., 2010). In the context of CF, Xue and colleagues challenged ELX-02 on the four most common CFTR PTCs (p.G542X, p.R553X, p.R1162X, and p.W1282X). Several models were exploited including a dual-luciferase reporter, Fischer rat thyroid cells stably expressing the p.G542X variant, primary human bronchial epithelial cells derived from lung explants (CFTR genotype G542X/delF508), and a CF mouse model expressing the human p.G542X CFTR variant (Xue et al., 2014). ELX-02 was superior to gentamicin and other aminoglycosides in recovering CFTR expression and activity, it exhibited favorable pharmacokinetic properties and lower cytotoxicity than gentamicin. These promising results were then confirmed by two randomized, double-blind, placebo-controlled phase 1a studies in healthy volunteers (Leubitz et al., 2019) and ELX-02 is currently being evaluated in phase 2 clinical trials for CF patients harboring the p.G542X variant (NCT04135495 and NCT04126473).

Overall, an effective readthrough-inducing therapy for CF patients carrying nonsense variants has not been established yet, but novel compounds hold promise for the development of new personalized therapeutic approaches. Nevertheless, the high threshold of functional rescue needed to reach a therapeutic benefit would likely require the combination of various approaches such as readthrough-inducing compounds, NMD inhibitors and corrector/potentiators (Pranke et al., 2018; Sharma et al., 2021).

4.2 Duchenne muscular dystrophy

A paradigmatic disease model for PTC suppression therapies based on readthrough has been Duchenne muscular dystrophy (DMD, OMIM #310200) (Babbs et al., 2020; Shimizu-Motohashi et al., 2019), for which the readthrough-inducing treatment received EMA approval.

DMD is an X-linked degenerative disorder (1:3500 male births) caused by mutations in the *DMD* gene encoding the dystrophin protein, which is involved, as part of the transmembrane scaffolding dystrophin-associated protein complex (DAPC), in the connection of actin cytoskeleton of a muscle fiber to the extracellular matrix (Hoffman, Brown, & Kunkel, 1987; Nowak & Davies, 2004). The consequence of dystrophin alterations (DMD gene homepage, https://databases.lovd.nl/shared/genes/DMD) is a progressive muscle fiber damage, leading DMD-affected patients to inability to walk by age 12, with associated cardiac and respiratory issues (Fairclough, Bareja, & Davies, 2011). Nonsense mutations in the *DMD* gene accounts for ~15% of all gene lesions and are responsible for a truncated/dysfunctional dystrophin protein giving rise the most severe clinical form.

First studies on DMD were conducted by using the aminoglycoside gentamicin as PTC suppressor, which rescued the desired range of normal dystrophin alleviating the DMD phenotype in the *mdx* mouse model (bearing the CAA > TAA in exon 23, corresponding to a mild DMD form) (Arakawa et al., 2003; Barton-Davis, Cordier, Shoturma, Leland, & Sweeney, 1999; Politano et al., 2003). However, the different aminoglycosides components of gentamicin gave rise to variable results, which prevented the evaluation of its effectiveness in patients (Shimizu-Motohashi et al., 2019). To overcome the nephrotoxicity and neurotoxicity effects of gentamicin, other promising aminoglycosides and analogs were developed such as negamycin, 3-epi-deoxynegamycin and leucyl-3-epi-deoxynegamycin, which can promote the suppression of PTC with production of full-length dystrophin (Taguchi et al., 2017). These potential therapeutic molecules are in progress for clinical trials. On the other hand, one of the most successful non-aminoglycoside molecule is PTC124 (ataluren), associated, as negamycin, with a low toxicity (Hirawat et al., 2007; Welch et al., 2007). Ataluren reached around 25% of wild-type dystrophin through oral administration and subsequent supplementation through intraperitoneal injection in *mdx* mouse (Welch et al., 2007). Prompted by these results in the *mdx* mouse, PTC124 entered in the first clinical trial phases. Currently, a 10-year phase 3 trial with PTC124 on 270 previously treated participants with nonsense mutations permitted the evaluation of tolerability and safety at different concentrations (NCT01247207).

In the last years, new therapeutics approaches have been explored to suppress nonsense mutations for DMD, as the incorporation of unnatural amino acids (UAAs) (Shi et al., 2022). Different elements should be considered to investigate this system, namely, (i) an engineered aminoacyl-tRNA-synthetase (aaRS)-tRNA pair accurately recognizing UAAs, able to insert UAAs at the PTC, (ii) the type of nonsense codon, and (iii) the appropriate UAAs among 150 types, which has been revealed to be pyrrolysine (Nozawa et al., 2009; Wan, Thrarp, & Liu, 2014). This approach resulted in suppression of TAA PTCs in differentiated myoblasts arising from *mdx* mouse and a DMD patient cells. In vitro studies revealed that dystrophin was restored after a treatment of 8 days, as well as in patient cells by co-administration of Nε-2-azidoethyloxycarbonyl-L-lysine (NAEK). In vivo studies in transgenic mdx mice further validated the efficacy of this system, detecting a significant rescue of dystrophin protein up to 8 weeks leading to alleviation of DMD symptoms (Shi et al., 2022).

4.3 Coagulation factor deficiencies

Coagulation factors are involved in a finely tuned series of events (the coagulation cascade) that ultimately lead to the formation of a clot (Monroe & Hoffman, 2006). Thus, a defect in one of these components may lead to a coagulation disorder with different phenotype severity as a function of the residual levels of the affected protein (Shee, 2005).

In the mutational pattern of coagulation factor disorders, nonsense mutations are relatively frequent, particularly in the X-linked hemophilia A (HA, OMIM #306700) and B (HB, OMIM #306900), with 10–13% of nucleotide changes resulting in a PTC (http://www.factorviii-db.org; http://www.factorix.org) (Li, Miller, Payne, & Craig Hooper, 2013; McVey et al., 2020; Payne, Miller, Kelly, Michael Soucie, & Craig Hooper, 2013; Rallapalli, Kemball-Cook, Tuddenham, Gomez, & Perkins, 2013). Severe disease forms in almost all other bleeding disorders are caused by nonsense mutations associated with alterations of serine proteases, cofactors and inhibitors as well as of proteins involved in primary hemostasis and clot formation/stabilization (HGMD, http://www.hgmd.cf.ac.uk/ac/index.php).

However, notable exceptions have been reported and characterized for factor (F)VII and FX, in particular for alterations affecting their carboxyl-terminal region (Branchini, 2021), in which missense/nonsense mutations are predicted to have a detrimental impact on secretion of homologous coagulation factors (Branchini et al., 2014, 2013; Katsumi et al., 1998; Kurachi, Pantazatos, & Kurachi, 1997; Mariani et al., 2005). Nonsense variants affecting the carboxyl-terminal region of FX revealed a slight impact on protein secretion/function (Branchini et al., 2015), thus predicting associated asymptomatic phenotypes. On the other hand, the homozygous *F7* p.R462X nonsense variant, predicting a potentially null genetic condition incompatible with life (Rosen et al., 1997), is associated with an asymptomatic phenotype due to the production of very low amounts of a gain-of-function FVII variant lacking the last five carboxyl-terminal residues (Branchini et al., 2012).

Notably, the low therapeutic threshold sufficient to ameliorate the clinical phenotype in coagulation factor disorders (in the range of 2–5% of normal protein) (Shee, 2005) is compatible with the predicted extent of rescue expected from readthrough, as well as from other strategies targeting mutations at different levels (Andersen et al., 2019; Balestra & Branchini, 2019; Barbon et al., 2016; Cavallari et al., 2012; Chollet, Skarpen, Iversen, Sandset, & Skretting, 2015; Lombardi, Ferrarese, et al., 2020; Lombardi, Testa, Pinotti, & Branchini, 2020; Lombardi et al., 2021; Malhotra et al., 2008; Pignani et al., 2018, 2019; Roth et al., 2012). This notion drove attempts for readthrough as therapy in patients with nonsense mutations in *F8* (HA) and *F9* (HB) (James et al., 2005) as well as *F7* (FVII deficiency, OMIM #22750) (Pinotti et al., 2006) genes, or in HB mouse models (Yang et al., 2007). These studies indicated that the overall response to readthrough was compatible with subtherapeutic levels or moderate phenotype amelioration for a few nonsense variants.

The interplay between sequence and protein contexts, as well as the influence of PTC localization, has been investigated in HB models. In a study on a wide panel of *F9* nonsense mutations, the two most responsive variants (p.W240X and p.R384X) were paradigmatically indicative of the influence of nucleotide/protein contexts (Branchini et al., 2017). Indeed, the relevant functional rescue observed was related to (i) favorable nucleotide contexts (UGAC and UGAU, respectively), and re-insertion of the original residue (W240X), as confirmed by the low or negligible activity of overlapping natural or recombinant variants (Jayandharan et al., 2005; Thompson, Schoof, Weinmann, & Chen, 1992); or (ii) gain-of-function features (R384X) of the most probable readthrough-derived hyperactive R384W missense variant, in accordance with the features of the so-called FIX "Padua" position (Samelson-Jones et al., 2021; Simioni et al., 2009). Similar results were obtained on *F8* nonsense mutations (Liu, Zhang, Zhu, & Zhang, 2019; Martorell, Cortina, Parra, Barquinero, & Vidal, 2020).

Interestingly, differently from PTCs located in the coding region corresponding to the mature protein, which influences the output of the resulting full-length protein, the occurrence of readthrough in regions removed during processing is predicted to be more tolerated. This was demonstrated by investigating three *F9* PTCs affecting the signal and pro-peptide region of FIX, which provided experimental evidence for a position-specific effect of PTCs to shape the observed protein output (Ferrarese et al., 2018). These two regions are crucial for directing precursor

FIX to the endoplasmic reticulum as well as for the key post-translational modification γ-carboxylation (Kaufman, 1998), and are cleaved intracellularly during FIX processing/maturation. Interestingly, rescue of PTCs at key cleavage sites for FIX processing, impaired also by overlapping missense mutations (Liddell, Lillicrap, Peake, & Bloom, 1989; Montejo, Magallón, Tizzano, & Solera, 1999; Wulff, Schröder, Wehnert, & Herrmann, 1995), was prevented due to sequence constraints. The relevant functional rescue observed for the responsive PTC variant affecting the signal peptide was driven by the synthesis of a full-length FIX protein displaying wild-type features, as also demonstrated by expression of the corresponding readthrough-derived missense variants, due to removal of this region tolerating amino acid changes.

The combination of nucleotide and protein contexts as determinants of readthrough susceptibility has been also investigated for homozygous nonsense mutations in rare autosomal coagulation factor disorders, namely, FVII (Branchini et al., 2016) and FX (Ferrarese et al., 2019) deficiency. Since the complete absence of FVII and FX is virtually incompatible with life (Dewerchin et al., 2000; Rosen et al., 1997), the possibility to rescue protein synthesis and function of these serine proteases is of great clinical significance.

Two *F7* homozygous nonsense variants, affecting the N-terminal region of FVII and predicted to be lethal due to the production of a truncated protein devoid of the catalytic domain, were identified in patients affected by severe (p.S112X) (Giansily-Blaizot et al., 2003) or unexpectedly moderate (p.C132X) (Chafa, Fischer, Reghis, & Tapon-Bretaudiere, 2005) FVII deficiency (McVey, Boswell, Mumford, Kemball-Cook, & Tuddenham, 2001; Millar et al., 2000). Readthrough induction and expression of predicted missense variants depicted two different scenarios. While the 112 position (surface-exposed), at which insertion of the original residue is virtually not predicted, tolerated amino acid substitutions, readthrough at the 132 position (part of a structural intrachain disulfide bridge) (Eigenbrot, 2002) was driven only by insertion of the (predicted) original residue. Although this event would limit responsiveness to readthrough, the underlying different susceptibility of PTC nucleotide contexts (132X-UGAC > 112X-UGAA), maybe through an additive compensatory effect, led to a higher functional rescue after correction of the C132X PTC.

An example of unsuccessful readthrough induction due to protein constraints has been provided for the very rare, and virtually unique, *F10* p.W461X nonsense mutation, found in a compound heterozygote patient with severe FX deficiency (OMIM #227600). Nonsense and missense variants showed negligible effects of readthrough induction as well as a very low functional profile. This protein output can be interpreted on the basis of the unfavorable UAG sequence context, not predicting re-insertion of the original tryptophan, in combination with specific protein constraints related to FX function (Branchini, 2021; Branchini et al., 2015; Ferrarese et al., 2019). In addition, the amino acid stretch containing the 461 residue is involved in crucial protein-protein interactions with prothrombin (Yegneswaran, Mesters, & Griffin, 2003), impaired by substitutions of the original tryptophan, and thus producing FX variants unable to convert prothrombin to thrombin, the central enzyme in the coagulation cascade.

4.4 Lysosomal storage disorders

Lysosomal storage disorders (LSDs) are a group of rare inherited metabolic diseases associated with deficiencies in key metabolic enzymes and characterized by the accumulation of undegraded substrates within lysosomes in multiple tissues and organs (Parenti, Andria, & Ballabio, 2015).

A correction approach based on drug-induced readthrough could be particularly relevant for LSDs patients, due to the relatively low threshold necessary to potentially alleviate the disease phenotype (Oussoren et al., 2013). In addition, nonsense mutations are relatively frequent in LSDs, particularly in Mucopolysaccharidosis I-Hurler (MPS I-H), in which a high proportion of patients (50–70%) is estimated to be affected by a nonsense mutation (Keeling et al., 2014).

MPS I-H (OMIM #607014) is the severe form of α-L-iduronidase (IDUA) deficiency and is associated with glycosaminoglycan accumulation (Coutinho, Lacerda, & Alves, 2012). In this disease, the p.Q70X and p.W402X are the two most common IDUA nonsense mutations in Caucasians (Bunge et al., 1998; Scott et al., 1995). Constructs bearing either the p.W402X (UGG > UAG) or p.Q70X (CAG > UAG) mutations were challenged with gentamicin in cell-free systems (Keeling et al., 2001). Higher effects were observed on production of full-length proteins in vitro for the p.Q70X variant (~10% of the normal protein) as a consequence of a readthrough-susceptible sequence context (UAGC) (Hein et al., 2004) in accordance with theoretical scores (Manuvakhova et al., 2000), and due to the predicted insertion of the wild-type residue. Differently, the sequence context of the p.W402X PTC (UAGG), challenged with different aminoglycosides, was identified to be low-score in in vitro assays with a reporter system. Moreover, re-introduction of the original residue (tryptophan) is not predicted in the subset of amino acids introduced by readthrough over UAG PTCs, thus resulting only in a partial rescue (~3%), as also observed in a second independent study where authors also predicted *in silico* the impact of the inserted amino acid(s) to interpret the unsuccessful output of readthrough (Matalonga et al., 2015). The first evidences on frequent MPS I-H nonsense mutations prompted attempts in a knock-in mouse model for the W392X PTC (Gunn et al., 2014; Mattis et al., 2012), corresponding to the human p. W402X variant, as well as in the presence of NMD attenuators to increase the mRNA levels potentially available as readthrough substrates (Keeling et al., 2013).

Additional complexity may emerge when considering oligomeric proteins. Here, the amino acid changes introduced by readthrough may exert unpredicted effects on the quaternary structure of the protein and thus on its biosynthesis/function. In this view, Fabry disease (FD, OMIM #301500) provides a paradigmatic model for oligomeric proteins. FD is a rare X-linked LSD caused by deficiency of the homodimeric α-galactosidase A (α-gal A) enzyme (Garman & Garboczi, 2004; Meikle, Hopwood, Clague, & Carey, 1999; Zarate & Hopkin, 2008), and failure in the metabolism of glycosphingolipids. Depending on residual enzymatic activity, FD patients may display either a severe/classic (<1% activity or complete loss-of-function) or a late-onset (residual activity) phenotype (Germain, 2010). Among the FD-causing alterations in the *GLA* gene, the relatively frequent nonsense mutations (Fabry mutation database, http://fabry-database.org/) show the highest association with the classic FD form.

A first attempt to correct FD nonsense mutations was provided in patients' fibroblasts bearing the p.R227X (UGA) variant (Matalonga et al., 2015), which resulted in low responsiveness and rescue. These negative results can be explained in light of the protein context, since R227 is involved in the formation of the α-gal A active site, where amino acid substitutions are unlikely to be compatible with rescue of enzyme function. A second attempt was made for a wide panel of *GLA* nonsense variants by using G418 as the readthrough-inducing agent (Lombardi, Ferrarese, et al., 2020; Lombardi, Testa, et al., 2020). Interestingly, the most responsive variant revealed a low functional response although expression of the predicted readthrough-mediated missense variants showed detectable levels, normal activity, and proper lysosomal localization. However, the observed discrepancy in protein/functional levels of the α-gal A homodimer after readthrough was not compatible with amino acid insertions as the sole determinant. Noticeably, co-expression studies with wild-type α-gal A resulted in decreased activity of the wild-type enzyme, suggesting potential dominant-negative effects. The hypothesis was that part of the amino acid changes introduced by readthrough would lead to the assembly into lysosomes of a fraction of functional homo/heterodimers as well as heterodimers with lower or no function, resulting from the interaction between wild-type and missense monomeric variants.

5. Conclusions

Among the molecular steps of protein synthesis, translation termination provides different layers of "meaning" of stop codons, ranging from normal termination to regulated recoding of those located at the canonical 3′ end of mRNA, including recoding events leading to suppression of premature termination signals by insertion of subset(s) of amino acids.

In this view, the ability of some drugs to magnify readthrough of PTCs have posed the rationale to take advantage of this recoding process as a strategy for genetic disorders caused by nonsense mutations. However, insertion of amino acids during translational readthrough is not a random event, but it is related to the involvement of near-cognate aa-tRNAs able to base pair with the PTC and drive the elongation of the growing polypeptide by adding one among a restricted subset of amino acids. This aspect is of primary importance to predict readthrough-favorable features and efficacy of this approach to achieve relevant protein/functional rescue, and thus the potential translatability of readthrough into a personalized and mutation-specific, and thus patient-oriented, therapeutic strategy.

Author contributions

The authors have equally contributed to conception, literature review, preparation and writing of the manuscript.

Funding

Authors would like to acknowledge the support provided by the Grifols Martin Villar Hemostasis Award (A.B.), the Italian Association of Hemophilia Centers (AICE; A.B.) and the "Fondo Ateneo per la Ricerca (FAR)" from the University of Ferrara (A.B. and M.P.).

Conflicts of interest

The authors declare no conflict of interest.

References

Agirrezabala, X., Schreiner, E., Trabuco, L. G., Lei, J., Ortiz-Meoz, R. F., Schulten, K., et al. (2011). Structural insights into cognate versus near-cognate discrimination during decoding. *The EMBO Journal*, *30*(8), 1497–1507. https://doi.org/10.1038/emboj.2011.58.

Allmang, C., & Krol, A. (2006). Selenoprotein synthesis: UGA does not end the story. *Biochimie*, *88*(11), 1561–1571. https://doi.org/10.1016/j.biochi.2006.04.015.

Amrani, N., Ganesan, R., Kervestin, S., Mangus, D. A., Ghosh, S., & Jacobson, A. (2004). A faux 3'-UTR promotes aberrant termination and triggers nonsense-mediated mRNA decay. *Nature*, *432*(7013), 112–118. https://doi.org/10.1038/nature03060.

Andersen, E., Chollet, M. E., Baroni, M., Pinotti, M., Bernardi, F., Skarpen, E., et al. (2019). The effect of the chemical chaperone 4-phenylbutyrate on secretion and activity of the p. Q160R missense variant of coagulation factor FVII. *Cell & Bioscience*, *9*, 69. https://doi.org/10.1186/s13578-019-0333-8.

Arakawa, M., Shiozuka, M., Nakayama, Y., Hara, T., Hamada, M., Kondo, S., et al. (2003). Negamycin restores dystrophin expression in skeletal and cardiac muscles of mdx mice. *Journal of Biochemistry*, *134*(5), 751–758. https://doi.org/10.1093/jb/mvg203.

Arribere, J. A., Cenik, E. S., Jain, N., Hess, G. T., Lee, C. H., Bassik, M. C., et al. (2016). Translation readthrough mitigation. *Nature*, *534*(7609), 719–723. https://doi.org/10.1038/nature18308.

Atkinson, J., & Martin, R. (1994). Mutations to nonsense codons in human genetic disease: Implications for gene therapy by nonsense suppressor tRNAs. *Nucleic Acids Research*, *22*(8), 1327–1334. https://doi.org/10.1093/nar/22.8.1327.

Babbs, A., Chatzopoulou, M., Edwards, B., Squire, S. E., Wilkinson, I. V. L., Wynne, G. M., et al. (2020). From diagnosis to therapy in Duchenne muscular dystrophy. *Biochemical Society Transactions*. https://doi.org/10.1042/BST20190282.

Balestra, D., & Branchini, A. (2019). Molecular mechanisms and determinants of innovative correction approaches in coagulation factor deficiencies. *International Journal of Molecular Sciences*, *20*(12). https://doi.org/10.3390/ijms20123036.

Ballut, L., Marchadier, B., Baguet, A., Tomasetto, C., Séraphin, B., & Le Hir, H. (2005). The exon junction core complex is locked onto RNA by inhibition of eIF4AIII ATPase activity. *Nature Structural & Molecular Biology*, *12*(10), 861–869. https://doi.org/10.1038/nsmb990.

Baradaran-Heravi, A., Balgi, A. D., Zimmerman, C., Choi, K., Shidmoossavee, F. S., Tan, J. S., et al. (2016). Novel small molecules potentiate premature termination codon readthrough by aminoglycosides. *Nucleic Acids Research*, *44*(14), 6583–6598. https://doi.org/10.1093/nar/gkw638.

Baradaran-Heravi, A., Bauer, C. C., Pickles, I. B., Hosseini-Farahabadi, S., Balgi, A. D., Choi, K., et al. (2022). Nonselective TRPC channel inhibition and suppression of aminoglycoside-induced premature termination codon readthrough by the small molecule AC1903. *The Journal of Biological Chemistry*, *298*(2), 101546. https://doi.org/10.1016/j.jbc.2021.101546.

Barbon, E., Pignani, S., Branchini, A., Bernardi, F., Pinotti, M., & Bovolenta, M. (2016). An engineered tale-transcription factor rescues transcription of factor VII impaired by promoter mutations and enhances its endogenous expression in hepatocytes. *Scientific Reports*, *6*, 28304. https://doi.org/10.1038/srep28304.

Barton-Davis, E. R., Cordier, L., Shoturma, D. I., Leland, S. E., & Sweeney, H. L. (1999). Aminoglycoside antibiotics restore dystrophin function to skeletal muscles of mdx mice. *The Journal of Clinical Investigation*, *104*(4), 375–381. https://doi.org/10.1172/JCI7866.

Behm-Ansmant, I., Kashima, I., Rehwinkel, J., Saulière, J., Wittkopp, N., & Izaurralde, E. (2007). mRNA quality control: An ancient machinery recognizes and degrades mRNAs with nonsense codons. *FEBS Letters*, *581*(15), 2845–2853. https://doi.org/10.1016/j.febslet.2007.05.027.

Beier, H., & Grimm, M. (2001). Misreading of termination codons in eukaryotes by natural nonsense suppressor tRNAs. *Nucleic Acids Research*, *29*(23), 4767–4782. https://doi.org/10.1093/nar/29.23.4767.

Ben-Shem, A., de Loubresse, N. G., Melnikov, S., Jenner, L., Yusupova, G., & Yusupov, M. (2011). The structure of the eukaryotic ribosome at 3.0 Å. *Resolution.*, *334*, 6.

Beryozkin, A., Samanta, A., Gopalakrishnan, P., Khateb, S., Banin, E., Sharon, D., et al. (2022). Translational read-through drugs (TRIDs) are able to restore protein expression and Ciliogenesis in fibroblasts of patients with retinitis pigmentosa caused by a premature termination codon in FAM161A. *International Journal of Molecular Sciences*, *23*(7), 3541. https://doi.org/10.3390/ijms23073541.

Beznosková, P., Wagner, S., Jansen, M. E., von der Haar, T., & Valášek, L. S. (2015). Translation initiation factor eIF3 promotes programmed stop codon readthrough. *Nucleic Acids Research*, *43*(10), 5099–5111. https://doi.org/10.1093/nar/gkv421.

Bezzerri, V., Lentini, L., Api, M., Busilacchi, E. M., Cavalieri, V., Pomilio, A., et al. (2022). Novel translational read-through-inducing drugs as a therapeutic option for Shwachman-diamond syndrome. *Biomedicine*, *10*(4), 886. https://doi.org/10.3390/biomedicines10040886.

Bidou, L., Allamand, V., Rousset, J.-P., & Namy, O. (2012). Sense from nonsense: Therapies for premature stop codon diseases. *Trends in Molecular Medicine*, *18*(11), 679–688. https://doi.org/10.1016/j.molmed.2012.09.008.

Bidou, L., Bugaud, O., Belakhov, V., Baasov, T., & Namy, O. (2017). Characterization of new-generation aminoglycoside promoting premature termination codon readthrough in cancer cells. *RNA Biology*, *14*(3), 378–388. https://doi.org/10.1080/15476286.2017.1285480.

Bidou, L., Hatin, I., Perez, N., Allamand, V., Panthier, J.-J., & Rousset, J.-P. (2004). Premature stop codons involved in muscular dystrophies show a broad spectrum of readthrough efficiencies in response to gentamicin treatment. *Gene Therapy*, *11*(7), 619–627. https://doi.org/10.1038/sj.gt.3302211.

Blanchard, S. C., Gonzalez, R. L., Kim, H. D., Chu, S., & Puglisi, J. D. (2004). TRNA selection and kinetic proofreading in translation. *Nature Structural & Molecular Biology*, *11*(10), 1008–1014. https://doi.org/10.1038/nsmb831.

Blanchet, S., Cornu, D., Argentini, M., & Namy, O. (2014). New insights into the incorporation of natural suppressor tRNAs at stop codons in Saccharomyces cerevisiae. *Nucleic Acids Research*, *42*(15), 10061–10072. https://doi.org/10.1093/nar/gku663.

Blanchet, S., Cornu, D., Hatin, I., Grosjean, H., Bertin, P., & Namy, O. (2018). Deciphering the reading of the genetic code by near-cognate tRNA. *Proceedings of the National Academy of Sciences of the United States of America*, *115*(12), 3018–3023. https://doi.org/10.1073/pnas.1715578115.

Bonetti, B., Fu, L., Moon, J., & Bedwell, D. M. (1995). The efficiency of translation termination is determined by a synergistic interplay between upstream and downstream sequences inSaccharomyces cerevisiae. *Journal of Molecular Biology*, *251*(3), 334–345. https://doi.org/10.1006/jmbi.1995.0438.

Borgatti, M., Altamura, E., Salvatori, F., D'Aversa, E., & Altamura, N. (2020). Screening readthrough compounds to suppress nonsense mutations: Possible application to β-thalassemia. *Journal of Clinical Medicine, 9*(2), 289. https://doi.org/10.3390/jcm9020289.

Branchini, A. (2021). The carboxyl-terminal region of coagulation serine proteases: A matter of cut and change. *Journal of Thrombosis and Haemostasis, 19*(4), 917–919. https://doi.org/10.1111/jth.15237.

Branchini, A., Baroni, M., Burini, F., Puzzo, F., Nicolosi, F., Mari, R., et al. (2015). The carboxyl-terminal region is NOT essential for secreted and functional levels of coagulation factor X. *Journal of Thrombosis and Haemostasis, 13*(8), 1468–1474. https://doi.org/10.1111/jth.13034.

Branchini, A., Baroni, M., Pfeiffer, C., Batorova, A., Giansily-Blaizot, M., Schved, J. F., et al. (2014). Coagulation factor VII variants resistant to inhibitory antibodies. *Thrombosis and Haemostasis, 112*(5), 972–980. https://doi.org/10.1160/TH14-03-0198.

Branchini, A., Campioni, M., Mazzucconi, M. G., Biondo, F., Mari, R., Bicocchi, M. P., et al. (2013). Replacement of the Y450 (c234) phenyl ring in the carboxyl-terminal region of coagulation factor IX causes pleiotropic effects on secretion and enzyme activity. *FEBS Letters, 587*(19), 3249–3253. https://doi.org/10.1016/j.febslet.2013.08.019.

Branchini, A., Ferrarese, M., Campioni, M., Castaman, G., Mari, R., Bernardi, F., et al. (2017). Specific factor IX mRNA and protein features favor drug-induced readthrough over recurrent nonsense mutations. *Blood, 129*(16), 2303–2307. https://doi.org/10.1182/blood-2016-09-738641.

Branchini, A., Ferrarese, M., Lombardi, S., Mari, R., Bernardi, F., & Pinotti, M. (2016). Differential functional readthrough over homozygous nonsense mutations contributes to the bleeding phenotype in coagulation factor VII deficiency. *Journal of Thrombosis and Haemostasis, 14*(10), 1994–2000. https://doi.org/10.1111/jth.13443.

Branchini, A., Rizzotto, L., Mariani, G., Napolitano, M., Lapecorella, M., Giansily-Blaizot, M., et al. (2012). Natural and engineered carboxy-terminal variants: Decreased secretion and gain-of-function result in asymptomatic coagulation factor VII deficiency. *Haematologica, 97*(5), 705–709. https://doi.org/10.3324/haematol.2011.049403.

Brown, A., Shao, S., Murray, J., Hegde, R. S., & Ramakrishnan, V. (2015). Structural basis for stop codon recognition in eukaryotes. *Nature, 524*(7566), 493–496. https://doi.org/10.1038/nature14896.

Bunge, S., Clements, P. R., Byers, S., Kleijer, W. J., Brooks, D. A., & Hopwood, J. J. (1998). Genotype-phenotype correlations in mucopolysaccharidosis type I using enzyme kinetics, immunoquantification and in vitro turnover studies. *Biochimica et Biophysica Acta, 1407*(3), 249–256. https://doi.org/10.1016/s0925-4439(98)00046-5.

Campbell, K. C. M., Meech, R. P., Klemens, J. J., Gerberi, M. T., Dyrstad, S. S. W., Larsen, D. L., et al. (2007). Prevention of noise- and drug-induced hearing loss with d-methionine. *Hearing Research, 226*(1–2), 92–103. https://doi.org/10.1016/j.heares.2006.11.012.

Cassan, M., & Rousset, J. P. (2001). UAG readthrough in mammalian cells: Effect of upstream and downstream stop codon contexts reveal different signals. *BMC Molecular Biology, 2*, 3. https://doi.org/10.1186/1471-2199-2-3.

Cavallari, N., Balestra, D., Branchini, A., Maestri, I., Chuamsunrit, A., Sasanakul, W., et al. (2012). Activation of a cryptic splice site in a potentially lethal coagulation defect accounts for a functional protein variant. *Biochimica et Biophysica Acta, 1822*(7), 1109–1113. https://doi.org/10.1016/j.bbadis.2012.03.001.

Chafa, O., Fischer, A.-M., Reghis, A., & Tapon-Bretaudiere, J. (2005). Homozygous nonsense mutation (Cys72- ->stop) in the human F7 gene: A not life-threatening mutation despite the absence of circulating factor VII. *Journal of Thrombosis and Haemostasis, 3*(1), 175–177. https://doi.org/10.1111/j.1538-7836.2004.01089.x.

Chen, C.-Y. A., & Shyu, A.-B. (2011). Mechanisms of deadenylation-dependent decay: Mechanisms of deadenylation-dependent decay. *Wiley Interdisciplinary Reviews: RNA*, *2*(2), 167–183. https://doi.org/10.1002/wrna.40.

Chiu, S.-Y., Serin, G., Ohara, O., & Maquat, L. E. (2003). Characterization of human Smg5/7a: A protein with similarities to Caenorhabditis elegans SMG5 and SMG7 that functions in the dephosphorylation of Upf1. *RNA (New York, N.Y.)*, *9*(1), 77–87. https://doi.org/10.1261/rna.2137903.

Chollet, M. E., Skarpen, E., Iversen, N., Sandset, P. M., & Skretting, G. (2015). The chemical chaperone sodium 4-phenylbutyrate improves the secretion of the protein CA267T mutant in CHO-K1 cells trough the GRASP55 pathway. *Cell & Bioscience*, *5*, 57. https://doi.org/10.1186/s13578-015-0048-4.

Clancy, J. P., Bebök, Z., Ruiz, F., King, C., Jones, J., Walker, L., et al. (2001). Evidence that systemic gentamicin suppresses premature stop mutations in patients with cystic fibrosis. *American Journal of Respiratory and Critical Care Medicine*, *163*(7), 1683–1692. https://doi.org/10.1164/ajrccm.163.7.2004001.

Cobucci-Ponzano, B., Rossi, M., & Moracci, M. (2012). Translational recoding in archaea. *Extremophiles*, *16*(6), 793–803. https://doi.org/10.1007/s00792-012-0482-8.

Cooper, D. N., Stenson, P. D., & Chuzhanova, N. A. (2006). The human gene mutation database (HGMD) and its exploitation in the study of mutational mechanisms. *Current Protocols in Bioinformatics*. Chapter 1, Unit 1.13. https://doi.org/10.1002/0471250953.bi0113s12.

Cosson, B., Couturier, A., Chabelskaya, S., Kiktev, D., Inge-Vechtomov, S., Philippe, M., et al. (2002). Poly(a)-binding protein acts in translation termination via eukaryotic release factor 3 interaction and does not influence [PSI+] propagation. *Molecular and Cellular Biology*, *22*(10), 3301–3315. https://doi.org/10.1128/MCB.22.10.3301-3315.2002.

Coutinho, M. F., Lacerda, L., & Alves, S. (2012). Glycosaminoglycan storage disorders: A review. *Biochemistry Research International*, *2012*, 471325. https://doi.org/10.1155/2012/471325.

Crawford, D. K., Alroy, I., Sharpe, N., Goddeeris, M. M., & Williams, G. (2020). ELX-02 generates protein via premature stop codon read-through without inducing native stop codon read-through proteins. *The Journal of Pharmacology and Experimental Therapeutics*, *374*(2), 264–272. https://doi.org/10.1124/jpet.120.265595.

Cridge, A. G., Crowe-McAuliffe, C., Mathew, S. F., & Tate, W. P. (2018). Eukaryotic translational termination efficiency is influenced by the 3′ nucleotides within the ribosomal mRNA channel. *Nucleic Acids Research*, *46*(4), 1927–1944. https://doi.org/10.1093/nar/gkx1315.

Demeshkina, N., Jenner, L., Westhof, E., Yusupov, M., & Yusupova, G. (2012). A new understanding of the decoding principle on the ribosome. *Nature*, *484*(7393), 256–259. https://doi.org/10.1038/nature10913.

Demeshkina, N., Jenner, L., Westhof, E., Yusupov, M., & Yusupova, G. (2013). New structural insights into the decoding mechanism: Translation infidelity via a G·U pair with Watson-crick geometry. *FEBS Letters*, *587*(13), 1848–1857. https://doi.org/10.1016/j.febslet.2013.05.009.

des Georges, A., Hashem, Y., Unbehaun, A., Grassucci, R. A., Taylor, D., Hellen, C. U. T., et al. (2014). Structure of the mammalian ribosomal pre-termination complex associated with eRF1.eRF3.GDPNP. *Nucleic Acids Research*, *42*(5), 3409–3418. https://doi.org/10.1093/nar/gkt1279.

Dever, T. E., & Green, R. (2012). The elongation, termination, and recycling phases of translation in eukaryotes. *Cold Spring Harbor Perspectives in Biology*, *4*(7), a013706. https://doi.org/10.1101/cshperspect.a013706.

Dewerchin, M., Liang, Z., Moons, L., Carmeliet, P., Castellino, F. J., Collen, D., et al. (2000). Blood coagulation factor X deficiency causes partial embryonic lethality and fatal neonatal bleeding in mice. *Thrombosis and Haemostasis*, *83*(2), 185–190.
Du, L., Damoiseaux, R., Nahas, S., Gao, K., Hu, H., Pollard, J. M., et al. (2009). Nonaminoglycoside compounds induce readthrough of nonsense mutations. *Journal of Experimental Medicine*, *206*(10), 2285–2297. https://doi.org/10.1084/jem.20081940.
Du, M., Jones, J. R., Lanier, J., Keeling, K. M., Lindsey, J. R., Tousson, A., et al. (2002). Aminoglycoside suppression of a premature stop mutation in a Cftr −/− mouse carrying a human CFTR-G542X transgene. *Journal of Molecular Medicine (Berlin, Germany)*, *80*(9), 595–604. https://doi.org/10.1007/s00109-002-0363-1.
Du, L., Jung, M. E., Damoiseaux, R., Completo, G., Fike, F., Ku, J.-M., et al. (2013). A new series of small molecular weight compounds induce read through of all three types of nonsense mutations in the ATM gene. *Molecular Therapy: The Journal of the American Society of Gene Therapy*, *21*(9), 1653–1660. https://doi.org/10.1038/mt.2013.150.
Du, M., Keeling, K. M., Fan, L., Liu, X., & Bedwell, D. M. (2009). Poly-L-aspartic acid enhances and prolongs gentamicin-mediated suppression of the CFTR-G542X mutation in a cystic fibrosis mouse model. *The Journal of Biological Chemistry*, *284*(11), 9.
Du, M., Liu, X., Welch, E. M., Hirawat, S., Peltz, S. W., & Bedwell, D. M. (2008). PTC124 is an orally bioavailable compound that promotes suppression of the human CFTR-G542X nonsense allele in a CF mouse model. *Proceedings of the National Academy of Sciences of the United States of America*, *105*(6), 2064–2069. https://doi.org/10.1073/pnas.0711795105.
Eigenbrot, C. (2002). Structure, function, and activation of coagulation factor VII. *Current Protein & Peptide Science*, *3*(3), 287–299. https://doi.org/10.2174/1389203023380675.
Eswarappa, S. M., Potdar, A. A., Koch, W. J., Fan, Y., Vasu, K., Lindner, D., et al. (2014). Programmed translational readthrough generates antiangiogenic VEGF-ax. *Cell*, *157*(7), 1605–1618. https://doi.org/10.1016/j.cell.2014.04.033.
Fairclough, R. J., Bareja, A., & Davies, K. E. (2011). Progress in therapy for Duchenne muscular dystrophy. *Experimental Physiology*, *96*(11), 1101–1113. https://doi.org/10.1113/expphysiol.2010.053025.
Fan-Minogue, H., & Bedwell, D. M. (2007). Eukaryotic ribosomal RNA determinants of aminoglycoside resistance and their role in translational fidelity. *RNA*, *14*(1), 148–157. https://doi.org/10.1261/rna.805208.
Fazzari, M., Frasca, A., Bifari, F., & Landsberger, N. (2019). Aminoglycoside drugs induce efficient read-through of CDKL5 nonsense mutations, slightly restoring its kinase activity. *RNA Biology*, *16*(10), 1414–1423. https://doi.org/10.1080/15476286.2019.1632633.
Ferrarese, M., Baroni, M., Della Valle, P., Spiga, I., Poloniato, A., D'Angelo, A., et al. (2019). Missense changes in the catalytic domain of coagulation factor X account for minimal function preventing a perinatal lethal condition. *Haemophilia: The Official Journal of the World Federation of Hemophilia*, *25*(4), 685–692. https://doi.org/10.1111/hae.13761.
Ferrarese, M., Testa, M. F., Balestra, D., Bernardi, F., Pinotti, M., & Branchini, A. (2018). Secretion of wild-type factor IX upon readthrough over F9 pre-peptide nonsense mutations causing hemophilia B. *Human Mutation*, *39*(5), 702–708. https://doi.org/10.1002/humu.23404.
Fischer, N., Neumann, P., Bock, L. V., Maracci, C., Wang, Z., Paleskava, A., et al. (2016). The pathway to GTPase activation of elongation factor SelB on the ribosome. *Nature*, *540*(7631), 80–85. https://doi.org/10.1038/nature20560.
Fleming, I., & Cavalcanti, A. R. O. (2019). Selection for tandem stop codons in ciliate species with reassigned stop codons. *PLoS One*, *14*(11), e0225804. https://doi.org/10.1371/journal.pone.0225804.

Garman, S. C., & Garboczi, D. N. (2004). The molecular defect leading to Fabry disease: Structure of human alpha-galactosidase. *Journal of Molecular Biology*, *337*(2), 319–335. https://doi.org/10.1016/j.jmb.2004.01.035.

Geggier, P., Dave, R., Feldman, M. B., Terry, D. S., Altman, R. B., Munro, J. B., et al. (2010). Conformational sampling of aminoacyl-tRNA during selection on the bacterial ribosome. *Journal of Molecular Biology*, *399*(4), 576–595. https://doi.org/10.1016/j.jmb.2010.04.038.

Germain, D. P. (2010). Fabry disease. *Orphanet Journal of Rare Diseases*, *5*, 30. https://doi.org/10.1186/1750-1172-5-30.

Giansily-Blaizot, M., Aguilar-Martinez, P., Briquel, M.-E., d'Oiron, R., De Maistre, E., Epelbaum, S., et al. (2003). Two novel cases of cerebral haemorrhages at the neonatal period associated with inherited factor VII deficiency, one of them revealing a new nonsense mutation (Ser52Stop). *Blood Coagulation & Fibrinolysis: An International Journal in Haemostasis and Thrombosis*, *14*(2), 217–220. https://doi.org/10.1097/00001721-200302000-00016.

Gonzalez-Hilarion, S., Beghyn, T., Jia, J., Debreuck, N., Berte, G., Mamchaoui, K., et al. (2012). Rescue of nonsense mutations by amlexanox in human cells. *Orphanet Journal of Rare Diseases*, 7(1), 58. https://doi.org/10.1186/1750-1172-7-58.

Gunn, G., Dai, Y., Du, M., Belakhov, V., Kandasamy, J., Schoeb, T. R., et al. (2014). Long-term nonsense suppression therapy moderates MPS I-H disease progression. *Molecular Genetics and Metabolism*, *111*(3), 374–381. https://doi.org/10.1016/j.ymgme.2013.12.007.

Guthrie, O. W. (2008). Aminoglycoside induced ototoxicity. *Toxicology*, *249*(2–3), 91–96. https://doi.org/10.1016/j.tox.2008.04.015.

Haas, M., Vlcek, V., Balabanov, P., Salmonson, T., Bakchine, S., Markey, G., et al. (2015). European medicines agency review of ataluren for the treatment of ambulant patients aged 5 years and older with Duchenne muscular dystrophy resulting from a nonsense mutation in the dystrophin gene. *Neuromuscular Disorders*, *25*(1), 5–13. https://doi.org/10.1016/j.nmd.2014.11.011.

Harrell, L. (2002). Predominance of six different hexanucleotide recoding signals 3′ of read-through stop codons. *Nucleic Acids Research*, *30*(9), 2011–2017. https://doi.org/10.1093/nar/30.9.2011.

Has, C., Sayar, S. B., Zheng, S., Chacón-Solano, E., Condrat, I., Yadav, A., et al. (2022). Read-through for nonsense mutations in type XVII collagen–deficient junctional epidermolysis bullosa. *The Journal of Investigative Dermatology*, *142*(4), 1227–1230. https://doi.org/10.1016/j.jid.2021.09.018.

Hatfield, D. L., Tsuji, P. A., Carlson, B. A., & Gladyshev, V. N. (2014). Selenium and selenocysteine: Roles in cancer, health, and development. *Trends in Biochemical Sciences*, *39*(3), 112–120. https://doi.org/10.1016/j.tibs.2013.12.007.

Heier, C. R., & DiDonato, C. J. (2009). Translational readthrough by the aminoglycoside geneticin (G418) modulates SMN stability in vitro and improves motor function in SMA mice in vivo. *Human Molecular Genetics*, *18*(7), 1310–1322. https://doi.org/10.1093/hmg/ddp030.

Hein, L. K., Bawden, M., Muller, V. J., Sillence, D., Hopwood, J. J., & Brooks, D. A. (2004). Alpha-L-iduronidase premature stop codons and potential read-through in mucopolysaccharidosis type I patients. *Journal of Molecular Biology*, *338*(3), 453–462. https://doi.org/10.1016/j.jmb.2004.03.012.

Hellen, C. U. T. (2018). Translation termination and ribosome recycling in eukaryotes. *Cold Spring Harbor Perspectives in Biology*, *10*(10). https://doi.org/10.1101/cshperspect.a032656.

Hirawat, S., Welch, E. M., Elfring, G. L., Northcutt, V. J., Paushkin, S., Hwang, S., et al. (2007). Safety, tolerability, and pharmacokinetics of PTC124, a nonaminoglycoside nonsense mutation suppressor, following single- and multiple-dose administration to healthy male and female adult volunteers. *Journal of Clinical Pharmacology*, *47*(4), 430–444. https://doi.org/10.1177/0091270006297140.

Hirsh, D. (1971). Tryptophan transfer RNA as the UGA suppressor. *Journal of Molecular Biology*, *58*(2), 439–458. https://doi.org/10.1016/0022-2836(71)90362-7.

Hoffman, E. P., Brown, R. H., & Kunkel, L. M. (1987). Dystrophin: The protein product of the Duchenne muscular dystrophy locus. *Cell*, *51*(6), 919–928. https://doi.org/10.1016/0092-8674(87)90579-4.

Hopfield, J. J. (1974). Kinetic proofreading: A new mechanism for reducing errors in biosynthetic processes requiring high specificity. *Proceedings of the National Academy of Sciences of the United States of America*, *71*(10), 4135–4139. https://doi.org/10.1073/pnas.71.10.4135.

Howard, M., Frizzell, R. A., & Bedwell, D. M. (1996). Aminoglycoside antibiotics restore CFTR function by overcoming premature stop mutations. *Nature Medicine*, *2*(4), 467–469.

Huang, S., Bhattacharya, A., Ghelfi, M. D., Li, H., Fritsch, C., Chenoweth, D. M., et al. (2022). Ataluren binds to multiple protein synthesis apparatus sites and competitively inhibits release factor-dependent termination. *Nature Communications*, *13*(1), 2413. https://doi.org/10.1038/s41467-022-30080-6.

Hwang, J., Sato, H., Tang, Y., Matsuda, D., & Maquat, L. E. (2010). UPF1 association with the cap-binding protein, CBP80, promotes nonsense-mediated mRNA decay at two distinct steps. *Molecular Cell*, *39*(3), 396–409. https://doi.org/10.1016/j.molcel.2010.07.004.

Isken, O., & Maquat, L. E. (2007). Quality control of eukaryotic mRNA: Safeguarding cells from abnormal mRNA function. *Genes & Development*, *21*(15), 1833–1856. https://doi.org/10.1101/gad.1566807.

Jackson, R. J., Hellen, C. U. T., & Pestova, T. V. (2012). Termination and post-termination events in eukaryotic translation. *Advances in Protein Chemistry and Structural Biology*, *86*, 45–93. https://doi.org/10.1016/B978-0-12-386497-0.00002-5.

Jacobs, G. H., Rackham, O., Stockwell, P. A., Tate, W., & Brown, C. M. (2002). Transterm: A database of mRNAs and translational control elements. *Nucleic Acids Research*, *30*(1), 310–311. https://doi.org/10.1093/nar/30.1.310.

James, P. D., Raut, S., Rivard, G. E., Poon, M.-C., Warner, M., McKenna, S., et al. (2005). Aminoglycoside suppression of nonsense mutations in severe hemophilia. *Blood*, *106*(9), 3043–3048. https://doi.org/10.1182/blood-2005-03-1307.

Jayandharan, G. R., Shaji, R. V., Baidya, S., Nair, S. C., Chandy, M., & Srivastava, A. (2005). Molecular characterization of factor IX gene mutations in 53 patients with haemophilia B in India. *Thrombosis and Haemostasis*, *94*(4), 883–886.

Kandasamy, J., Atia-Glikin, D., Shulman, E., Shapira, K., Shavit, M., Belakhov, V., et al. (2012). Increased selectivity toward cytoplasmic versus mitochondrial ribosome confers improved efficiency of synthetic aminoglycosides in fixing damaged genes: A strategy for treatment of genetic diseases caused by nonsense mutations. *Journal of Medicinal Chemistry*, *55*(23), 10630–10643. https://doi.org/10.1021/jm3012992.

Karki, P., Carney, T. D., Maracci, C., Yatsenko, A. S., Shcherbata, H. R., & Rodnina, M. V. (2021). Tissue-specific regulation of translational readthrough tunes functions of the traffic jam transcription factor. *Nucleic Acids Research*. https://doi.org/10.1093/nar/gkab1189.

Kashima, I., Yamashita, A., Izumi, N., Kataoka, N., Morishita, R., Hoshino, S., et al. (2006). Binding of a novel SMG-1-Upf1-eRF1-eRF3 complex (SURF) to the exon junction complex triggers Upf1 phosphorylation and nonsense-mediated mRNA decay. *Genes & Development*, *20*(3), 355–367. https://doi.org/10.1101/gad.1389006.

Katsumi, A., Kojima, T., Senda, T., Yamazaki, T., Tsukamoto, H., Sugiura, I., et al. (1998). The carboxyl-terminal region of protein C is essential for its secretion. *Blood*, *91*(10), 3784–3791.

Kaufman, R. J. (1998). Post-translational modifications required for coagulation factor secretion and function. *Thrombosis and Haemostasis*, *79*(6), 1068–1079.

Keeling, K. M., & Bedwell, D. M. (2002). Clinically relevant aminoglycosides can suppress disease-associated premature stop mutations in the IDUA and P53 cDNAs in a mammalian translation system. *Journal of Molecular Medicine (Berlin, Germany)*, *80*(6), 367–376. https://doi.org/10.1007/s00109-001-0317-z.

Keeling, K. M., Brooks, D. A., Hopwood, J. J., Li, P., Thompson, J. N., & Bedwell, D. M. (2001). Gentamicin-mediated suppression of hurler syndrome stop mutations restores a low level of alpha-L-iduronidase activity and reduces lysosomal glycosaminoglycan accumulation. *Human Molecular Genetics*, *10*(3), 291–299. https://doi.org/10.1093/hmg/10.3.291.

Keeling, K. M., Wang, D., Conard, S. E., & Bedwell, D. M. (2012). Suppression of premature termination codons as a therapeutic approach. *Critical Reviews in Biochemistry and Molecular Biology*, *47*(5), 444–463. https://doi.org/10.3109/10409238.2012.694846.

Keeling, K. M., Wang, D., Dai, Y., Murugesan, S., Chenna, B., Clark, J., et al. (2013). Attenuation of nonsense-mediated mRNA decay enhances in vivo nonsense suppression. *PLoS One*, *8*(4), e60478. https://doi.org/10.1371/journal.pone.0060478.

Keeling, K. M., Xue, X., Gunn, G., & Bedwell, D. M. (2014). Therapeutics based on stop codon readthrough. *Annual Review of Genomics and Human Genetics*, *15*, 371–394. https://doi.org/10.1146/annurev-genom-091212-153527.

Kerem, E. (2020). ELX-02: An investigational read-through agent for the treatment of nonsense mutation-related genetic disease. *Expert Opinion on Investigational Drugs*, *29*(12), 1347–1354. https://doi.org/10.1080/13543784.2020.1828862.

Kerem, E., Hirawat, S., Armoni, S., Yaakov, Y., Shoseyov, D., Cohen, M., et al. (2008). Effectiveness of PTC124 treatment of cystic fibrosis caused by nonsense mutations: A prospective phase II trial. *Lancet (London, England)*, *372*(9640), 719–727. https://doi.org/10.1016/S0140-6736(08)61168-X.

Kerem, E., Konstan, M. W., De Boeck, K., Accurso, F. J., Sermet-Gaudelus, I., Wilschanski, M., et al. (2014). Ataluren for the treatment of nonsense-mutation cystic fibrosis: A randomised, double-blind, placebo-controlled phase 3 trial. *The Lancet Respiratory Medicine*, *2*(7), 539–547. https://doi.org/10.1016/S2213-2600(14)70100-6.

Khajavi, M., Inoue, K., & Lupski, J. R. (2006). Nonsense-mediated mRNA decay modulates clinical outcome of genetic disease. *European Journal of Human Genetics: EJHG*, *14*(10), 1074–1081. https://doi.org/10.1038/sj.ejhg.5201649.

Kim, V. N., Kataoka, N., & Dreyfuss, G. (2001). Role of the nonsense-mediated decay factor hUpf3 in the splicing-dependent exon-exon junction complex. *Science (New York, N.Y.)*, *293*(5536), 1832–1836. https://doi.org/10.1126/science.1062829.

Klauer, A. A., & van Hoof, A. (2012). Degradation of mRNAs that lack a stop codon: A decade of nonstop progress: Degradation of mRNAs that lack a stop codon. *Wiley Interdisciplinary Reviews: RNA*, *3*(5), 649–660. https://doi.org/10.1002/wrna.1124.

Kryukov, G. V., Castellano, S., Novoselov, S. V., Lobanov, A. V., Zehtab, O., Guigó, R., et al. (2003). Characterization of mammalian selenoproteomes. *Science (New York, N.Y.)*, *300*(5624), 1439–1443. https://doi.org/10.1126/science.1083516.

Kurachi, S., Pantazatos, D. P., & Kurachi, K. (1997). The carboxyl-terminal region of factor IX is essential for its secretion. *Biochemistry*, *36*(14), 4337–4344. https://doi.org/10.1021/bi962002v.

Kurosaki, T., & Maquat, L. E. (2016). Nonsense-mediated mRNA decay in humans at a glance. *Journal of Cell Science*, *129*(3), 461–467. https://doi.org/10.1242/jcs.181008.

Kuzmiak, H. A., & Maquat, L. E. (2006). Applying nonsense-mediated mRNA decay research to the clinic: Progress and challenges. *Trends in Molecular Medicine*, *12*(7), 306–316. https://doi.org/10.1016/j.molmed.2006.05.005.

Labunskyy, V. M., Hatfield, D. L., & Gladyshev, V. N. (2014). Selenoproteins: Molecular pathways and physiological roles. *Physiological Reviews*, *94*(3), 739–777. https://doi.org/10.1152/physrev.00039.2013.

Le Hir, H., Izaurralde, E., Maquat, L. E., & Moore, M. J. (2000). The spliceosome deposits multiple proteins 20-24 nucleotides upstream of mRNA exon-exon junctions. *The EMBO Journal*, *19*(24), 6860–6869. https://doi.org/10.1093/emboj/19.24.6860.

Lee, H.-L. R., & Dougherty, J. P. (2012). Pharmaceutical therapies to recode nonsense mutations in inherited diseases. *Pharmacology & Therapeutics*, *136*(2), 227–266. https://doi.org/10.1016/j.pharmthera.2012.07.007.

Lejeune, F., Ishigaki, Y., Li, X., & Maquat, L. E. (2002). The exon junction complex is detected on CBP80-bound but not eIF4E-bound mRNA in mammalian cells: Dynamics of mRNP remodeling. *The EMBO Journal*, *21*(13), 3536–3545. https://doi.org/10.1093/emboj/cdf345.

Lejeune, F., Li, X., & Maquat, L. E. (2003). Nonsense-mediated mRNA decay in mammalian cells involves decapping, deadenylating, and exonucleolytic activities. *Molecular Cell*, *12*(3), 675–687. https://doi.org/10.1016/s1097-2765(03)00349-6.

Leubitz, A., Frydman-Marom, A., Sharpe, N., van Duzer, J., Campbell, K. C. M., & Vanhoutte, F. (2019). Safety, tolerability, and pharmacokinetics of single ascending doses of ELX-02, a potential treatment for genetic disorders caused by nonsense mutations, in healthy volunteers. *Clinical Pharmacology in Drug Development*, *8*(8), 984–994. https://doi.org/10.1002/cpdd.647.

Li, T., Miller, C. H., Payne, A. B., & Craig Hooper, W. (2013). The CDC hemophilia B mutation project mutation list: A new online resource. *Molecular Genetics & Genomic Medicine*, *1*(4), 238–245. https://doi.org/10.1002/mgg3.30.

Liang, H., Cavalcanti, A. R., & Landweber, L. F. (2005). Conservation of tandem stop codons in yeasts. *Genome Biology*, *8*.

Liddell, M. B., Lillicrap, D. P., Peake, I. R., & Bloom, A. L. (1989). Defective propeptide processing and abnormal activation underlie the molecular pathology of factor IX Troed-y-Rhiw. *British Journal of Haematology*, *72*(2), 208–215. https://doi.org/10.1111/j.1365-2141.1989.tb07684.x.

Linde, L., & Kerem, B. (2008). Introducing sense into nonsense in treatments of human genetic diseases. *Trends in Genetics: TIG*, *24*(11), 552–563. https://doi.org/10.1016/j.tig.2008.08.010.

Liu, X., Zhang, Y., Zhang, B., Gao, H., & Qiu, C. (2020). Nonsense suppression induced readthrough of a novel PAX6 mutation in patient-derived cells of congenital aniridia. *Molecular Genetics & Genomic Medicine*, *8*(5), e1198. https://doi.org/10.1002/mgg3.1198.

Liu, Z., Zhang, Y., Zhu, M., & Zhang, B. (2019). Identification of candidate nonsense mutations of FVIII for ribosomal readthrough therapy. *Haematologica*, *104*(12), e573–e576. https://doi.org/10.3324/haematol.2018.205104.

Lombardi, S., Ferrarese, M., Marchi, S., Pinton, P., Pinotti, M., Bernardi, F., et al. (2020). Translational readthrough of GLA nonsense mutations suggests dominant-negative effects exerted by the interaction of wild-type and missense variants. *RNA Biology*, *17*(2), 254–263. https://doi.org/10.1080/15476286.2019.1676115.

Lombardi, S., Leo, G., Merlin, S., Follenzi, A., McVey, J. H., Maestri, I., et al. (2021). Dissection of pleiotropic effects of variants in and adjacent to F8 exon 19 and rescue of mRNA splicing and protein function. *American Journal of Human Genetics*, *108*(8), 1512–1525. https://doi.org/10.1016/j.ajhg.2021.06.012.

Lombardi, S., Testa, M. F., Pinotti, M., & Branchini, A. (2020). Molecular insights into determinants of translational readthrough and implications for nonsense suppression approaches. *International Journal of Molecular Sciences*, *21*(24), E9449. https://doi.org/10.3390/ijms21249449.

Lopez-Novoa, J. M., Quiros, Y., Vicente, L., Morales, A. I., & Lopez-Hernandez, F. J. (2011). New insights into the mechanism of aminoglycoside nephrotoxicity: An integrative point of view. *Kidney International*, *79*(1), 33–45. https://doi.org/10.1038/ki.2010.337.

Loughran, G., Chou, M.-Y., Ivanov, I. P., Jungreis, I., Kellis, M., Kiran, A. M., et al. (2014). Evidence of efficient stop codon readthrough in four mammalian genes. *Nucleic Acids Research*, *42*(14), 8928–8938. https://doi.org/10.1093/nar/gku608.

Loughran, G., Jungreis, I., Tzani, I., Power, M., Dmitriev, R. I., Ivanov, I. P., et al. (2018). Stop codon readthrough generates a C-terminally extended variant of the human vitamin D receptor with reduced calcitriol response. *Journal of Biological Chemistry*, *293*(12), 4434–4444. https://doi.org/10.1074/jbc.M117.818526.

Lueck, J. D., Yoon, J. S., Perales-Puchalt, A., Mackey, A. L., Infield, D. T., Behlke, M. A., et al. (2019). Engineered transfer RNAs for suppression of premature termination codons. *Nature Communications*, *10*(1), 822. https://doi.org/10.1038/s41467-019-08329-4.

Lykke-Andersen, J., Shu, M. D., & Steitz, J. A. (2001). Communication of the position of exon-exon junctions to the mRNA surveillance machinery by the protein RNPS1. *Science (New York, N.Y.)*, *293*(5536), 1836–1839. https://doi.org/10.1126/science.1062786.

Malhotra, J. D., Miao, H., Zhang, K., Wolfson, A., Pennathur, S., Pipe, S. W., et al. (2008). Antioxidants reduce endoplasmic reticulum stress and improve protein secretion. *Proceedings of the National Academy of Sciences of the United States of America*, *105*(47), 18525–18530. https://doi.org/10.1073/pnas.0809677105.

Manuvakhova, M., Keeling, K., & Bedwell, D. M. (2000). Aminoglycoside antibiotics mediate context-dependent suppression of termination codons in a mammalian translation system. *RNA*, *6*(7), 1044–1055. https://doi.org/10.1017/S1355838200000716.

Mariani, G., Herrmann, F. H., Dolce, A., Batorova, A., Etro, D., Peyvandi, F., et al. (2005). Clinical phenotypes and factor VII genotype in congenital factor VII deficiency. *Thrombosis and Haemostasis*, *93*(3), 481–487. https://doi.org/10.1160/TH04-10-0650.

Martínez-Santamaría, L., Maseda, R., de Arriba, M. D. C., Membrilla, J. A., Sigüenza, A. I., Mascías, J., et al. (2022). Evaluation of systemic gentamicin as translational readthrough therapy for a patient with epidermolysis bullosa simplex with muscular dystrophy owing to PLEC1 pathogenic nonsense variants. *JAMA Dermatology*, *158*(4), 439–443. https://doi.org/10.1001/jamadermatol.2022.0112.

Martorell, L., Cortina, V., Parra, R., Barquinero, J., & Vidal, F. (2020). Variable readthrough responsiveness of nonsense mutations in hemophilia a. *Haematologica*, *105*(2), 508–518. https://doi.org/10.3324/haematol.2018.212118.

Matalonga, L., Arias, Á., Tort, F., Ferrer-Cortés, X., Garcia-Villoria, J., Coll, M. J., et al. (2015). Effect of readthrough treatment in fibroblasts of patients affected by lysosomal diseases caused by premature termination codons. *Neurotherapeutics: The Journal of the American Society for Experimental NeuroTherapeutics*, *12*(4), 874–886. https://doi.org/10.1007/s13311-015-0368-4.

Matheisl, S., Berninghausen, O., Becker, T., & Beckmann, R. (2015). Structure of a human translation termination complex. *Nucleic Acids Research*, *43*(18), 8615–8626. https://doi.org/10.1093/nar/gkv909.

Mattis, V. B., Rai, R., Wang, J., Chang, C.-W. T., Coady, T., & Lorson, C. L. (2006). Novel aminoglycosides increase SMN levels in spinal muscular atrophy fibroblasts. *Human Genetics*, *120*(4), 589–601. https://doi.org/10.1007/s00439-006-0245-7.

Mattis, V. B., Tom Chang, C.-W., & Lorson, C. L. (2012). Analysis of a read-through promoting compound in a severe mouse model of spinal muscular atrophy. *Neuroscience Letters*, *525*(1), 72–75. https://doi.org/10.1016/j.neulet.2012.07.024.

McCaughan, K. K., Brown, C. M., Dalphin, M. E., Berry, M. J., & Tate, W. P. (1995). Translational termination efficiency in mammals is influenced by the base following the stop codon. *Proceedings of the National Academy of Sciences*, *92*(12), 5431–5435. https://doi.org/10.1073/pnas.92.12.5431.

McDonald, C. M., Campbell, C., Torricelli, R. E., Finkel, R. S., Flanigan, K. M., Goemans, N., et al. (2017). Ataluren in patients with nonsense mutation Duchenne muscular dystrophy (ACT DMD): A multicentre, randomised, double-blind, placebo-controlled, phase 3 trial. *The Lancet*, *390*(10101), 1489–1498. https://doi.org/10.1016/S0140-6736(17)31611-2.

McVey, J. H., Boswell, E., Mumford, A. D., Kemball-Cook, G., & Tuddenham, E. G. (2001). Factor VII deficiency and the FVII mutation database. *Human Mutation*, *17*(1), 3–17. https://doi.org/10.1002/1098-1004(2001)17:1<3::AID-HUMU2>3.0.CO;2-V.

McVey, J. H., Rallapalli, P. M., Kemball-Cook, G., Hampshire, D. J., Giansily-Blaizot, M., Gomez, K., et al. (2020). The European Association for Haemophilia and Allied Disorders (EAHAD) coagulation factor variant databases: Important resources for haemostasis clinicians and researchers. *Haemophilia: The Official Journal of the World Federation of Hemophilia*, *26*(2), 306–313. https://doi.org/10.1111/hae.13947.

Meikle, P. J., Hopwood, J. J., Clague, A. E., & Carey, W. F. (1999). Prevalence of lysosomal storage disorders. *JAMA*, *281*(3), 249–254. https://doi.org/10.1001/jama.281.3.249.

Millar, D. S., Kemball-Cook, G., McVey, J. H., Tuddenham, E. G., Mumford, A. D., Attock, G. B., et al. (2000). Molecular analysis of the genotype-phenotype relationship in factor VII deficiency. *Human Genetics*, *107*(4), 327–342. https://doi.org/10.1007/s004390000373.

Monroe, D. M., & Hoffman, M. (2006). What does it take to make the perfect clot? *Arteriosclerosis, Thrombosis, and Vascular Biology*, *26*(1), 41–48. https://doi.org/10.1161/01.ATV.0000193624.28251.83.

Montejo, J. M., Magallón, M., Tizzano, E., & Solera, J. (1999). Identification of twenty-one new mutations in the factor IX gene by SSCP analysis. *Human Mutation*, *13*(2), 160–165. https://doi.org/10.1002/(SICI)1098-1004(1999)13:2<160::AID-HUMU9>3.0.CO;2-C.

Moosajee, M., Tracey-White, D., Smart, M., Weetall, M., Torriano, S., Kalatzis, V., et al. (2016). Functional rescue of REP1 following treatment with PTC124 and novel derivative PTC-414 in human choroideremia fibroblasts and the nonsense-mediated zebrafish model. *Human Molecular Genetics*, *25*(16), 3416–3431. https://doi.org/10.1093/hmg/ddw184.

Mort, M., Ivanov, D., Cooper, D. N., & Chuzhanova, N. A. (2008). A meta-analysis of nonsense mutations causing human genetic disease. *Human Mutation*, *29*(8), 1037–1047. https://doi.org/10.1002/humu.20763.

Namy, O., Rousset, J.-P., Napthine, S., & Brierley, I. (2004). Reprogrammed genetic decoding in cellular gene expression. *Molecular Cell*, *13*(2), 157–168. https://doi.org/10.1016/s1097-2765(04)00031-0.

Ng, M. Y., Li, H., Ghelfi, M. D., Goldman, Y. E., & Cooperman, B. S. (2021). Ataluren and aminoglycosides stimulate read-through of nonsense codons by orthogonal mechanisms. *Proceedings of the National Academy of Sciences of the United States of America*, *118*(2), e2020599118. https://doi.org/10.1073/pnas.2020599118.

Nowak, K. J., & Davies, K. E. (2004). Duchenne muscular dystrophy and dystrophin: Pathogenesis and opportunities for treatment. *EMBO Reports*, *5*(9), 872–876. https://doi.org/10.1038/sj.embor.7400221.

Nozawa, K., O'Donoghue, P., Gundllapalli, S., Araiso, Y., Ishitani, R., Umehara, T., et al. (2009). Pyrrolysyl-tRNA synthetase-tRNA(Pyl) structure reveals the molecular basis of orthogonality. *Nature*, *457*(7233). https://doi.org/10.1038/nature07611.

Nudelman, I., Glikin, D., Smolkin, B., Hainrichson, M., Belakhov, V., & Baasov, T. (2010). Repairing faulty genes by aminoglycosides: Development of new derivatives of geneticin (G418) with enhanced suppression of diseases-causing nonsense mutations. *Bioorganic & Medicinal Chemistry*, *18*(11), 3735–3746. https://doi.org/10.1016/j.bmc.2010.03.060.

Nudelman, I., Rebibo-Sabbah, A., Cherniavsky, M., Belakhov, V., Hainrichson, M., Chen, F., et al. (2009). Development of novel aminoglycoside (NB54) with reduced toxicity and enhanced suppression of disease-causing premature stop mutations. *Journal of Medicinal Chemistry*, *52*(9), 2836–2845. https://doi.org/10.1021/jm801640k.

Ogle, J. M. (2001). Recognition of cognate transfer RNA by the 30S ribosomal subunit. *Science*, *292*(5518), 897–902. https://doi.org/10.1126/science.1060612.

Ogle, J. M., Brodersen, D. E., Clemons, W. M., Tarry, M. J., Carter, A. P., & Ramakrishnan, V. (2001). Recognition of cognate transfer RNA by the 30S ribosomal subunit. *Science (New York, N.Y.)*, *292*(5518), 897–902. https://doi.org/10.1126/science.1060612.

Ogle, J. M., Murphy, F. V., Tarry, M. J., & Ramakrishnan, V. (2002). Selection of tRNA by the ribosome requires a transition from an open to a closed form. *Cell*, *111*(5), 721–732. https://doi.org/10.1016/s0092-8674(02)01086-3.

Okada-Katsuhata, Y., Yamashita, A., Kutsuzawa, K., Izumi, N., Hirahara, F., & Ohno, S. (2012). N- and C-terminal Upf1 phosphorylations create binding platforms for SMG-6 and SMG-5:SMG-7 during NMD. *Nucleic Acids Research*, *40*(3), 1251–1266. https://doi.org/10.1093/nar/gkr791.

Otani, Y., Taguchi, A., Hamada, K., Hayashi, Y., Yamaguchi, Y., & Baba, H. (2022). Influence of novel readthrough agents on myelin protein zero translation in the peripheral nervous system. *Neuropharmacology*, *211*, 109059. https://doi.org/10.1016/j.neuropharm.2022.109059.

Oussoren, E., Keulemans, J., van Diggelen, O. P., Oemardien, L. F., Timmermans, R. G., van der Ploeg, A. T., et al. (2013). Residual α-L-iduronidase activity in fibroblasts of mild to severe mucopolysaccharidosis type I patients. *Molecular Genetics and Metabolism*, *109*(4), 377–381. https://doi.org/10.1016/j.ymgme.2013.05.016.

Pape, T., Wintermeyer, W., & Rodnina, M. (1999). Induced fit in initial selection and proofreading of aminoacyl-tRNA on the ribosome. *The EMBO Journal*, *18*(13), 3800–3807. https://doi.org/10.1093/emboj/18.13.3800.

Parenti, G., Andria, G., & Ballabio, A. (2015). Lysosomal storage diseases: From pathophysiology to therapy. *Annual Review of Medicine*, *66*, 471–486. https://doi.org/10.1146/annurev-med-122313-085916.

Payne, A. B., Miller, C. H., Kelly, F. M., Michael Soucie, J., & Craig Hooper, W. (2013). The CDC hemophilia a mutation project (CHAMP) mutation list: A new online resource. *Human Mutation*, *34*(2), E2382–E2391. https://doi.org/10.1002/humu.22247.

Pignani, S., Todaro, A., Ferrarese, M., Marchi, S., Lombardi, S., Balestra, D., et al. (2018). The chaperone-like sodium phenylbutyrate improves factor IX intracellular trafficking and activity impaired by the frequent p.R294Q mutation. *Journal of Thrombosis and Haemostasis*, *16*(10), 2035–2043. https://doi.org/10.1111/jth.14236.

Pignani, S., Zappaterra, F., Barbon, E., Follenzi, A., Bovolenta, M., Bernardi, F., et al. (2019). Tailoring the CRISPR system to transactivate coagulation gene promoters in normal and mutated contexts. *Biochimica Et Biophysica Acta. Gene Regulatory Mechanisms*, *1862*(6), 619–624. https://doi.org/10.1016/j.bbagrm.2019.04.002.

Pinotti, M., Rizzotto, L., Chuansumrit, A., Mariani, G., Bernardi, F., & International Factor VII Deficiency Study Group. (2006). Gentamicin induces sub-therapeutic levels of coagulation factor VII in patients with nonsense mutations. *Journal of Thrombosis and Haemostasis*, *4*(8), 1828–1830. https://doi.org/10.1111/j.1538-7836.2006.02057.x.

Polikanov, Y. S., Starosta, A. L., Juette, M. F., Altman, R. B., Terry, D. S., Lu, W., et al. (2015). Distinct tRNA accommodation intermediates observed on the ribosome with the antibiotics hygromycin a and A201A. *Molecular Cell*, *58*(5), 832–844. https://doi.org/10.1016/j.molcel.2015.04.014.

Politano, L., Nigro, G., Nigro, V., Piluso, G., Papparella, S., Paciello, O., et al. (2003). Gentamicin administration in Duchenne patients with premature stop codon. Preliminary results. *Acta Myologica: Myopathies and Cardiomyopathies: Official Journal of the Mediterranean Society of Myology*, *22*(1), 15–21.

Popescu, A. C., Sidorova, E., Zhang, G., & Eubanks, J. H. (2010). Aminoglycoside-mediated partial suppression of MECP2 nonsense mutations responsible for Rett syndrome in vitro. *Journal of Neuroscience Research*, *88*(11), 2316–2324. https://doi.org/10.1002/jnr.22409.

Popp, M. W.-L., & Maquat, L. E. (2013). Organizing principles of mammalian nonsense-mediated mRNA decay. *Annual Review of Genetics*, *47*, 139–165. https://doi.org/10.1146/annurev-genet-111212-133424.

Porter, J. J., Heil, C. S., & Lueck, J. D. (2021). Therapeutic promise of engineered nonsense suppressor tRNAs. *Wiley Interdisciplinary Reviews. RNA*, *12*(4), e1641. https://doi.org/10.1002/wrna.1641.

Pranke, I., Bidou, L., Martin, N., Blanchet, S., Hatton, A., Karri, S., et al. (2018). Factors influencing readthrough therapy for frequent cystic fibrosis premature termination codons. *ERJ Open Research*, *4*(1). https://doi.org/10.1183/23120541.00080-2017.

Prokhorova, I., Altman, R. B., Djumagulov, M., Shrestha, J. P., Urzhumtsev, A., Ferguson, A., et al. (2017). Aminoglycoside interactions and impacts on the eukaryotic ribosome. *Proceedings of the National Academy of Sciences*, *114*(51), E10899–E10908. https://doi.org/10.1073/pnas.1715501114.

Rallapalli, P. M., Kemball-Cook, G., Tuddenham, E. G., Gomez, K., & Perkins, S. J. (2013). An interactive mutation database for human coagulation factor IX provides novel insights into the phenotypes and genetics of hemophilia B. *Journal of Thrombosis and Haemostasis*, *11*(7), 1329–1340. https://doi.org/10.1111/jth.12276.

Rebibo-Sabbah, A., Nudelman, I., Ahmed, Z. M., Baasov, T., & Ben-Yosef, T. (2007). In vitro and ex vivo suppression by aminoglycosides of PCDH15 nonsense mutations underlying type 1 usher syndrome. *Human Genetics*, *122*(3–4), 373–381. https://doi.org/10.1007/s00439-007-0410-7.

Reeves, M. A., & Hoffmann, P. R. (2009). The human selenoproteome: Recent insights into functions and regulation. *Cellular and Molecular Life Sciences: CMLS*, *66*(15), 2457–2478. https://doi.org/10.1007/s00018-009-0032-4.

Richardson, R., Smart, M., Tracey-White, D., Webster, A. R., & Moosajee, M. (2017). Mechanism and evidence of nonsense suppression therapy for genetic eye disorders. *Experimental Eye Research*, *155*, 24–37. https://doi.org/10.1016/j.exer.2017.01.001.

Rodnina, M. V., Korniy, N., Klimova, M., Karki, P., Peng, B.-Z., Senyushkina, T., et al. (2019). Translational recoding: Canonical translation mechanisms reinterpreted. *Nucleic Acids Research*. https://doi.org/10.1093/nar/gkz783.

Rodnina, M. V., & Wintermeyer, W. (2001). Ribosome fidelity: TRNA discrimination, proofreading and induced fit. *Trends in Biochemical Sciences*, *26*(2), 124–130. https://doi.org/10.1016/S0968-0004(00)01737-0.

Rosen, E. D., Chan, J. C., Idusogie, E., Clotman, F., Vlasuk, G., Luther, T., et al. (1997). Mice lacking factor VII develop normally but suffer fatal perinatal bleeding. *Nature*, *390*(6657), 290–294. https://doi.org/10.1038/36862.

Roth, S. D., Schüttrumpf, J., Milanov, P., Abriss, D., Ungerer, C., Quade-Lyssy, P., et al. (2012). Chemical chaperones improve protein secretion and rescue mutant factor VIII in mice with hemophilia a. *PLoS One*, 7(9), e44505. https://doi.org/10.1371/journal.pone.0044505.

Roy, B., Leszyk, J. D., Mangus, D. A., & Jacobson, A. (2015). Nonsense suppression by near-cognate tRNAs employs alternative base pairing at codon positions 1 and 3. *Proceedings of the National Academy of Sciences*, *112*(10), 3038–3043. https://doi.org/10.1073/pnas.1424127112.

Rozov, A., Demeshkina, N., Khusainov, I., Westhof, E., Yusupov, M., & Yusupova, G. (2016). Novel base-pairing interactions at the tRNA wobble position crucial for accurate reading of the genetic code. *Nature Communications*, 7, 10457. https://doi.org/10.1038/ncomms10457.

Rozov, A., Westhof, E., Yusupov, M., & Yusupova, G. (2016). The ribosome prohibits the G•U wobble geometry at the first position of the codon-anticodon helix. *Nucleic Acids Research*, *44*(13), 6434–6441. https://doi.org/10.1093/nar/gkw431.

Ruusala, T., Ehrenberg, M., & Kurland, C. G. (1982). Is there proofreading during polypeptide synthesis? *The EMBO Journal*, *1*(6), 741–745.

Salas-Marco, J., & Bedwell, D. M. (2004). GTP hydrolysis by eRF3 facilitates stop codon decoding during eukaryotic translation termination. *Molecular and Cellular Biology*, *24*(17), 7769–7778. https://doi.org/10.1128/MCB.24.17.7769-7778.2004.

Samelson-Jones, B. J., Finn, J. D., Raffini, L. J., Merricks, E. P., Camire, R. M., Nichols, T. C., et al. (2021). Evolutionary insights into coagulation factor IX Padua and other high-specific-activity variants. *Blood Advances*, *5*(5), 1324–1332. https://doi.org/10.1182/bloodadvances.2019000405.

Sánchez-Alcudia, R., Pérez, B., Ugarte, M., & Desviat, L. R. (2012). Feasibility of nonsense mutation readthrough as a novel therapeutical approach in propionic acidemia. *Human Mutation*, *33*(6), 973–980. https://doi.org/10.1002/humu.22047.

Schiffelers, R. (2001). Liposome-encapsulated aminoglycosides in pre-clinical and clinical studies. *Journal of Antimicrobial Chemotherapy*, *48*(3), 333–344. https://doi.org/10.1093/jac/48.3.333.

Schilff, M., Sargsyan, Y., Hofhuis, J., & Thoms, S. (2021). Stop codon context-specific induction of translational readthrough. *Biomolecules*, *11*(7), 1006. https://doi.org/10.3390/biom11071006.

Schuller, A. P., & Green, R. (2018). Roadblocks and resolutions in eukaryotic translation. *Nature Reviews. Molecular Cell Biology*, *19*(8), 526–541. https://doi.org/10.1038/s41580-018-0011-4.

Schwarz, N., Carr, A.-J., Lane, A., Moeller, F., Chen, L. L., Aguilà, M., et al. (2015). Translational read-through of the RP2 Arg120stop mutation in patient iPSC-derived retinal pigment epithelium cells. *Human Molecular Genetics*, *24*(4), 972–986. https://doi.org/10.1093/hmg/ddu509.

Scott, H. S., Bunge, S., Gal, A., Clarke, L. A., Morris, C. P., & Hopwood, J. J. (1995). Molecular genetics of mucopolysaccharidosis type I: Diagnostic, clinical, and biological implications. *Human Mutation*, *6*(4), 288–302. https://doi.org/10.1002/humu.1380060403.

Sermet-Gaudelus, I., Boeck, K. D., Casimir, G. J., Vermeulen, F., Leal, T., Mogenet, A., et al. (2010). Ataluren (PTC124) induces cystic fibrosis transmembrane conductance regulator protein expression and activity in children with nonsense mutation cystic fibrosis. *American Journal of Respiratory and Critical Care Medicine*, *182*(10), 1262–1272. https://doi.org/10.1164/rccm.201001-0137OC.

Sermet-Gaudelus, I., Renouil, M., Fajac, A., Bidou, L., Parbaille, B., Pierrot, S., et al. (2018). Correction to: In vitro prediction of stop-codon suppression by intravenous gentamicin in patients with cystic fibrosis: A pilot study. *BMC Medicine*, *16*(1), 159. https://doi.org/10.1186/s12916-018-1138-z.

Shao, S., Murray, J., Brown, A., Taunton, J., Ramakrishnan, V., & Hegde, R. S. (2016). Decoding mammalian ribosome-mRNA states by translational GTPase complexes. *Cell*, *167*(5), 1229–1240. https://doi.org/10.1016/j.cell.2016.10.046.

Sharma, J., Du, M., Wong, E., Mutyam, V., Li, Y., Chen, J., et al. (2021). A small molecule that induces translational readthrough of CFTR nonsense mutations by eRF1 depletion. *Nature Communications*, *12*(1), 4358. https://doi.org/10.1038/s41467-021-24575-x.

Shee, C. D. (2005). Oxford textbook of medicine. *Journal of the Royal Society of Medicine*, *98*(7), 333–334.

Shi, N., Yang, Y., Zhang, H., Lu, J., Lin, H., Yang, X., et al. (2022). Restoration of dystrophin expression in mice by suppressing a nonsense mutation through the incorporation of unnatural amino acids. *Nature. Nature Biomedical Engineering*, *6*(2). https://doi.org/10.1038/s41551-021-00774-1.

Shimizu-Motohashi, Y., Komaki, H., Motohashi, N., Takeda, S., Yokota, T., & Aoki, Y. (2019). Restoring dystrophin expression in Duchenne muscular dystrophy: Current status of therapeutic approaches. *Journal of. Personalized Medicine*, *9*(1). https://doi.org/10.3390/jpm9010001.

Shteinberg, M., Haq, I. J., Polineni, D., & Davies, J. C. (2021). Cystic fibrosis. *Lancet (London, England)*, *397*(10290), 2195–2211. https://doi.org/10.1016/S0140-6736(20)32542-3.

Simioni, P., Tormene, D., Tognin, G., Gavasso, S., Bulato, C., Iacobelli, N. P., et al. (2009). X-linked thrombophilia with a mutant factor IX (factor IX Padua). *The New England Journal of Medicine*, *361*(17), 1671–1675. https://doi.org/10.1056/NEJMoa0904377.

Stenson, P. D., Ball, E. V., Mort, M., Phillips, A. D., Shiel, J. A., Thomas, N. S. T., et al. (2003). Human gene mutation database (HGMD): 2003 update. *Human Mutation*, *21*(6), 577–581. https://doi.org/10.1002/humu.10212.

Stiebler, A. C., Freitag, J., Schink, K. O., Stehlik, T., Tillmann, B. A. M., Ast, J., et al. (2014). Ribosomal readthrough at a short UGA stop codon context triggers dual localization of metabolic enzymes in fungi and animals. *PLoS Genetics*, *10*(10), e1004685. https://doi.org/10.1371/journal.pgen.1004685.

Taguchi, A., Hamada, K., Shiozuka, M., Kobayashi, M., Murakami, S., Takayama, K., et al. (2017). Structure-activity relationship study of Leucyl-3-epi-deoxynegamycin for potent premature termination codon readthrough. *ACS Medicinal Chemistry Letters*, *8*(10), 1060–1065. https://doi.org/10.1021/acsmedchemlett.7b00269.

Tate, W. P., Cridge, A. G., & Brown, C. M. (2018). 'Stop' in protein synthesis is modulated with exquisite subtlety by an extended RNA translation signal. *Biochemical Society Transactions*, *46*(6), 1615–1625. https://doi.org/10.1042/BST20180190.

Thompson, A. R., Schoof, J. M., Weinmann, A. F., & Chen, S. H. (1992). Factor IX mutations: Rapid, direct screening methods for 20 new families with hemophilia B. *Thrombosis Research*, *65*(2), 289–295. https://doi.org/10.1016/0049-3848(92)90249-a.

Tomkova, M., McClellan, M., Kriaucionis, S., & Schuster-Böckler, B. (2018). DNA replication and associated repair pathways are involved in the mutagenesis of methylated cytosine. *DNA Repair*, *62*, 1–7. https://doi.org/10.1016/j.dnarep.2017.11.005.

Tomkova, M., McClellan, M., Kriaucionis, S., & Schuster-Boeckler, B. (2016). 5-hydroxymethylcytosine marks regions with reduced mutation frequency in human DNA. *eLife*, *5*, e17082. https://doi.org/10.7554/eLife.17082.

Tork, S. (2004). The major 5′ determinant in stop codon read-through involves two adjacent adenines. *Nucleic Acids Research*, *32*(2), 415–421. https://doi.org/10.1093/nar/gkh201.

Trzaska, C., Amand, S., Bailly, C., Leroy, C., Marchand, V., Duvernois-Berthet, E., et al. (2020). 2,6-Diaminopurine as a highly potent corrector of UGA nonsense mutations. *Nature. Communications*, *11*(1). https://doi.org/10.1038/s41467-020-15140-z.

Vecsler, M., Ben Zeev, B., Nudelman, I., Anikster, Y., Simon, A. J., Amariglio, N., et al. (2011). Ex vivo treatment with a novel synthetic aminoglycoside NB54 in primary fibroblasts from Rett syndrome patients suppresses MECP2 nonsense mutations. *PLoS One*, *6*(6), e20733. https://doi.org/10.1371/journal.pone.0020733.

Wan, W., Thrarp, J. M., & Liu, W. R. (2014). Pyrrolysyl-tRNA synthetase: An ordinary enzyme but an outstanding genetic code expansion tool. *Biochimica et Biophysica Acta*, *1844*(6). https://doi.org/10.1016/j.bbapap.2014.03.002.

Welch, E. M., Barton, E. R., Zhuo, J., Tomizawa, Y., Friesen, W. J., Trifillis, P., et al. (2007). PTC124 targets genetic disorders caused by nonsense mutations. *Nature*, *447*(7140), 87–91. https://doi.org/10.1038/nature05756.

Westhof, E., Yusupov, M., & Yusupova, G. (2014). Recognition of Watson-crick base pairs: Constraints and limits due to geometric selection and tautomerism. *F1000Prime Reports*, *6*. https://doi.org/10.12703/P6-19.

Wilschanski, M., Miller, L. L., Shoseyov, D., Blau, H., Rivlin, J., Aviram, M., et al. (2011). Chronic ataluren (PTC124) treatment of nonsense mutation cystic fibrosis. *The European Respiratory Journal*, *38*(1), 59–69. https://doi.org/10.1183/09031936.00120910.

Wilschanski, M., Yahav, Y., Yaacov, Y., Blau, H., Bentur, L., Rivlin, J., et al. (2003). Gentamicin-induced correction of CFTR function in patients with cystic fibrosis and CFTR stop mutations. *The New England Journal of Medicine*, *349*(15), 1433–1441. https://doi.org/10.1056/NEJMoa022170.

Wohlgemuth, I., Pohl, C., & Rodnina, M. V. (2010). Optimization of speed and accuracy of decoding in translation. *The EMBO Journal*, *29*(21), 3701–3709. https://doi.org/10.1038/emboj.2010.229.

Wulff, K., Schröder, W., Wehnert, M., & Herrmann, F. H. (1995). Twenty-five novel mutations of the factor IX gene in haemophilia B. *Human Mutation*, *6*(4), 346–348. https://doi.org/10.1002/humu.1380060410.

Xie, J., Talaska, A. E., & Schacht, J. (2011). New developments in aminoglycoside therapy and ototoxicity. *Hearing Research*, *281*(1–2), 28–37. https://doi.org/10.1016/j.heares.2011.05.008.

Xue, X., Mutyam, V., Tang, L., Biswas, S., Du, M., Jackson, L. A., et al. (2014). Synthetic aminoglycosides efficiently suppress cystic fibrosis transmembrane conductance regulator nonsense mutations and are enhanced by ivacaftor. *American Journal of Respiratory Cell and Molecular Biology*, *50*(4), 805–816. https://doi.org/10.1165/rcmb.2013-0282OC.

Yamashita, A., Izumi, N., Kashima, I., Ohnishi, T., Saari, B., Katsuhata, Y., et al. (2009). SMG-8 and SMG-9, two novel subunits of the SMG-1 complex, regulate remodeling of the mRNA surveillance complex during nonsense-mediated mRNA decay. *Genes & Development*, *23*(9), 1091–1105. https://doi.org/10.1101/gad.1767209.

Yang, C., Feng, J., Song, W., Wang, J., Tsai, B., Zhang, Y., et al. (2007). A mouse model for nonsense mutation bypass therapy shows a dramatic multiday response to geneticin. *Proceedings of the National Academy of Sciences of the United States of America*, *104*(39), 15394–15399. https://doi.org/10.1073/pnas.0610878104.

Yarus, M. (1992). Proofreading, NTPases and translation: Constraints on accurate biochemistry. *Trends in Biochemical Sciences*, *17*(4), 130–133. https://doi.org/10.1016/0968-0004(92)90320-9.

Yegneswaran, S., Mesters, R. M., & Griffin, J. H. (2003). Identification of distinct sequences in human blood coagulation factor Xa and prothrombin essential for substrate and cofactor recognition in the prothrombinase complex. *The Journal of Biological Chemistry*, *278*(35), 33312–33318. https://doi.org/10.1074/jbc.M305906200.

Zaher, H. S., & Green, R. (2010). Hyperaccurate and error-prone ribosomes exploit distinct mechanisms during tRNA selection. *Molecular Cell*, *39*(1), 110–120. https://doi.org/10.1016/j.molcel.2010.06.009.

Zarate, Y. A., & Hopkin, R. J. (2008). Fabry's disease. *Lancet (London, England)*, *372*(9647), 1427–1435. https://doi.org/10.1016/S0140-6736(08)61589-5.

Zhang, Y., Baranov, P. V., Atkins, J. F., & Gladyshev, V. N. (2005). Pyrrolysine and selenocysteine use dissimilar decoding strategies. *Journal of Biological Chemistry*, *280*(21), 20740–20751. https://doi.org/10.1074/jbc.M501458200.

CHAPTER TWO

The disturbance of protein synthesis/degradation homeostasis is a common trait of age-related neurodegenerative disorders

Fabio Di Domenico* and Chiara Lanzillotta
Department of Biochemical Sciences "A. Rossi Fanelli", Sapienza University of Rome, Rome, Italy
*Corresponding author: e-mail address: fabio.didomenico@uniroma1.it

Contents

Abstract

Protein homeostasis or "proteostasis" represent the process that regulates the balance of the intracellular functional and "healthy" proteins. Proteostasis is fundamental to preserve physiological metabolic processes in the cell and it allow to respond to any given stimulus as the expression of components of the proteostasis network is customized according to the proteomic demands of different cellular environments. In conditions that promote unfolding/misfolding of proteins chaperones act as signaling molecules inducing extreme measures to either fix the problem or destroy unfolded proteins. When the chaperone machinery fails under pathological insults unfolded proteins induce the endoplasmic reticulum (ER) stress activating the unfolded protein response (UPR) machinery. The activation of the UPR restores ER proteostasis primarily through the transcriptional remodeling of ER protein folding, trafficking, and degradation pathways, such as the ubiquitin proteasome system (UPS). If these mechanisms do not manage to clear the aberrant proteins, proteasome overload and become defective, and misfolded proteins may form aggregates thus extending the UPR mechanism. These aggregates are then attempted to be cleared by macroautophagy. Impaired proteostasis promote the accumulation of misfolded proteins that exacerbate the

Advances in Protein Chemistry and Structural Biology, Volume 132
ISSN 1876-1623
https://doi.org/10.1016/bs.apcsb.2022.05.008

damage to chaperones, surveillance systems and/or degradative activities. Remarkably, the removal of toxic misfolded proteins is critical for all cells, but it is especially significant in neurons since these cannot be readily replaced. In neurons, the maintenance of efficient proteostasis is essential to healthy aging since the dysregulation of the proteostasis network can lead to neurodegenerative disease. Each of these brain pathologies is characterized by the repeated misfolding of one of more peculiar proteins, which evade both the protein folding machinery and cellular degradation mechanisms and begins to form aggregates that nucleate out into large fibrillar aggregates. In this chapter we describe the mechanisms, associated with faulty proteostasis, that promote the formation of protein aggregates, amyloid fibrils, intracellular, and extracellular inclusions in the most common nondegenerative disorders also referred to as protein misfolding disorders.

1. Introduction

The human proteome expresses in the cell about ten thousand of diverse proteins at any condition and the most of these need to fold in a specific and unique structure to allow a multitude of cellular functions (Klaips, Jayaraj, & Hartl, 2018). Although the amino acid sequence (Anfinsen, 1973) governs the native conformation of protein, molecular chaperones, and other factors are required to assist the folding process (Anfinsen, 1973). Besides, proteins to function properly are required to preserve flexibility in their structure or comprise substantial unstructured regions, which are at high risk of misfolding and aggregation (Chiti & Dobson, 2017). About 30% of the newly produced proteins may unfold and possibly aggregate under stress conditions. Lastly, if proteins become critically misfolded or functionally defective, they need to be degraded to prevent their detrimental effect within the cell. Protein homeostasis or "proteostasis" represent the process that regulates the dynamic equilibrium of the intracellular pool of functional and "healthy" proteins. Maintaining an integral proteome involves the regulation of the initial synthesis and folding of a protein, its conformational maintenance, the control of abundance and subcellular localization and clearance by degradation mechanisms (Lanzillotta & Di Domenico, 2021). Proteostasis is fundamental to preserve physiological metabolic processes in the cell, but it is also crucial to allow all the dynamic changes required to respond to any given stimulus. A wide and complex network of signaling pathways is involved in defending cells and organisms against proteotoxic stress (Hipp, Kasturi, & Hartl, 2019). The proteostasis network is estimated to comprises about two thousand components that act in concert to maintain cellular functionality. All the factors involved in proteostasis can be divided into three major groups:

(i) protein synthesis that contains the components of the translational machinery; (ii) folding and conformational surveillance that contains chaperons and co-chaperones but also of protein quality control systems components; and (iii) protein degradation that contains components of the ubiquitin–proteasome system and of the autophagy–lysosome system (Klaips et al., 2018). Despite common factors required for protein synthesis and maintenance, the expression of several components of the proteostasis network is customized according to the proteomic demands of different cellular environments (Labbadia & Morimoto, 2015). Furthermore, the induction and regulation of the proteostasis network machinery can be changed transiently or permanently different conditions, including development, aging, or exposure to stressors. The scrupulous control of such machinery is vital in lowering the toxic related with the accumulation of misfolded and/or damaged proteins. In this context, defects of one or more elements of proteostasis network could reduce its function and affect the ability of cells to correctly fold newly synthetized protein or degrade misfolded proteins. Subsequently, the accumulation of misfolded proteins could exacerbate the damage by overwhelming molecular chaperones and/or degradative activities (Di Domenico, Head, Butterfield, & Perluigi, 2014; Hohn, Tramutola, & Cascella, 2020; Scheper, Nijholt, & Hoozemans, 2011). Remarkably, the removal of toxic misfolded proteins is critical for all cells, but it is especially significant in post-mitotic cells, such as neurons since these cannot be readily replaced (Kurtishi, Rosen, Patil, Alves, & Moller, 2019). In neurons, the maintenance of efficient proteostasis is essential to healthy aging since the dysregulation of the proteostasis network can lead to neurodegenerative disease (Ciechanover & Brundin, 2003; Di Domenico, Barone, Perluigi, & Butterfield, 2017; Hipp et al., 2019; Hoozemans et al., 2005; Klaips et al., 2018; Orr & Oddo, 2013). Therefore, the successful maintenance of proteostasis is directly related to brain integrity and cognition abilities preservation. In this chapter, we describe the main pathways involved in proteostasis reporting the challenges they confront during the development of common neurodegenerative disorders.

2. Proteostasis: The balance between synthesis, folding and degradation

The synthesis of proteins is regulated by specific factors and pathways, whose function must be accustomed to the protein folding capacity of the cell to avoid the accumulation of misfolded proteins. Indeed, in conditions that favor lifespan extension, such as caloric restriction, a general decrease in protein translation occurs promoting the improved proteostasis capacity of the cell (Taylor & Dillin, 2011). Translation mitigation is also important in alleviating the overload of the proteostasis network during sustained stress. One of the main factors involved is the initiation factor 2α (eIF2α), which upon activation mediates the repression of translation (Costa-Mattioli et al., 2007). When a protein fails to fold correctly the cell utilizes extensive protection measures to maintain proper function, as described in Fig. 1. To ensure efficient folding and prevent aggregation, cells express several classes of molecular chaperones that guide the nascent polypeptide chain along a productive folding pathway, facilitating the formation of the most stable conformation and avoiding or sometime reversing misfolding and

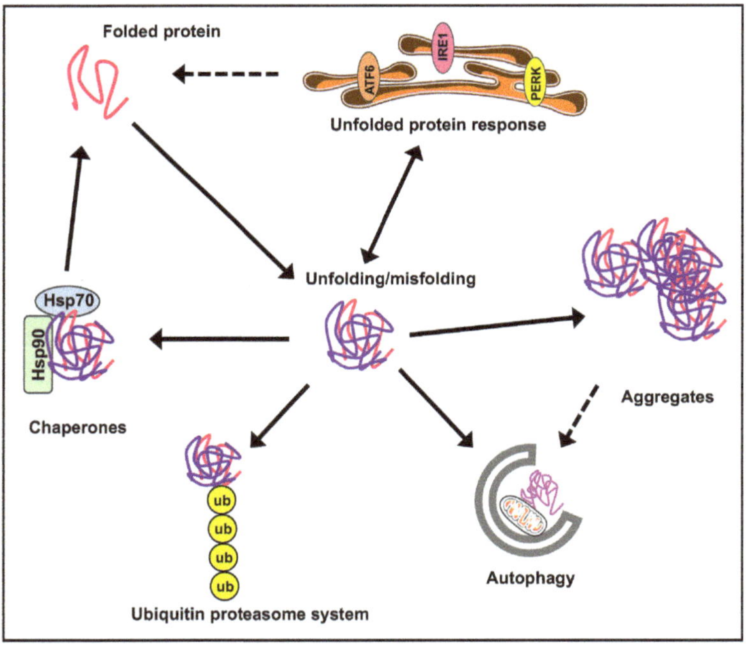

Fig. 1 Molecular processes involved in regulating the proteostasis network. In case of protein unfolding/misfolding molecular chaperones bind to hydrophobic regions within unfolded chains, to protect these regions from aggregation, thus rescuing folding. When the chaperone machinery fails proteins the development of ER stress activates the unfolded protein response, which restores ER proteostasis through the transcriptional remodeling of ER protein folding, trafficking, and degradation through the ubiquitin proteasome system. If a proteasomal overload occurs, misfolded proteins may form aggregates, which are attempted to be cleared by macroautophagy.

aggregation (Chiti & Dobson, 2017; Cuanalo-Contreras, Mukherjee, & Soto, 2013). The protein-folding networks consist of cytoplasmic and endoplasmic reticulum (ER)-resident chaperones that guarantee proper folding of native proteins in a crowded cellular environment. Furthermore, many organelles possess their own sets of chaperones to ensure that proteins remain folded correctly once they have reached their final location (Harding, Zhang, & Ron, 1999; Roussel et al., 2013). In conditions that promote unfolding/misfolding of proteins, chaperones act as signaling molecules inducing extreme measures to either fix the problem or terminate the damaged protein (Hoozemans & Scheper, 2012). The main causes of misfolding and aggregation are nonspecific interactions between exposed hydrophobic regions, which segregate to avoid unfavorable contacts with water and polar/charged moieties (Hohn et al., 2020). Chaperones preferentially bind to hydrophobic regions within unfolded chains, to protect these regions from nonspecific aggregation, thus acting as regulatory machineries accountable for the quality control of protein folding. When the chaperone machinery fails under pathological insults unfolded proteins are accumulated in the ER as effect of induced ER stress activating the unfolded protein response (UPR) (Hetz & Papa, 2018; Walter & Ron, 2011). The activation of UPR restores ER proteostasis through the transcriptional remodeling of ER folding capacity or the induction of degradation pathways, such as the ubiquitin proteasome system (UPS). If these mechanisms do not manage to clear irregular proteins due to proteasome overload, defective proteasome structure or to excess reactive oxygen species (ROS) in the cytoplasm, misfolded proteins may form aggregates (Hetz & Papa, 2018), which are attempted to be cleared by macroautophagy (Scheper et al., 2011; Wang & Kaufman, 2012). The foster inefficient clearance of aggregates promotes the increase of oxidative stress that impinge on proteasome and ER-associated degradation (ERAD) function resulting in chronic stress and apoptosis (Lanzillotta et al., 2018). More than 40 human diseases have been directly associated with the deposition of toxic aggregates in different tissue and organs. If aberrant protein deposition occurs in the brain this contribute to the progression of neurodegenerative diseases such as Alzheimer's disease and Parkinson's disease (Bell, Meier, Ingram, & Abisambra, 2016; Kurtishi et al., 2019; Metaxakis, Ploumi, & Tavernarakis, 2018; Schmidt, Gan, Komander, & Dewson, 2021). Each of these brain pathologies is characterized by the repeated misfolding of one of more peculiar proteins, which evade both the protein folding machinery and cellular degradation mechanisms and begins to form aggregates that nucleate out into large fibrillar aggregates (Chiti & Dobson, 2017). Further, the majority of these diseases are age-dependent, since aging promote the decline of the proteostasis network (Chondrogianni et al., 2014; Hipp et al., 2019).

2.1 Protein re-folding and surveillance

2.1.1 Molecular chaperones

Molecular chaperones are defined as proteins that assists in the folding, assembly and conformational maintenance other proteins without becoming part of its final structure (Hartl, 1996).

Molecular chaperones aided by protein disulfide isomerases (PDI) help nascent peptides avoid alternative, non-functional, low free energy states (Wilkinson & Gilbert, 2004). Although many proteins may independently fold into their native state, chaperones act to prevent protein misfolding and subsequent aggregation (Barral, Broadley, Schaffar, & Hartl, 2004). Chaperones are mostly ATP-dependent and function by binding and freeing polypeptides as they fold, mitigating the surface that could be negatively influenced by other cellular species (Saibil, 2013). In addition to that, PDIs guarantee that disulfide bridges between cysteine residues form in properly (Wilkinson & Gilbert, 2004). Chaperone proteins may even bind nascent polypeptide as they are translated and conduct the protein to final location within the cell. Membrane proteins, which must not only be inserted into a membrane but also fold properly within it take advantage of chaperone assistance. Chaperones are bind and aid numerous proteins under specific physiological and pathological, however if a protein fails to fold appropriately, despite the support of a chaperones, the chaperone becomes stuck on the protein. This causes the induction of second messengers or molecular inducers that promote the expression of further chaperone proteins or favor the degradation of defective proteins (Barral et al., 2004; Hetz, 2012; Hetz & Glimcher, 2011). Molecular chaperones include, among others, heat shock proteins (HSP), a group of proteins activated by heat shock mainly characterized by HSP70, HSP90, DNAJ/HSP40, chaperonin/HSP60, and by the small HSP (sHSP) families (Barral et al., 2004). Chaperones are able to independently assist protein folding but, to finely regulate client–substrate interactions, folding, disaggregation, degradation, and trafficking within the cell they function in combinations with different cochaperones (Yerbury et al., 2016). The best studied members of HSPs are the HSP70 and HSP90 families mostly described by the cytosolic/nuclear HSC70, HSP90, and inducible HSP70 (*HSPA1A/B*). However, homologs of HSP70 and HSP90 are also integral to ER (BiP and GRP94) and mitochondrial (Mortalin and TRAP1) function (Gandhi et al., 2019). HSP70 and HSP90 are highly abundant proteins that in some cell can reach up to 1–2% of the total protein. ATP hydrolysis is necessary for the chaperone activity of HSP70 and HSP90, causing conformational changes that result in substrate binding. When protein refolding is ineffective, the cochaperones C terminus of HSC70-interacting protein (CHIP) and Bcl2-associated athanogene 1 (BAG1) interact with HSP70 and HSP90 complexes to promote substrate ubiquitylation, thereby redirecting HSP70 and HSP90 clients to proteasomal degradation (Saibil, 2013). HSP70 and HSP90 are essential for proteins surveillance endorsing refolding or elimination. HSP90 associates with many cochaperones, while HSP70 is mainly regulated by HSP40/DNAJ chaperones, which act by presenting misfolded proteins to HSP70 and by stimulating its ATPase activity (Barral et al., 2004). In addition, to the more common chaperone functions exerted by HSP70/HSP40 and HSP90, some proteins require the specialized functions provided by the HSP60/chaperonin family. HSP60 is essential for maturation and maintenance of the mitochondrial proteome and is therefore intimately linked to energy production.

2.1.2 The unfolded protein response

Within the cell, a variety of stress sensor molecules induce specialized mechanisms to respond to malfunction in protein synthesis, folding, assembling, and degradation processes, and to maintain or re-establish proteostasis. The UPR senses misfolded protein accumulation in the ER, (Hetz, 2021; Hetz, Chevet, & Oakes, 2015), whereas the ISR responds to stress conditions from both the lumen of the ER and the cytosol (Chou et al., 2017; Harding et al., 2003). The UPR activates a widespread transcriptional response, which corrects the ER protein folding ability according to its needs (Hetz, Zhang, & Kaufman, 2020; Walter & Ron, 2011). As such, the UPR represents a standard among intracellular control mechanisms that compensate organelle abundance in response to environmental or developmental signs. The UPR rises the amount of ER components, including chaperones, and protein-modifying enzymes, needed to fold proteins. The UPR also decreases proteins translation and loading into the ER and increases the targeting of unfolded/misfolded proteins for degradation. As described in Fig. 2, the induction of the UPR initiates by the detachment of glucose-regulating proteins (GRPs) from three types of ER transmembrane anchors, namely activating transcription factor 6 (ATF6), inositol-requiring

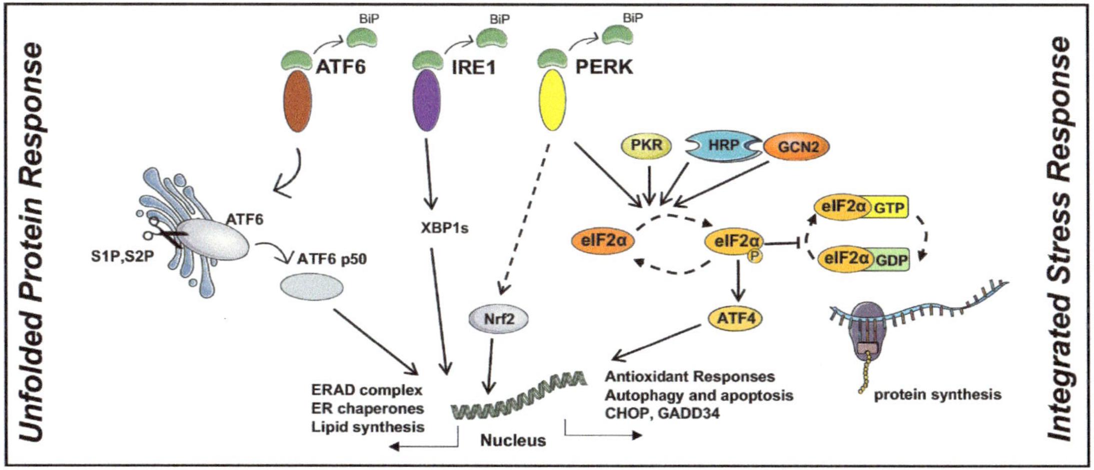

Fig. 2 Unfolded protein response and integrated stress response. Upon activation of the UPR, BiP releases the effectors of the three branches of the UPR. IRE1 dimerizes and autophosphorylates activating XBP1, which in turn is translocated into the nucleus to upregulate the transcription of other adaptive gene of the UPR. ATF6 is cleaved and translocate into the nucleus to enhance transcription of UPR-controlled genes. PERK is the only kinase of the UPR that overlap with ISR sensor kinases. PERK dimerizes, autophosphorylates, and targets Nrf2 and eIF2α. PKR, GCN2, and HRI (ISR sensors) once activated converge on the eIF2α phosphorylation, to elicit a translational and transcriptional stress response, thus inhibiting the initiation of translation. Phosphorylation of eIF2α also triggers the translation of specific mRNAs, including key transcription factors, such as ATF4, which activate pro-survival pathways.

protein 1 (IRE1), and PKR-like ER kinase (PERK) (Chakrabarti, Chen, & Varner, 2011; Hetz et al., 2015). Once separated from the membrane, GRPs bind to nascent proteins to assist their folding and secretion from the ER. The most abundant GRPs are Grp78 (known also as BiP) and Grp94. Once free, ATF6, IRE1, and PERK initiate their own signaling pathways. ATF6 is exported to the Golgi apparatus and processed by the S1P and S2P proteases. This process leads to the release of the cytosolic fragment domain of ATF6, which in the nucleus, simultaneously with X box-binding protein 1 (XBP1) a transcriptional activator of UPR controlled genes, upregulates the expression of CHOP and other genes involved in the regulation of ER size, protein-folding capacity, and the ERAD (Adachi et al., 2008; Yamamoto et al., 2007). IRE1 dimerizes and auto phosphorylates activating RNase domains. Phosphorylated and active IRE1 targets and cleaves XBP1, which encode for proteins belonging to ER chaperone family or to ERAD pathways. Under sustained ER stress, IRE1 mediates the activation of signaling cascades involved in cell death (Lee et al., 2012; Ron & Hubbard, 2008; Urano et al., 2000). The third branch of the UPR is initiated by PERK and its overall objective is to reduce translation of mRNA, through the phosphorylation of eIF2α, to constraint the input of nascent proteins in the ER (Bell et al., 2016; Cui, Li, Ron, & Sha, 2011; Lanzillotta et al., 2021). The phosphorylation of eIF2α at serine 51, block protein synthesis by repressing the action of the GTP exchange factor eIF2B by turning eIF2 into a non-competitive inhibitor (Costa-Mattioli & Walter, 2020; Zhu et al., 2019). This event allows the selected translation of mRNAs containing upstream short reading frames, highlighting the mRNA encoding for activating of transcription factor 4 (ATF4), which regulates the expression of multiple genes involved in proteostasis control, including chaperones, redox balance, autophagy, and amino acid synthesis (Lanzillotta & Di Domenico, 2021). eIF2α phosphorylation is also regulated by two phosphatase complexes, the constitutively expressed CReP and the stress-inducible subunit GADD34 (Pakos-Zebrucka et al., 2016). ATF4 regulates the levels of GADD34, mediating a feedforward loop to regulate translation initiation transiently and shut down the UPR. Under prolonged eIF2α phosphorylation, ATF4 triggers cell death by engaging the pro-apoptotic factor CHOP in addition to promoting the expression of members of the BCL-2 protein family (Wang & Kaufman, 2012). Furtherly, PERK can target and phosphorylate Nrf2 inducing the expression of proteins involved in the adaptation to oxidative stress (Lanzillotta et al., 2021). Upon activation of the UPR, PERK-mediated phosphorylation of Nrf2 lead to the Keap1/Nrf2 complex dissociation favoring Nrf2 translocation to the nucleus where it activates the transcription of antioxidant proteins.

2.1.3 The integrated stress response

The integrated stress response (ISR) is a cellular signaling network that combines the detection of cellular stresses with the inhibition of translation. Four different kinases are involved in the initiation of the ISR: PERK detects the accumulation of unfolded proteins in the lumen of the ER and it is a central ER stress sensor that is shared by UPR and ISR, protein kinase R (PKR) senses double-stranded RNA, general control nonderepressible 2 (GCN2) responds to amino acid deprivation, and eukaryotic translation initiation factor 2-alpha kinase 1 (EIF2AK1 or HRI) senses heme deficiency (Costa-Mattioli & Walter, 2020). Once activated, these kinases converge on the phosphorylation of eIF2α to stimulate a translational and transcriptional stress response as described above (Zhu et al., 2019). The ISR also modulates the two major forms of synaptic plasticity in the mammalian brain, protein synthesis-dependent long-term potentiation (LTP) and long-term depression (LTD) that are crucial for long-term memory formation (Costa-Mattioli et al., 2007; Huang et al., 2016; Zhu et al., 2011).

2.2 Protein degradation

Proteolysis plays a central role in protein renewal, which maintains the quality of proteins within the cell by degrading non-functional components (Hohn & Grune, 2014; Ruggiano, Foresti, & Carvalho, 2014). Two main complexes, the proteasome, and autophagy ensure proteolysis in eukaryotic cells UPS acts both in the cytosolic and nuclear compartment, while autophagy acts only in the cell cytoplasm (Grimm, Hohn, & Grune, 2012; Kwon & Ciechanover, 2017; Shin, Park, & Chung, 2020). The proteasome, in collaboration with a sophisticated ubiquitin system for target proteins tagging, selectively destroys short-lived proteins involved in the cell regulation, and anomalous proteins with irregular structures that upon deposition become harmful for the cell (Komander & Rape, 2012; Mulder, Witting, & Ovaa, 2020; Schmidt et al., 2021). On the contrary, autophagy leads to the lysosomal proteolysis of non-selectively targeted cytoplasmic constituents (Di Domenico, Zuliani, & Tramutola, 2019; Li, Liu, & Sun, 2017; Park, Kang, & Lee, 2020). However, recent data demonstrated that autophagy, by working in tandem with ubiquitin, perform also a selective-degradation process (Kwon & Ciechanover, 2017; Shin et al., 2020; Zhang, Chen, Zhao, Ponnusamy, & Liu, 2017). In this scenario, it is reasonable that UPS and autophagy are associated, using ubiquitin as a common label of proteins to be degraded. ERAD has a specific function in protein degradation since the ER is a major site for protein folding. When aberrant ER-resident proteins are processed by ERAD, they are released into the cytosol for proteasomal or autophagic clearance (Hetz, 2012). Even in the case of the UPR-ERAD, a central role is played by an HSP70, BiP (or HSPA5 or GRP-78), which has low intrinsic ATPase activity, enhanced by co-chaperones of the DNAJ-proteins (the same class of the HSP40, like ERdj4 or DNAJB9). Once activated, the UPR-ERAD converges on the proteasome or to autophagy.

2.2.1 The ubiquitin proteome system

The degradation of most intracellular proteins occurs through the UPS, which uses polyubiquitin chains to mark proteins for proteolysis and a great array of ubiquitin-activating, ubiquitin-conjugating, and ubiquitin-ligases enzymes to ensure protein turnover (Chondrogianni et al., 2014). The UPS degrades misfolded, oxidized, or damaged proteins, but it is also involved in eliminating proteins from different cellular processes, such as signal transduction, cell cycle regulation, and cell death (Hohn et al., 2020). Furthermore, it eventually regulates gene transcription. The 20S proteasome, which is the core particle where the proteolysis takes place, contains four stacked rings that form a barrel-shaped molecule with a central cavity. This structure is the functional core of all proteasomal forms, accompanied by regulatory components influencing specificity and activity of the 20S proteasome. Association of 19S structure with the 20S core particle results in the formation of 26S proteasomes. The 19S proteasome binds and unfolds ubiquitinated proteins and opens the entry gate of the 20S proteasome to let protein degradation in the central cavity (Ciechanover & Brundin, 2003). To label proteins for degradation, the covalent binding of target proteins with ubiquitin, a small protein with 76 amino acids, is requisite (Komander & Rape, 2012). This labeling generally involves three different enzymes: ubiquitin-activating enzyme (E1) hydrolyses ATP and forms a thioester-linked conjugate between itself and ubiquitin; ubiquitin-conjugating enzyme (E2) receives ubiquitin from E1 and forms a similar thioester intermediate with ubiquitin; and ubiquitin ligase (E3) binds both E2 and the substrate and transfers the ubiquitin to the substrate (Mulder et al., 2020). In few conditions, a fourth enzyme, ubiquitin chain elongation

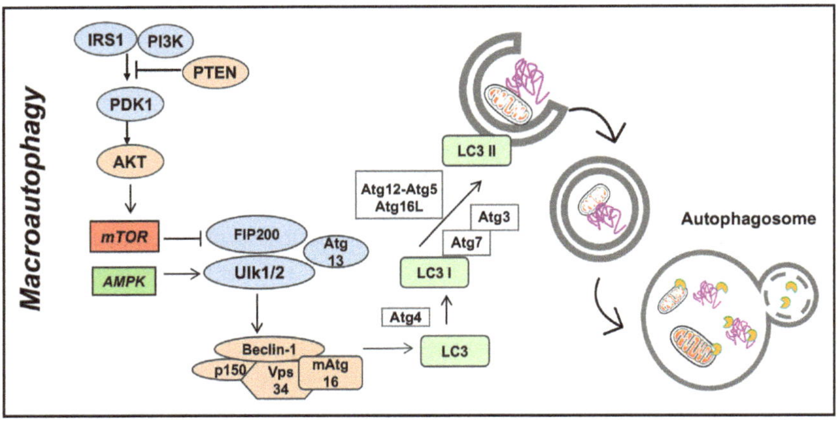

Fig. 3 Ubiquitin proteasome system. Proteins are directed to proteasomal degradation after being covalently modified with ubiquitin through the formation of an isopeptide bond. This conjugation involves three types of enzymes: E1 hydrolyzes ATP and forms a thioester-linked conjugate between itself and ubiquitin; E2 receives ubiquitin from E1 and forms a thioester intermediate with ubiquitin; and E3 binds both E2 and the substrate and transfers the ubiquitin to the substrate. Polyubiquitin chain is recognized by the 26S proteasome and protein degraded by its three proteolytic activities.

factor (E4), is necessary to extend a polyubiquitin chain. After polyubiquitination, substrate proteins are transferred for degradation to the 26S proteasome, where they are cleaved into oligopeptides by three proteolytic enzymes that exert chymotrypsin-like, caspase-like and trypsin- like activities (Schmidt et al., 2021) (Fig. 3). Besides, ubiquitin is freed by de-ubiquitinating enzymes to recycle it for new polyubiquitination reactions (Tramutola, Di Domenico, Barone, Perluigi, & Butterfield, 2016).

2.2.2 Autophagy

Autophagy is a degrading pathway essential for maintaining cellular homeostasis. Autophagy is considered as a recovery process that provides essential components to sustain principal metabolic functions during starvation or stress (Giorgi et al., 2021; Metaxakis et al., 2018; Murrow & Debnath, 2013). It is divided into three different processes according with the molecular mechanism involved, macroautophagy, microautophagy, and chaperone-mediated autophagy (CMA) (Perluigi, Di Domenico, & Butterfield, 2015). Among the three, macroautophagy is the best understood process and is often referred to as autophagy. The macroautophagy is critical in the cell to degrade protein aggregates and superfluous or damaged organelles, which are not suitable substrate for proteasomal degradation

(Garcia-Arencibia, Hochfeld, Toh, & Rubinsztein, 2010; Saha, Panigrahi, Patil, & Bhutia, 2018). In macroautophagy, portions of the cytoplasm are enveloped by a double-membrane that expands into a cytosolic vesicle named autophagosome. The outer membrane of the mature autophagosome is fused with lysosomal membrane forming an autolysosome that shares the contents of both and enables lysosomal acid hydrolases to degrade and recycle autophagosome cargo (Zheng et al., 2011). In microautophagy and CMA, cargo is directly taken up by the lysosome, either through the invagination of the lysosomal membrane or by unfolding and translocation of proteins with a specific signal sequence that is recognized by the LAMP2A receptor on the lysosome (Bandyopadhyay, Sridhar, Kaushik, Kiffin, & Cuervo, 2010). Autophagosome formation is often analyzed by the levels of LC3II which is degraded after formation of the autolysosome (Di Domenico, Zuliani, et al., 2019; Perluigi et al., 2015). The homeostatic role of autophagy includes both nonselective and selective degradation mechanisms (Kraft, Peter, & Hofmann, 2010; Kwon & Ciechanover, 2017). However, differently from the UPS, autophagy often mediates non-selective and bulk degradation of cytoplasmic contents or entire organelles, such as mitochondria (mitophagy), ER or ribosomes (Giorgi et al., 2021). As pictured in Fig. 4 macroautophagy is regulated by several protein kinases including mTOR that inhibits macroautophagy, AMPK and Ulk1 that promote macroautophagy (Heras-Sandoval, Perez-Rojas, Hernandez-Damian, & Pedraza-Chaverri, 2014; Kim, Kundu, Viollet, & Guan, 2011; Perluigi et al., 2015). Downstream activators of macroautophagy are the proteins RAB1α, Beclin1, and the ATG protein family. Under ordinary nutrient circumstances, active mTOR complex 1 (mTORC1) phosphorylates Ulk1 and sequesters it in a complex with Atg13 and FIP200, thereby inhibiting autophagy (Perluigi et al., 2015). Starvation, amino acid deprivation, or growth factors remove mTOR control allowing Ulk1 to promote autophagy. AMPK is a major positive regulator of autophagy. Under high intracellular AMP/ATP ratio, AMPK induces autophagy by Ulk1 activation and by mTORC1 inhibition (Kim et al., 2011). Once activated, Ulk1 initiate autophagosome nucleation by creating the class III phosphatidylinositol 3-kinase (PtdIns3K) complex. PtdIns3k, together with other Atg-proteins, has the function to recruit two ubiquitin-like conjugation systems, which are necessary for the expansion and elongation of the phagophore and with the formation of mature autophagosomes (Saha et al., 2018; Uddin et al., 2018). At last, autophagosome fuses with the lysosome forming the autophagolysosome where the

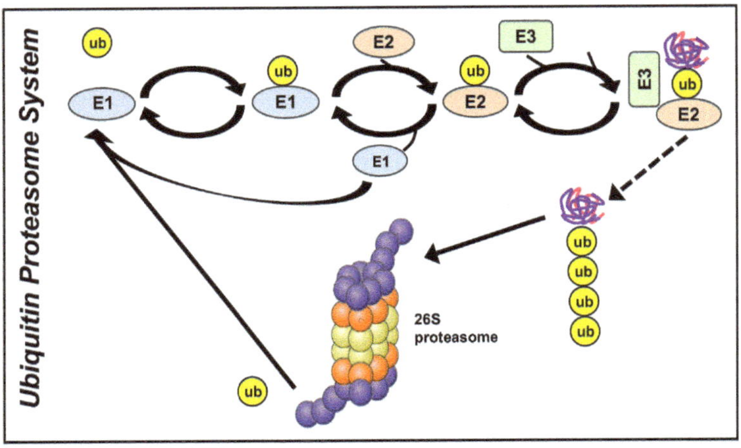

Fig. 4 Macroautophagy: Macroautophagy is finely regulated by a protein complex named Ulk1/2-Atg13-FIP200, which works as a sensor for cellular state. Under nutrient-rich conditions, mTOR is associated with the induction complex, keeping autophagy in an inactive state. During nutrient starvation mTOR is inactivated and dissociates from Ulk1/2 complex resulting in autophagy induction. AMPK, a key- role energy sensor, promotes autophagy under glucose starvation. Once activated, Ulk1/2 complex recruits PtdIns3K Vps34 (vacuolar protein sorting 34), p150, mAtg16 and Beclin-1 to participate at the autophagosome nucleation, creating the class III phosphatidylinositol 3-kinase (PtdIns3K) complex. PtdIns3k, together with other Atg-proteins, has the function to recruit the Atg12–Atg5-Atg16L1 and LC3 systems that coordinate the phagophore elongation phase, finally resulting in a mature autophagosome. After the formation of the autophagolysosome, cargo degradation is strictly dependent on the activation of lysosomal hydrolases, such as cathepsin B, D, and L.

acidic environment of the lysosome is maintained by the activity of the V_0-ATPase proton pump that allow cargo degradation. Lysosomal hydrolases, such as cathepsin B, D, and L, are involved in the cleavage of autophagic substrates and resulting molecules are transported back to the cytosol for nutrient recycling (Di Domenico, Zuliani, et al., 2019).

3. The proteostasis network in the brain

Understanding how the brain works is one of the major challenges of science, not only due to the convolution of neuronal networks and the diversity of neuronal characteristics but also for the plastic nature of the nervous system that implicates the remodeling of synaptic contacts and their abundancy. It is recognized that de novo protein synthesis is required to support the persistence of synaptic changes to convert short-term memories into long-term memories (Sossin & Costa-Mattioli, 2019). Neurons exhibit a unique degree of spatial compartmentalization and can preserve and modify their proteomes independently from other cells. During their extended lifetime, neurons are exposed to several stressors and harmful conditions that can disrupt proteostasis. The importance of maintaining the integrity of proteostasis is proven by the fact that most neurodegenerative disorders are characterized by the accumulation of abnormal protein aggregates in the form of soluble oligomers, fibrils, and large protein inclusions (Pritzkow et al., 2018). Past reports revealed that the presence of large fibrillar aggregate deposits does not always correlate with disease onset or severity and most toxic aggregate species may be soluble oligomers and small insoluble species (Chiti & Dobson, 2017; Leitman, Ron, Ogen-Shtern, & Lederkremer, 2013). Indeed, oligomers expose hydrophobic residues and unpaired polypeptide backbone structures, becoming highly interactive with other proteins, including critical factors of the proteostasis network (Kayed et al., 2003; Olzscha et al., 2011; Woerner et al., 2016). Brain proteostasis, due to the increased susceptibility of neurons to hypoxia, the absence of cell division, the effects of chronic disorders and the high cell components turnover and metabolic rate at the synapses, is more susceptible to perturbations than other organs. In this scenario, the successful maintenance of proteostasis is directly related to brain preservation. Neurons, motor neurons and glial cells are highly efficient to respond to misfolded protein species by the overexpression of specific chaperones and co-chaperones, paralleled by the potentiation of the degradative pathways (Giandomenico, Alvarez-Castelao, & Schuman, 2022; Hanus & Schuman, 2013; Yerbury et al., 2016). All these factors are extremely well-coordinated to protect against proteotoxicity and acts as the first line of defense to avoid brain damage and the development of protein misfolding diseases.

3.1 Evidence of aberrant proteostasis in age-related neurodegenerative diseases

Neurodegeneration is identified as the progressive, irreversible loss of neurons, which may affect either the peripheral or central nervous system (CNS). As the neuronal structures deteriorate, a gradual and progressive loss of cognitive and/or motor skills occurs causing mental impairment, functional loss, and debilitation (Kurtishi et al., 2019; Lanzillotta & Di Domenico, 2021). Compelling evidence suggests that neurodegenerative diseases exhibit common pathological features, among which one of the most representative is aberrant proteostasis as effect of peculiar protein aggregation and proteasomal and/or autophagic dysfunction. Neurons are vulnerable to the impairment of protein degradation pathways and the progressive decline of both proteasome activity and autophagic induction, observed in neurodegeneration, result in proteotoxicity, ER stress, and ultimately to neuronal loss (Gandhi et al., 2019; Hohn et al., 2020; Picca et al., 2020; Soto & Pritzkow, 2018). The presence of insoluble aggregates is one of the main characteristics of neuropathological disorders, such as Alzheimer's disease (AD), Parkinson's disease (PD), and amyotrophic lateral sclerosis (ALS) (Bell et al., 2016; Cai & Jeong, 2020; Di Domenico, Zuliani, et al., 2019; Gandhi et al., 2019). For each of these diseases, a key protein is repeatedly produced and misfolded, evading both the protein folding machinery and cellular degradation mechanisms, and begins to form aggregates that nucleate out into large fibrillar aggregates (Rochet & Lansbury, 2000). These fibrils are known as amyloid fibrils when they form extracellularly or intracellular inclusions when they form intracellularly (Chiti & Dobson, 2006). These misfolded proteins are one component of a cerebral plaque, along with other molecules, neurites, and reactive glia. Evidence to date indicates that, although they differ in amino acid sequence, the various proteins involved in neurodegenerative diseases share common structures in their aggregate forms. Genetic mutations to metastable proteins may increase their susceptibility to aggregate, causing the proteostasis network to be overloaded. Importantly, each protein aggregate, specific of a given neuropathological condition, impairs a unique cellular signaling network, thereby contributing to proteostasis damage and selective neuronal impairment (Chiti & Dobson, 2017; Hetz et al., 2015; Hetz & Glimcher, 2011; Labbadia & Morimoto, 2015). The identification of affected signaling pathways is fundamental for the understanding of the pathological mechanisms associated with neurodegenerative diseases. However, there is evidence that expression of aggregation-prone proteins can be dealt by the proteostasis network in early life but as the organisms age the capacity to deal with this condition of added proteotoxic stress is impaired, facilitating aggregation (Ben-Zvi, Miller, & Morimoto, 2009; Hohn et al., 2020; Taylor & Dillin, 2011). In sporadic neurodegenerative diseases characterized by the absence of genetic mutations associated with the increased propensity of a single protein to aggregate, the failure of the proteostasis network is likely a requisite for aggregation to occur. In this scenario, the build-up of persistent chronic stress leads to the prolonged activation of protein folding and surveillance systems, which from protective become harmful by promoting the sustained reduction of protein synthesis and the activation of cell death pathways.

3.1.1 Alzheimer's disease

AD is the most common neurodegenerative disease and the most common form of cognitive decline accounting for 60–80% of dementia cases. The greatest known risk factor is increasing age, and the majority of people with Alzheimer's are 65 and older (Querfurth & LaFerla, 2010). The classical symptoms of AD include mental illness and amnesia, long-term memory loss, confusion, personality change and difficulty with language. AD leads to many changes in the brain, including diffuse brain atrophy, specifically in the frontal lobe and hippocampus, with degeneration of cholinergic neurons. Age is the greatest risk factor for AD. Many molecular lesions have been detected in Alzheimer's disease, but the principal theme that emerge from the data is that an accumulation of misfolded proteins in the aging brain, which results in oxidative and inflammatory damage and in turn leads to energy failure and synaptic dysfunction (Haass & Selkoe, 2007). The clinical manifestation of AD is misfolded proteins that form cerebral plaques. Aβ form β-amyloid plaques, while hyperphosphorylated tau proteins form aggregates of neurofibrillary tangles (NFTs). Aβ peptides are natural products of metabolism consisting of 36 to 43 amino acids. Monomers of Aβ 1–40 are much more prevalent than the aggregation prone and damaging Aβ 1–42 species. β-amyloid peptides originate from proteolysis of the amyloid precursor protein (APP) by the sequential enzymatic actions of beta-site amyloid precursor protein–cleaving enzyme 1 (BACE-1), a β-secretase, and γ-secretase, a protein complex with presenilin 1 at its catalytic core. An imbalance between production and clearance that lead to the aggregation of peptides causes Aβ to accumulate, and this excess may be the initiating factor in Alzheimer's disease (Busciglio et al., 2002). Tau promotes assembly and stability of microtubules and vesicle transport. Hyperphosphorylated tau is insoluble, lacks affinity for microtubules, and self-associates into paired helical filament (PHF) structures. The activity of kinases and phosphatases regulate the extent of tau phosphorylation (Iqbal et al., 2005). Like Aβ oligomers, intermediate aggregates of abnormal tau molecules are cytotoxic and impair cognition (Oddo et al., 2006). It is now increasingly recognized that the brains of AD patients can display pathologic inclusions containing α-synuclein and TDP-43 in addition to the habitually found amyloid and NFT pathology. Approximately 20–50% of AD cases present neuronal cells with TDP-43 along with tau inclusions as separate entities (Hu et al., 2008; Nag et al., 2015).

Molecular chaperones play an indispensable role in neurodegenerative disorder progression, as for AD, since HSPs bind to APP reducing Aβ40 and Aβ42 production. HSP70 and HSP90 degrade tau NFTs through the proteasome system (Dickey et al., 2009; Yang, Turner, & Gaut, 1998). Overexpression of HSP70 decreases the amount of soluble and insoluble tau in animal models, reduces insoluble tau phosphorylation leading to tau stability and enhances tau binding to microtubules declining its toxicity (Dou et al., 2003; Petrucelli et al., 2004; Sarkar, Kuret, & Lee, 2008). Heat shock cognate 70 (HSC70) together with HSP70 or HSP90 directly binds to tau, whether it is phosphorylated or not, facilitating microtubule polymerization and controlling tau aggregation (Matts et al., 2011). Hence, HSPs stimulation may represents possible therapeutic approach for treating AD. Immunohistochemical studies and expression analyses in AD brain tissue showed the increased expression levels of HSP27 and HSP70 associated with activated glia and stressed neurons (Hanus & Schuman, 2013). In addition, protein levels of HSP27, HSP32, HSP60, HSP70, and HSP90 were found to be increased in hippocampus and inferior parietal lobule from amnestic mild cognitive impairment (MCI) patients, suggesting that alteration in the chaperone protein systems might contribute to the pathogenesis and progression of AD (Di Domenico et al., 2010). Interestingly, although HSPs are neuroprotective against and induced in AD brain, they were also found post-translationally modified (e.g., oxidized) suggesting the potential alteration of their protective function (Butterfield, Gu, Di Domenico, & Robinson, 2014).

Several studies imply that Aβ interferes with ER functioning. Analyses of AD post-mortem brain revealed altered levels of ER stress markers, like Grp78/BiP, IRE1α, PERK, eIF2α phosphorylation and PDIs (Hoozemans et al., 2005, 2009). Aβ affects ER functioning, leading to ER stress in the early stages of AD. Exogenous Aβ induces ER stress in primary neuronal cultures, while tau has been also shown to block ERAD, resulting in accumulation of misfolded proteins in the ER lumen (Hoozemans et al., 2009; Hoozemans & Scheper, 2012; Milisav, Suput, & Ribaric, 2015). Mitochondrial defects and free radical production are also strong contributors to ER stress, inducing the oxidation of ER proteins possibly by altering the activity of the resident chaperones (Gerakis & Hetz, 2018). Recent studies in AD brain and animal models suggest that abnormal PERK activation has detrimental effects to neuronal plasticity and cognition. Indeed, repression of protein translation through eIF2α phosphorylation is a well-known negative regulator of synaptic plasticity and learning and memory- related process (Bell et al., 2016; Lanzillotta et al., 2021). Targeted deletion of PERK or GCN2 improves synaptic plasticity and memory function in the APP/PS1 AD mouse model. Similar results were obtained when PERK was targeted in the AD model 5xFAD (Milisav et al., 2015). In the hippocampus of the 3xTg-AD mouse model increased levels of CHOP/GADD153 precede the increase in BACE and Aβ levels (Hetz & Soto, 2006) suggesting that chronic ER stress in AD neurons may result in synaptic dysfunction and cognitive impairment through the repression of synaptic protein expression. In the long term, sustained PERK signaling may also translate into neuronal loss through apoptosis. Studies in human AD brain demonstrated altered ER proteostasis. Markers of ER stress correlate with the progression of AD and are associated with the activation of UPR anchors and their downstream effectors. BiP expression is induced in the hippocampal and cortical areas, showing an increased staining in healthier neurons (Hoozemans et al., 2005, 2009; Hoozemans & Scheper, 2012). PERK is increasingly phosphorylated along with its main target, eIF2α, in the hippocampal area where they colocalize with abnormally phosphorylated tau. Phosphorylated IRE1α was detected in AD brain tissues and directly correlates with Braak stage. The proapoptotic UPR transcription factor CHOP is also upregulated. XBP1s, on the other hand, appears to follow a more complex kinetic with reports of a moderate increase or no modulation (Hetz & Saxena, 2017; Hoozemans & Scheper, 2012; Wang & Kaufman, 2012).

Many researchers investigated the involvement of impaired UPS in the pathogenesis of AD. Ubiquitin accumulates especially in early AD cases leading to changes in the composition of the proteasome and inhibiting the proteasome activity by the downregulation of ubiquitin-activating and ubiquitin-conjugating enzymes and the accumulation of mutated ubiquitin (Bence, Sampat, & Kopito, 2001; Hol, van Leeuwen, & Fischer, 2005; Pasinetti, 2001). The increased oxidation of ubiquitin carboxy-terminal hydrolase L1 (UCH-L1), a deubiquitinating enzyme accountable for the hydrolysis of ubiquitin, is shown AD promoting the downregulation of ubiquitin hydrolysis. Proteases are responsible for Aβ degradation (Tseng, Green, Chan, Blurton-Jones, & LaFerla, 2008), so age-dependent proteasome dysfunction may have a role in Aβ accumulation in AD brains, especially after demonstrating that the proteasome degrades both presenilin 1 and 2, the catalytic subunits of the γ-secretase intramembrane protease complex (Checler et al., 2000). It is suggested that Aβ production will increase upon the exacerbation of the proteasome function by increasing γ-secretase activity (Oddo, 2008). However, in AD pathogenesis, it is still unclear whether Aβ aggregation causes UPS dysfunction, or the opposite (Upadhya & Hegde, 2007). Aβ inhibits the proteasome in vitro, however, it is still unknown if AD patients follow a similar pattern. In contrast, tau is degraded by multiple proteases, including calpains, caspases, cathepsins, and thrombin (Oddo, 2008). Recent studies support a relation between the UPS and tau, it was suggested an inhibitory effect of the proteasome on tau proteins. Interestingly, aggregated forms of tau proteins isolated from AD brain cells were able to inhibit the proteasome activity, whereas non-aggregated isolated forms could not. Thus, tau turnover is somehow dependent on the activity of the proteasome and its different states of aggregation (Tramutola et al., 2016; Weng & He, 2021).

In the last two decades the mTOR/autophagy axis has been extensively analyzed in AD brains and in brains from AD mouse models, demonstrating its aberrant regulation during pathology progression (Li et al., 2017). Evidence that autophagy deficiency was involved in AD was shown by Nixon's group that found immature autophagic vacuoles accumulated in dystrophic neuritis in AD brains (Nixon, 2007; Nixon & Yang, 2011). Similar results were obtained on AP/PS-1double transgenic mice that showed autophagic vacuoles accumulated in neuronal dendrites and soma before Aβ plaques appearance. Further, several laboratories demonstrated changes in the expressions of Atgs, indicating autophagy dysfunction in AD (Uddin et al., 2018). Markers of autophagy initiation, autophagosome formation (e.g., LC3 II/I) and autophagic flux were found reduced in MCI and AD brain (Di Domenico, Zuliani, et al., 2019). mTOR through the control of autophagy is considered a key regulator of Aβ generation and clearance, and tangles formation (Lipton & Sahin, 2014; Switon, Kotulska, Janusz-Kaminska, Zmorzynska, & Jaworski, 2017). Several studies suggested that reduced autophagy in AD brain and animal models, leading to the accumulation of protein aggregates, is likely caused by the hyperactivation of the PI3K/Akt/mTOR axis. Studies from our group and others showed a significant increase in Aβ 1–42 levels, associated with a reduction in autophagy (Beclin-1 and LC3) and a hyperactivation of PI3K/Akt/mTOR pathway in Preclinical AD (PCAD), MCI, and AD subjects (Caccamo, Majumder, Richardson, Strong, & Oddo, 2010; Di Domenico, Tramutola, et al., 2019; Di Domenico, Zuliani, et al., 2019; Oddo, 2012; Spilman et al., 2010; Tramutola et al., 2015). In addition, mTOR hyperactivity correlated with Braak stages and/or cognitive severity of AD patients (Fig. 5).

Furtherly, it was shown that genetic reduction of Beclin-1 in APP transgenic mice, promoted intraneuronal, and extracellular Aβ deposition, likely contributing to synaptodendritic degeneration and neuronal loss. Moreover, macroautophagy inhibition by 3-metyhladenine associated with

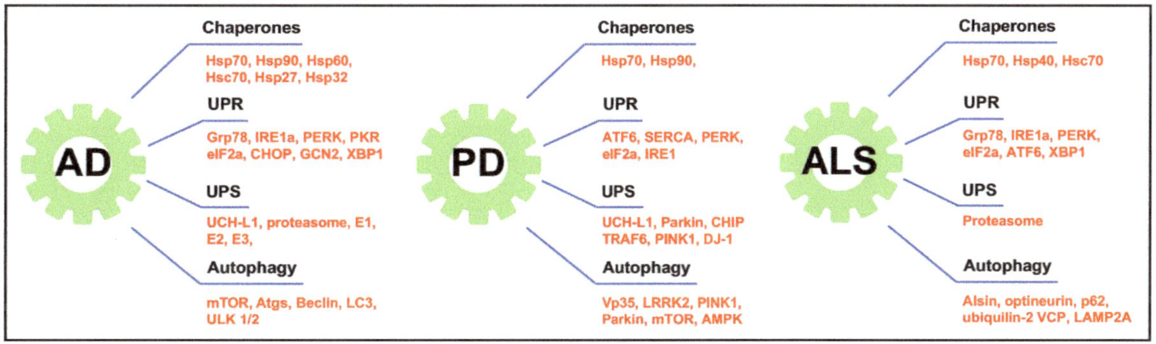

Fig. 5 Components of the proteostasis system found to be altered in selected neurodegenerative diseases.

tau increased aggregation and cytotoxicity, demonstrating a pivotal role of autophagy in preventing the formation of tau tangles (Caccamo et al., 2010; Mueed et al., 2018). In agreement, several studies revealed a tight association between reduced autophagosome formation and increased p-tau levels in AD. On the other hand, autophagy inducers (e.g., rapamycin) facilitated the degradation of insoluble forms of tau and protect against its toxicity in tau transgenic and AD mouse models (Caccamo et al., 2010; Di Domenico, Tramutola, et al., 2019; Di Domenico, Zuliani, et al., 2019; Tramutola et al., 2018; Van Skike & Galvan, 2018).

3.1.2 Parkinson's disease

PD is a common progressive neurodegenerative disorder that affects 1–3% of the aging population (Kalia & Lang, 2015; Pringsheim, Jette, Frolkis, & Steeves, 2014). PD patients suffer from a selective loss of dopaminergic neurons of the sustantia nigra pars compacta (SNpc), associated with a dramatic decrease in dopamine at the level of the nigro-striatal circuit, resulting in the loss of controlled movements (Dauer & Przedborski, 2003). Initial symptoms include shaking and movement- related problems, while dyskinesia becomes characteristic of more advanced stages of the disease. The canonical motor symptoms of PD include tremor, rigidity, bradykinesia, and postural instability (Shulman, De Jager, & Feany, 2011). The predominant proportion of PD cases are sporadic, while 10–15% are familiar with mutations in a variety of genes including SNCA/PARK1, Parkin (PRKN), leucine-rich repeat kinase 2 (LRRK2), Parkinson's disease protein 7 (PARK7/DJ1), PTEN-induced putative kinase 1 (PINK1) and Parkinson's disease protein 9 (PARK9/ATP13A2) among others (Corti, Lesage, & Brice, 2011). The histopathological hallmark of PD is the accumulation of proteinaceous inclusions called Lewy bodies (LBs), formed by misfolded and post-translationally modified α-synuclein (encoded by the *SNCA* gene) (Spillantini et al., 1997) and ubiquitinated proteins. α-synuclein is abundant in the human brain, found in high levels in neurons of the neocortex, hippocampus, substantia nigra, thalamus, and cerebellum. Cytosolic α-synuclein is natively unbound and unfolded. When bound to membranes or vesicles, α-synuclein acquire an α-helical structure. When misfolded, α-synuclein get a β-sheet-rich conformation, begins to oligomerize and then to aggregate into fibrils and insoluble LBs (Sulzer, 2010). The identification of α-synuclein fibrils as a major constituent of LBs in both sporadic and familial PD cases supports the association with PD pathology (Spillantini et al., 1997). Interestingly, oligomeric species of α-synuclein are considered as one of the main dangerous species in the induction and progression of PD, like the observations described in other diseases. In addition to α-synuclein, Lewy bodies are composed of ubiquitin, calbindin, complement proteins, actin, tubulin, MAP1 and 2, Pael-R, and synphilin-1. According to the Braak staging, Lewy body formation begins initially in the medulla oblongata and olfactory bulb and progresses rostrally over the years to the SNpc and basal forebrain leading to dopaminergic neurodegeneration (Galvan & Wichmann, 2008). Phosphorylation of α-synuclein is believed to be involved in the initiation of a-synuclein misfolding, while ubiquitination is believed to lead to mislocalization and escape from degradation.

Many studies have reported that molecular chaperones modulate and control the toxicity and degradation of α-synuclein and other PD-relevant proteins (e.g., Parkin, PINK1, DJ-1, and LRRK2). Regarding α-synuclein, several studies revealed that HSP70 may have a neuroprotective role in PD as observed in Drosophila and yeast models of PD (Auluck, Chan, Trojanowski, Lee, & Bonini, 2002; Flower, Chesnokova, Froelich, Dixon, & Witt, 2005). The neuroprotective effect of HSP70 is done by several mechanisms, including refolding of misfolded protein, guiding misfolded protein to degradation by UPS and CMA, and disaggregation of already formed α- synuclein protein (Friesen, De Snoo, Rajendran, Kalia, & Kalia, 2017; Nillegoda & Bukau, 2015). In agreement, the mutation of the ATPase function of HSP70 leads to increasing the toxicity of α-synuclein (Klucken, Shin, Hyman, & McLean, 2004; Klucken, Shin, Masliah, Hyman, & McLean, 2004). Modulating the activity of the molecular chaperones offers a potential therapeutic intervention for the management of PD. HSP90 inhibitors lead to dissociation of HSF-1 and move to the nucleus where they induce HSP70 (Putcha et al., 2010; Taldone & Chiosis, 2009). HSF-1 activator (e.g., carbenoxolone and celastrol) can also induce HSP70 expression (Thakur & Nehru, 2014).

Chronic ER stress contributes to degeneration of PD dopaminergic neurons of the *substantia nigra* (Hoozemans et al., 2007). α synuclein could directly interfere with the function of the UPR by an abnormal physical association with ATF6, affecting its activation status in response to ER stress (Credle et al., 2015). This model was also validated in iPSC-derived dopaminergic neurons from PD patients (Chung et al., 2013). αSyn expression results in altered translocation of ATF6 into COPII-coated vesicles, a process required for its activation. Impaired ATF6 signaling was associated with decreased ERAD function and increased pro-apoptotic signaling

(Milisav et al., 2015). Abnormal ER calcium homeostasis has been linked with the UPR, because most ER chaperones require calcium as a cofactor for their optimal activity (Carreras-Sureda, Pihan, & Hetz, 2018). Abnormal calcium signaling has been also linked with the manifestation of ER stress in PD. Recently, it was shown that α synuclein aggregates led to the activation of the ER calcium pump SERCA, disrupting calcium metabolism in neurons (Betzer et al., 2018). Abnormal rises of calcium in the cytosol then can be uptake by the mitochondria, sensitizing cells to apoptosis and enhancing ROS production (Giorgi et al., 2015). Studies on PD post-mortem brain demonstrated the immunohistochemical localization of phosphorylated PERK and eIF2α in the SN (Mercado et al., 2018). Further, the increased phosphorylation of PERK, eIF2α, and IRE1 activity that were closely associated with glial cytoplasmic inclusions containing α synuclein (Bell et al., 2016). Overexpression of ER chaperone Grp78/BiP was neuroprotective in mice (Gorbatyuk et al., 2012), in contrast to the removal of ATF6, which enhanced the susceptibility to PD-inducing neurotoxins (Egawa et al., 2011; Hashida et al., 2012). Salubrinal, an inducer of eIF2α phosphorylation was neuroprotective in αSYN transgenic mice (Boyce et al., 2005).

Accumulating evidence from many studies (Choi et al., 2004; Hol et al., 2005; Lim & Tan, 2007) shows that impairment of UPS plays an essential role in PD pathology. Four proteins, identified in patients with PD, were found to have a role in the dysfunction of UPS, Parkin, CHIP, UCHL1, and TRAF6 (Shulman et al., 2011). Parkin, a PD associated protein, is a ligase ubiquitin E3 that will mark the proteins poorly folded for ubiquitination (Myeku & Duff, 2018). Besides, PINK1 is generally a mitochondrial Ser/Thr kinase that triggers and activates Parkin via phosphorylation. Ubiquitin is phosphorylated in serine 65 by PINK1, then binds with Parkin and activates Parkin in the mitochondria to ubiquitinate the proteins faultily folded, this leads eventually to the deterioration of proteins by the proteasome (Kane et al., 2014; Kazlauskaite & Muqit, 2015). Mutations and malfunctions of *Parkin* and *PINK1* are associated with the early-onset familial and sporadic PD (Pickrell & Youle, 2015). It has also been demonstrated that Parkin ubiquitinates p62, an important shuttle protein that regulates the quality of the protein (Song et al., 2016). On the other hand, CHIP has many functions like targeting oligomer form of α-synuclein for degradation, increased ubiquitination of Pael receptor and regulates the LRRK2 ubiquitination and degradation. TRAF6 normally promotes K63-linked ubiquitination of targeted proteinin, while in PD patients, it is up-regulated and promotes K6, K27, and K29- linked ubiquitination of α-synuclein and DJ-1 (Romani-Aumedes et al., 2014). UCH-L1 is one of the de-ubiquitinating enzymes that are found abundantly in the brain and one of its main functions is hydrolysis the bonds between ubiquitin molecules to produce the ubiquitin monomer, an important component of UPS (Choi et al., 2004; Maraganore et al., 2004). A missense mutation in *UCH-L1* is linked to early-onset familial PD.

Normally, autophagy in all its different types can remove α-synuclein aggregates. Remarkably, the level of α-synuclein increases when lysosomes are inhibited highlighting the pathogenic relationship between degradation of α-synuclein and autophagy (Arotcarena, Teil, & Dehay, 2019). Studies on post-mortem brains from PD patients revealed a significant increase in autophagosome aggregates secondary to lysosomes depletion (Dehay et al., 2010). Vacuolar protein sorting-associated protein (Vps) 35 is a part promoting WASH complex (Zavodszky et al., 2014). The mutation of the *VPS35* gene, which encode for a component of the retromer complex that has an important role in endosomal trafficking has been recently identified as rare autosomal dominant PD mutations (Elmatboly et al., 2020). In a *C. elegans* model, a mutation in *LRRK2* led to the inhibition of autophagy causing age-linked dysfunction and increased neuronal (Di Domenico et al., 2012; Saha et al., 2015) (Fig. 5). Mutation in PINK and Parkin leads to disruption of the mitophagy pathway and progression of the familial form of PD (Klein et al., 2005). Genome-wide association studies (GWASs) of Parkinson's disease, identified many genes related to lysosomal function, such as *LIMP-2, TMEM175, SMPD1, GBA, VPS13C, ATP13A2,* and *ATP6AP2* (Elmatboly et al., 2020; Shulman et al., 2011). As mentioned earlier, autophagy is regulated by two important metabolic sensors, mTORC1, and AMPK, and the control of their activation holds a potential way for PD treatment. Curcumin and piperine both can activate the autophagy pathway by inhibition of the mTORC1 pathway which leads to degradation of α-synuclein and ameliorates pathology of PD (Jiang et al., 2013; Liu et al., 2016). Resveratrol, sestrin2, and calcitriol induce autophagy by stimulating AMPK pathway, hence degradation of α-synuclein in animal models (Hou et al., 2015; Wu et al., 2011). Many drugs that increase the level or mimic Beclin-1 activities exert anti-PD pathology properties in animal models (Lan et al., 2012).

3.1.3 Amyotrophic lateral sclerosis

ALS is a typical neurodegenerative disorder characterized by the loss of both the cerebral motor cortex or brainstem (upper) motor neurons and the cranial nerves and ventral horns of the spinal cord (lower) motor neurons. ALS is characterized by rapidly progressive weakness, muscle atrophy, fasciculation, spasticity, dysarthria, dysphagia, and dyspnea (Gandhi et al., 2019). Patients with ALS progressively lose almost all motor function (Pasinelli & Brown, 2006). The sensory neurons are typically not affected, and most ALS patients do not experience cognitive malfunctions, although a subset of patients with frontotemporal degeneration do develop dementia (Mitchell & Borasio, 2007). About 90% of ALS cases appear as sporadic (sALS) forms, and only 10% of cases are caused by inherited mutations linked to familial (fALS) forms. The two types of ALS are clinically indistinguishable. Up to now, more than 30 genes have been found altered in fALS (Cook & Petrucelli, 2019; Mathis, Goizet, Soulages, Vallat, & Masson, 2019; Mejzini et al., 2019). In inherited forms of ALS several genes that are mutated lead to the formation of abnormally folded proteins which aggregate and form inclusion bodies as well as actively impair proteostasis mechanisms. These are mainly SOD1, TDP-43, FUS, and C9ORF72 (Andersen & Al-Chalabi, 2011). Furthermore, dominant mutations in genes encoding for components of the protein quality control machinery related to vesicle transport, autophagy, ER homeostasis and UPS such as tubulin 4A (TUB4A), Dynein, Dynactin, sequestosome-1 (p62), Optineurin, valosin-containing protein (VCP or p97), vesicle-associated membrane protein associated protein B (VAPB), TANK1 binding kinase 1 (TBK1) and Ubiquilin-2 are associated with ALS (Ruegsegger & Saxena, 2016). Notably, most of the ALS-associated mutations have been identified to directly target proteostasis network components and contribute to familial ALS, thereby providing direct evidence for the importance of proteostasis in ALS. Historically, the SOD1 is the first gene associated with fALS and accounts for 15% of all fALS cases (Cristofani et al., 2020). The exact role of SOD1 in ALS is still unclear, although mutations in the gene lead to misfolding of the translated protein and its consequent aggregation. SOD1 aggregation is the common pathological trait in both familial and sporadic ALS. cytoplasmic aggregates are believed to damage mitochondria, proteasomes, chaperones, and other proteins (Kiernan et al., 2011; Pasinelli & Brown, 2006). Ubiquitylated inclusions (UBI) are the most common histopathological hallmarks, found in LMNs of the brain stem and spinal cord and in UMNs of the temporal and frontal lobes of the neocortex of almost 100% of ALS patients. UBIs are composed of ubiquitin, peripherin, SOD1, and dorfin. TDP-43 is also found in UBI inclusion bodies (Gandhi et al., 2019).

ER morphology alterations occur both in ALS patients and ALS mouse models (Lautenschlaeger, Prell, & Grosskreutz, 2012), possibly because of protein accumulation in ER causing ER stress (Sasaki, 2010). Also, the Golgi apparatus is affected in ALS (Fujita, Okamoto, Sakurai, Gonatas, & Hirano, 2000). Mutant SOD1 inclusions in ER are positive for Grp78/BiP and calnexin (Wate et al., 2005), while several ER chaperones are upregulated in ALS patients and mice (Medinas, Gonzalez, Falcon, & Hetz, 2017). Remarkably, mutant SOD1 specifically binds Derlin-1, which controls the ERAD machinery, and triggers ER stress-induced apoptosis. ER stress in ALS may also result from altered ER calcium homeostasis or by ER-mitochondria calcium cycle unbalance (Grosskreutz, Van Den Bosch, & Keller, 2010; Nishitoh et al., 2008). Studies in mutant SOD1 mice described ER stress as one of the earliest pathological signatures in the gene expression profile of vulnerable and resistant motoneurons. Basal levels of ER stress and proteostasis alterations are an intrinsic feature of patient iPSC-derived human motoneurons (Rozas, Bargsted, Martinez, Hetz, & Medinas, 2017). UPR sensors, namely ATF6 and phospho-PERK, are elevated in ALS mice and cell models (Cristofani et al., 2020). In the spinal cord of ALS patients and mice, IRE1 is increased and its phosphorylated form correlated with spliced XBP1 (Ruegsegger & Saxena, 2016). Forty different target genes, associated with ERAD and regulated by XBP1 or ATF6, are altered in spinal cord specimens from ALS patients (Montibeller & de Belleroche, 2018). Among these genes, co-chaperones of the DNAJ family modulating BiP/Grp78 were increased in this dataset (Kampinga & Bergink, 2016).

Evidence suggests that SOD1, TDP-43, and its ALS associated fragments, as well as, other ALS-proteins, are processed via the proteasome (Cristofani et al., 2020). However, the large amounts of misfolded proteins formed when specific gene mutations occur may overwhelm the UPS capability to degrade them efficiently. This process is accentuated during aging where the chaperone and UPS activities are reduced (Ciechanover, 2015; Ciechanover & Brundin, 2003; Hegde, Smith, Duke, Pourquoi, & Vaz, 2019) Mutant SOD1 is mainly cleared by the UPS, and its pharmacological inhibition results in an accumulation of ubiquitin-positive SOD1 aggregates (Cicardi et al., 2019; Crippa et al., 2010, 2016). Recent studies showed that cytosolic accumulation of TDP-43 is turned over mainly by UPS, and UPS inhibition results in an accumulation and mislocalization of TDP-43 and fragments. Ubiquitination of TDP-43 is one of the major features of pathological TDP-43 that has been found in the brain and spinal cord of patients with ALS. Parkin is shown to ubiquitinate TDP-43 via the ubiquitin lysines, K-48, and K-63. This facilitates the TDP-43s cytoplasmic accumulation into inclusions without any detectable evidence of its protein degradation (Hohn et al., 2020).

The dysfunction of different autophagy steps was identified in both ALS animal models and patients. ALS-associated genes can impair autophagy activity and lead to the aggregation of proteins and organelles damage (Medinas et al., 2017). The impairment of Alsin, optineurin, p62/SQTSM, and ubiquilin-2 and VCP, all involved in autophagosome formation, delivery of autophagy cargo to the autophagosome and in autophagosome maturation, were observed in different ALS models. The initiation of autophagy has been observed in SOD1G93A motoneurons (Ruegsegger & Saxena, 2016) as well as in spinal cord samples from ALS patients (Morimoto et al., 2007; Sasaki, 2011). The appearance of autophagic dysfunction due to mutant SOD1-associated inclusions, actively impair proteostasis by sequestering molecular chaperones, such as Hsp70 and Hsp40, and proteasomal subunits. Notably, autophagy is induced in double knock-out/transgenic mice with mutant SOD1 and XBP1 blockage, suggesting that it may serve to protect when UPR/ERAD fails (Matus, Valenzuela, Medinas, & Hetz, 2013; Medinas et al., 2017; Rozas et al., 2017). Defective autophagy might be related to the TDP-43-related toxicity in ALS. Impaired endocytosis has been linked with an increase in TDP-43 aggregation, whereas enhancing endocytosis can reverse the TDP-43 toxicity and the motor neuron dysfunction (Hohn et al., 2020). Another study has also shown that the endocytosis and the *endo*-lysosomal pathway are markedly disturbed by TDP-43 expression that probably alters the expression of key endocytic components (Cristofani et al., 2020). In ALS, CMA has been involved in TDP-43 metabolism (Huang et al., 2014). TDP-43 contains a KFERQ-like domain, the consensus sequence that allows the interaction with HSC70. Mutation in this domain blocks the ubiquitin-dependent binding of TDP-43 with HSC70. LAMP2A down-regulation induces the intracellular accumulation of the ALS-associated TDP-43 fragments (Huang et al., 2014) (Fig. 5).

References

Adachi, Y., Yamamoto, K., Okada, T., Yoshida, H., Harada, A., & Mori, K. (2008). ATF6 is a transcription factor specializing in the regulation of quality control proteins in the endoplasmic reticulum. *Cell Structure and Function*, *33*, 75–89.

Andersen, P. M., & Al-Chalabi, A. (2011). Clinical genetics of amyotrophic lateral sclerosis: What do we really know? *Nature Reviews. Neurology*, 7, 603–615.

Anfinsen, C. B. (1973). Principles that govern the folding of protein chains. *Science*, *181*, 223–230.
Arotcarena, M. L., Teil, M., & Dehay, B. (2019). Autophagy in synucleinopathy: The overwhelmed and defective machinery. *Cell*, *8*(6), 565.
Auluck, P. K., Chan, H. Y., Trojanowski, J. Q., Lee, V. M., & Bonini, N. M. (2002). Chaperone suppression of alpha-synuclein toxicity in a Drosophila model for Parkinson's disease. *Science*, *295*, 865–868.
Bandyopadhyay, U., Sridhar, S., Kaushik, S., Kiffin, R., & Cuervo, A. M. (2010). Identification of regulators of chaperone-mediated autophagy. *Molecular Cell*, *39*, 535–547.
Barral, J. M., Broadley, S. A., Schaffar, G., & Hartl, F. U. (2004). Roles of molecular chaperones in protein misfolding diseases. *Seminars in Cell & Developmental Biology*, *15*, 17–29.
Bell, M. C., Meier, S. E., Ingram, A. L., & Abisambra, J. F. (2016). PERK-opathies: An endoplasmic reticulum stress mechanism underlying neurodegeneration. *Current Alzheimer Research*, *13*, 150–163.
Bence, N. F., Sampat, R. M., & Kopito, R. R. (2001). Impairment of the ubiquitin-proteasome system by protein aggregation. *Science*, *292*, 1552–1555.
Ben-Zvi, A., Miller, E. A., & Morimoto, R. I. (2009). Collapse of proteostasis represents an early molecular event in Caenorhabditis elegans aging. *Proceedings of the National Academy of Sciences of the United States of America*, *106*, 14914–14919.
Betzer, C., Lassen, L. B., Olsen, A., Kofoed, R. H., Reimer, L., Gregersen, E., et al. (2018). Alpha-synuclein aggregates activate calcium pump SERCA leading to calcium dysregulation. *EMBO Reports*, *19*(5), 1–21.
Boyce, M., Bryant, K. F., Jousse, C., Long, K., Harding, H. P., Scheuner, D., et al. (2005). A selective inhibitor of eIF2alpha dephosphorylation protects cells from ER stress. *Science*, *307*, 935–939.
Busciglio, J., Pelsman, A., Wong, C., Pigino, G., Yuan, M., Mori, H., et al. (2002). Altered metabolism of the amyloid beta precursor protein is associated with mitochondrial dysfunction in Down's syndrome. *Neuron*, *33*, 677–688.
Butterfield, D. A., Gu, L., Di Domenico, F., & Robinson, R. A. (2014). Mass spectrometry and redox proteomics: Applications in disease. *Mass Spectrometry Reviews*, *33*, 277–301.
Caccamo, A., Majumder, S., Richardson, A., Strong, R., & Oddo, S. (2010). Molecular interplay between mammalian target of rapamycin (mTOR), amyloid-beta, and tau: Effects on cognitive impairments. *The Journal of Biological Chemistry*, *285*, 13107–13120.
Cai, Q., & Jeong, Y. Y. (2020). Mitophagy in Alzheimer's disease and other age-related neurodegenerative diseases. *Cell*, *9*(150), 1–28.
Carreras-Sureda, A., Pihan, P., & Hetz, C. (2018). Calcium signaling at the endoplasmic reticulum: Fine-tuning stress responses. *Cell Calcium*, *70*, 24–31.
Chakrabarti, A., Chen, A. W., & Varner, J. D. (2011). A review of the mammalian unfolded protein response. *Biotechnology and Bioengineering*, *108*, 2777–2793.
Checler, F., da Costa, C. A., Ancolio, K., Chevallier, N., Lopez-Perez, E., & Marambaud, P. (2000). Role of the proteasome in Alzheimer's disease. *Biochimica et Biophysica Acta*, *1502*, 133–138.
Chiti, F., & Dobson, C. M. (2006). Protein misfolding, functional amyloid, and human disease. *Annual Review of Biochemistry*, *75*, 333–366.
Chiti, F., & Dobson, C. M. (2017). Protein misfolding, amyloid formation, and human disease: A summary of progress over the last decade. *Annual Review of Biochemistry*, *86*, 27–68.
Choi, J., Levey, A. I., Weintraub, S. T., Rees, H. D., Gearing, M., Chin, L. S., et al. (2004). Oxidative modifications and down-regulation of ubiquitin carboxyl-terminal hydrolase L1 associated with idiopathic Parkinson's and Alzheimer's diseases. *The Journal of Biological Chemistry*, *279*, 13256–13264.

Chondrogianni, N., Petropoulos, I., Grimm, S., Georgila, K., Catalgol, B., Friguet, B., et al. (2014). Protein damage, repair and proteolysis. *Molecular Aspects of Medicine*, *35*, 1–71.
Chou, A., Krukowski, K., Jopson, T., Zhu, P. J., Costa-Mattioli, M., Walter, P., et al. (2017). Inhibition of the integrated stress response reverses cognitive deficits after traumatic brain injury. *Proceedings of the National Academy of Sciences of the United States of America*, *114*, E6420–E6426.
Chung, C. Y., Khurana, V., Auluck, P. K., Tardiff, D. F., Mazzulli, J. R., Soldner, F., et al. (2013). Identification and rescue of alpha-synuclein toxicity in Parkinson patient-derived neurons. *Science*, *342*, 983–987.
Cicardi, M. E., Cristofani, R., Crippa, V., Ferrari, V., Tedesco, B., Casarotto, E., et al. (2019). Autophagic and proteasomal mediated removal of mutant androgen receptor in muscle models of spinal and bulbar muscular atrophy. *Frontiers in Endocrinology*, *10*, 569.
Ciechanover, A. (2015). The unravelling of the ubiquitin system. *Nature Reviews. Molecular Cell Biology*, *16*, 322–324.
Ciechanover, A., & Brundin, P. (2003). The ubiquitin proteasome system in neurodegenerative diseases: Sometimes the chicken, sometimes the egg. *Neuron*, *40*, 427–446.
Cook, C., & Petrucelli, L. (2019). Genetic convergence brings clarity to the enigmatic red line in ALS. *Neuron*, *101*, 1057–1069.
Corti, O., Lesage, S., & Brice, A. (2011). What genetics tells us about the causes and mechanisms of Parkinson's disease. *Physiological Reviews*, *91*, 1161–1218.
Costa-Mattioli, M., Gobert, D., Stern, E., Gamache, K., Colina, R., Cuello, C., et al. (2007). eIF2alpha phosphorylation bidirectionally regulates the switch from short- to long-term synaptic plasticity and memory. *Cell*, *129*, 195–206.
Costa-Mattioli, M., & Walter, P. (2020). The integrated stress response: From mechanism to disease. *Science*, *368*(6489), 1–11.
Credle, J. J., Forcelli, P. A., Delannoy, M., Oaks, A. W., Permaul, E., Berry, D. L., et al. (2015). Alpha-Synuclein-mediated inhibition of ATF6 processing into COPII vesicles disrupts UPR signaling in Parkinson's disease. *Neurobiology of Disease*, *76*, 112–125.
Crippa, V., Carra, S., Rusmini, P., Sau, D., Bolzoni, E., Bendotti, C., et al. (2010). A role of small heat shock protein B8 (HspB8) in the autophagic removal of misfolded proteins responsible for neurodegenerative diseases. *Autophagy*, *6*, 958–960.
Crippa, V., Cicardi, M. E., Ramesh, N., Seguin, S. J., Ganassi, M., Bigi, I., et al. (2016). The chaperone HSPB8 reduces the accumulation of truncated TDP-43 species in cells and protects against TDP-43-mediated toxicity. *Human Molecular Genetics*, *25*, 3908–3924.
Cristofani, R., Crippa, V., Cicardi, M. E., Tedesco, B., Ferrari, V., Chierichetti, M., et al. (2020). A crucial role for the protein quality control system in motor neuron diseases. *Frontiers in Aging Neuroscience*, *12*, 191.
Cuanalo-Contreras, K., Mukherjee, A., & Soto, C. (2013). Role of protein misfolding and proteostasis deficiency in protein misfolding diseases and aging. *International Journal of Cell Biology*, *2013*, 638083.
Cui, W., Li, J., Ron, D., & Sha, B. (2011). The structure of the PERK kinase domain suggests the mechanism for its activation. *Acta Crystallographica. Section D, Biological Crystallography*, *67*, 423–428.
Dauer, W., & Przedborski, S. (2003). Parkinson's disease: Mechanisms and models. *Neuron*, *39*, 889–909.
Dehay, B., Bove, J., Rodriguez-Muela, N., Perier, C., Recasens, A., Boya, P., et al. (2010). Pathogenic lysosomal depletion in Parkinson's disease. *The Journal of Neuroscience*, *30*, 12535–12544.
Di Domenico, F., Barone, E., Perluigi, M., & Butterfield, D. A. (2017). The triangle of death in Alzheimer's disease brain: The aberrant cross-talk among energy metabolism, mammalian target of rapamycin signaling, and protein homeostasis revealed by redox proteomics. *Antioxidants & Redox Signaling*, *26*, 364–387.

Di Domenico, F., Head, E., Butterfield, D. A., & Perluigi, M. (2014). Oxidative stress and proteostasis network: Culprit and casualty of Alzheimer's-like neurodegeneration. *Advances in Geriatrics, 2014*, 14.

Di Domenico, F., Sultana, R., Ferree, A., Smith, K., Barone, E., Perluigi, M., et al. (2012). Redox proteomics analyses of the influence of co-expression of wild-type or mutated LRRK2 and tau on C. elegans protein expression and oxidative modification: Relevance to Parkinson disease. *Antioxidants & Redox Signaling, 17*, 1490–1506.

Di Domenico, F., Sultana, R., Tiu, G. F., Scheff, N. N., Perluigi, M., Cini, C., et al. (2010). Protein levels of heat shock proteins 27, 32, 60, 70, 90 and thioredoxin-1 in amnestic mild cognitive impairment: An investigation on the role of cellular stress response in the progression of Alzheimer disease. *Brain Research, 1333*, 72–81.

Di Domenico, F., Tramutola, A., Barone, E., Lanzillotta, C., Defever, O., Arena, A., et al. (2019). Restoration of aberrant mTOR signaling by intranasal rapamycin reduces oxidative damage: Focus on HNE-modified proteins in a mouse model of down syndrome. *Redox Biology, 23*, 1–14.

Di Domenico, F., Zuliani, I., & Tramutola, A. (2019). Shining a light on defective autophagy by proteomics approaches: Implications for neurodegenerative illnesses. *Expert Review of Proteomics, 16*, 951–964.

Dickey, C., Kraft, C., Jinwal, U., Koren, J., Johnson, A., Anderson, L., et al. (2009). Aging analysis reveals slowed tau turnover and enhanced stress response in a mouse model of tauopathy. *The American Journal of Pathology, 174*, 228–238.

Dou, F., Netzer, W. J., Tanemura, K., Li, F., Hartl, F. U., Takashima, A., et al. (2003). Chaperones increase association of tau protein with microtubules. *Proceedings of the National Academy of Sciences of the United States of America, 100*, 721–726.

Egawa, N., Yamamoto, K., Inoue, H., Hikawa, R., Nishi, K., Mori, K., et al. (2011). The endoplasmic reticulum stress sensor, ATF6alpha, protects against neurotoxin-induced dopaminergic neuronal death. *The Journal of Biological Chemistry, 286*, 7947–7957.

Elmatboly, A. M., Sherif, A. M., Deeb, D. A., Benmelouka, A., Bin-Jumah, M. N., Aleya, L., et al. (2020). The impact of proteostasis dysfunction secondary to environmental and genetic causes on neurodegenerative diseases progression and potential therapeutic intervention. *Environmental Science and Pollution Research International, 27*, 11461–11483.

Flower, T. R., Chesnokova, L. S., Froelich, C. A., Dixon, C., & Witt, S. N. (2005). Heat shock prevents alpha-synuclein-induced apoptosis in a yeast model of Parkinson's disease. *Journal of Molecular Biology, 351*, 1081–1100.

Friesen, E. L., De Snoo, M. L., Rajendran, L., Kalia, L. V., & Kalia, S. K. (2017). Chaperone-based therapies for disease modification in Parkinson's disease. *Parkinsons Disease, 2017*, 5015307.

Fujita, Y., Okamoto, K., Sakurai, A., Gonatas, N. K., & Hirano, A. (2000). Fragmentation of the Golgi apparatus of the anterior horn cells in patients with familial amyotrophic lateral sclerosis with SOD1 mutations and posterior column involvement. *Journal of the Neurological Sciences, 174*, 137–140.

Galvan, A., & Wichmann, T. (2008). Pathophysiology of parkinsonism. *Clinical Neurophysiology, 119*, 1459–1474.

Gandhi, J., Antonelli, A. C., Afridi, A., Vatsia, S., Joshi, G., Romanov, V., et al. (2019). Protein misfolding and aggregation in neurodegenerative diseases: A review of pathogeneses, novel detection strategies, and potential therapeutics. *Reviews in the Neurosciences, 30*, 339–358.

Garcia-Arencibia, M., Hochfeld, W. E., Toh, P. P., & Rubinsztein, D. C. (2010). Autophagy, a guardian against neurodegeneration. *Seminars in Cell & Developmental Biology, 21*, 691–698.

Gerakis, Y., & Hetz, C. (2018). Emerging roles of ER stress in the etiology and pathogenesis of Alzheimer's disease. *The FEBS Journal, 285*, 995–1011.

Giandomenico, S. L., Alvarez-Castelao, B., & Schuman, E. M. (2022). Proteostatic regulation in neuronal compartments. *Trends in Neurosciences*, *45*, 41–52.
Giorgi, C., Bouhamida, E., Danese, A., Previati, M., Pinton, P., & Patergnani, S. (2021). Relevance of autophagy and mitophagy dynamics and markers in neurodegenerative diseases. *Biomedicine*, *9*(2), 149.
Giorgi, C., Missiroli, S., Patergnani, S., Duszynski, J., Wieckowski, M. R., & Pinton, P. (2015). Mitochondria-associated membranes: Composition, molecular mechanisms, and physiopathological implications. *Antioxidants & Redox Signaling*, *22*, 995–1019.
Gorbatyuk, M. S., Shabashvili, A., Chen, W., Meyers, C., Sullivan, L. F., Salganik, M., et al. (2012). Glucose regulated protein 78 diminishes alpha-synuclein neurotoxicity in a rat model of Parkinson disease. *Molecular Therapy*, *20*, 1327–1337.
Grimm, S., Hohn, A., & Grune, T. (2012). Oxidative protein damage and the proteasome. *Amino Acids*, *42*, 23–38.
Grosskreutz, J., Van Den Bosch, L., & Keller, B. U. (2010). Calcium dysregulation in amyotrophic lateral sclerosis. *Cell Calcium*, *47*, 165–174.
Haass, C., & Selkoe, D. J. (2007). Soluble protein oligomers in neurodegeneration: Lessons from the Alzheimer's amyloid beta-peptide. *Nature Reviews. Molecular Cell Biology*, *8*, 101–112.
Hanus, C., & Schuman, E. M. (2013). Proteostasis in complex dendrites. *Nature Reviews. Neuroscience*, *14*, 638–648.
Harding, H. P., Zhang, Y., & Ron, D. (1999). Protein translation and folding are coupled by an endoplasmic-reticulum-resident kinase. *Nature*, *397*, 271–274.
Harding, H. P., Zhang, Y., Zeng, H., Novoa, I., Lu, P. D., Calfon, M., et al. (2003). An integrated stress response regulates amino acid metabolism and resistance to oxidative stress. *Molecular Cell*, *11*, 619–633.
Hartl, F. U. (1996). Molecular chaperones in cellular protein folding. *Nature*, *381*, 571–579.
Hashida, K., Kitao, Y., Sudo, H., Awa, Y., Maeda, S., Mori, K., et al. (2012). ATF6alpha promotes astroglial activation and neuronal survival in a chronic mouse model of Parkinson's disease. *PLoS One*, *7*, e47950.
Hegde, A. N., Smith, S. G., Duke, L. M., Pourquoi, A., & Vaz, S. (2019). Perturbations of ubiquitin-proteasome-mediated proteolysis in aging and Alzheimer's disease. *Frontiers in Aging Neuroscience*, *11*, 324.
Heras-Sandoval, D., Perez-Rojas, J. M., Hernandez-Damian, J., & Pedraza-Chaverri, J. (2014). The role of PI3K/AKT/mTOR pathway in the modulation of autophagy and the clearance of protein aggregates in neurodegeneration. *Cellular Signalling*, *26*, 2694–2701.
Hetz, C. (2012). The unfolded protein response: Controlling cell fate decisions under ER stress and beyond. *Nature Reviews. Molecular Cell Biology*, *13*, 89–102.
Hetz, C. (2021). Adapting the proteostasis capacity to sustain brain healthspan. *Cell*, *184*, 1545–1560.
Hetz, C., Chevet, E., & Oakes, S. A. (2015). Proteostasis control by the unfolded protein response. *Nature Cell Biology*, *17*, 829–838.
Hetz, C., & Glimcher, L. H. (2011). Protein homeostasis networks in physiology and disease. *Current Opinion in Cell Biology*, *23*, 123–125.
Hetz, C., & Papa, F. R. (2018). The unfolded protein response and cell fate control. *Molecular Cell*, *69*, 169–181.
Hetz, C., & Saxena, S. (2017). ER stress and the unfolded protein response in neurodegeneration. *Nature Reviews. Neurology*, *13*, 477–491.
Hetz, C. A., & Soto, C. (2006). Emerging roles of the unfolded protein response signaling in physiology and disease. *Current Molecular Medicine*, *6*, 1.

Hetz, C., Zhang, K., & Kaufman, R. J. (2020). Mechanisms, regulation and functions of the unfolded protein response. *Nature Reviews. Molecular Cell Biology*, *21*, 421–438.
Hipp, M. S., Kasturi, P., & Hartl, F. U. (2019). The proteostasis network and its decline in ageing. *Nature Reviews. Molecular Cell Biology*, *20*, 421–435.
Hohn, T. J., & Grune, T. (2014). The proteasome and the degradation of oxidized proteins: Part III-redox regulation of the proteasomal system. *Redox Biology*, *2*, 388–394.
Hohn, A., Tramutola, A., & Cascella, R. (2020). Proteostasis failure in neurodegenerative diseases: Focus on oxidative stress. *Oxidative Medicine and Cellular Longevity*, *2020*, 5497046.
Hol, E. M., van Leeuwen, F. W., & Fischer, D. F. (2005). The proteasome in Alzheimer's disease and Parkinson's disease: Lessons from ubiquitin B + 1. *Trends in Molecular Medicine*, *11*, 488–495.
Hoozemans, J. J. M., & Scheper, W. (2012). Endoplasmic reticulum: The unfolded protein response is tangled in neurodegeneration. *International Journal of Biochemistry & Cell Biology*, *44*, 1295–1298.
Hoozemans, J. J., van Haastert, E. S., Eikelenboom, P., de Vos, R. A., Rozemuller, J. M., & Scheper, W. (2007). Activation of the unfolded protein response in Parkinson's disease. *Biochemical and Biophysical Research Communications*, *354*, 707–711.
Hoozemans, J. J., van Haastert, E. S., Nijholt, D. A., Rozemuller, A. J., Eikelenboom, P., & Scheper, W. (2009). The unfolded protein response is activated in pretangle neurons in Alzheimer's disease hippocampus. *The American Journal of Pathology*, *174*, 1241–1251.
Hoozemans, J. J., Veerhuis, R., Van Haastert, E. S., Rozemuller, J. M., Baas, F., Eikelenboom, P., et al. (2005). The unfolded protein response is activated in Alzheimer's disease. *Acta Neuropathologica*, *110*, 165–172.
Hou, Y. S., Guan, J. J., Xu, H. D., Wu, F., Sheng, R., & Qin, Z. H. (2015). Sestrin2 protects dopaminergic cells against rotenone toxicity through AMPK-dependent autophagy activation. *Molecular and Cellular Biology*, *35*, 2740–2751.
Hu, W. T., Josephs, K. A., Knopman, D. S., Boeve, B. F., Dickson, D. W., Petersen, R. C., et al. (2008). Temporal lobar predominance of TDP-43 neuronal cytoplasmic inclusions in Alzheimer disease. *Acta Neuropathologica*, *116*, 215–220.
Huang, C. C., Bose, J. K., Majumder, P., Lee, K. H., Huang, J. T., Huang, J. K., et al. (2014). Metabolism and mis-metabolism of the neuropathological signature protein TDP-43. *Journal of Cell Science*, *127*, 3024–3038.
Huang, W., Placzek, A. N., Viana Di Prisco, G., Khatiwada, S., Sidrauski, C., Krnjevic, K., et al. (2016). Translational control by eIF2alpha phosphorylation regulates vulnerability to the synaptic and behavioral effects of cocaine. *eLife*, *5*, e12052.
Iqbal, K., Alonso Adel, C., Chen, S., Chohan, M. O., El-Akkad, E., Gong, C. X., et al. (2005). Tau pathology in Alzheimer disease and other tauopathies. *Biochimica et Biophysica Acta*, *1739*, 198–210.
Jiang, T. F., Zhang, Y. J., Zhou, H. Y., Wang, H. M., Tian, L. P., Liu, J., et al. (2013). Curcumin ameliorates the neurodegenerative pathology in A53T alpha-synuclein cell model of Parkinson's disease through the downregulation of mTOR/p70S6K signaling and the recovery of macroautophagy. *Journal of Neuroimmune Pharmacology*, *8*, 356–369.
Kalia, L. V., & Lang, A. E. (2015). Parkinson's disease. *Lancet*, *386*, 896–912.
Kampinga, H. H., & Bergink, S. (2016). Heat shock proteins as potential targets for protective strategies in neurodegeneration. *Lancet Neurology*, *15*, 748–759.
Kane, L. A., Lazarou, M., Fogel, A. I., Li, Y., Yamano, K., Sarraf, S. A., et al. (2014). PINK1 phosphorylates ubiquitin to activate Parkin E3 ubiquitin ligase activity. *The Journal of Cell Biology*, *205*, 143–153.
Kayed, R., Head, E., Thompson, J. L., McIntire, T. M., Milton, S. C., Cotman, C. W., et al. (2003). Common structure of soluble amyloid oligomers implies common mechanism of pathogenesis. *Science*, *300*, 486–489.

Kazlauskaite, A., & Muqit, M. M. (2015). PINK1 and Parkin—Mitochondrial interplay between phosphorylation and ubiquitylation in Parkinson's disease. *The FEBS Journal, 282*, 215–223.

Kiernan, M. C., Vucic, S., Cheah, B. C., Turner, M. R., Eisen, A., Hardiman, O., et al. (2011). Amyotrophic lateral sclerosis. *Lancet, 377*, 942–955.

Kim, J., Kundu, M., Viollet, B., & Guan, K. L. (2011). AMPK and mTOR regulate autophagy through direct phosphorylation of Ulk1. *Nature Cell Biology, 13*, 132–141.

Klaips, C. L., Jayaraj, G. G., & Hartl, F. U. (2018). Pathways of cellular proteostasis in aging and disease. *The Journal of Cell Biology, 217*, 51–63.

Klein, C., Djarmati, A., Hedrich, K., Schafer, N., Scaglione, C., Marchese, R., et al. (2005). PINK1, Parkin, and DJ-1 mutations in Italian patients with early-onset parkinsonism. *European Journal of Human Genetics, 13*, 1086–1093.

Klucken, J., Shin, Y., Hyman, B. T., & McLean, P. J. (2004). A single amino acid substitution differentiates Hsp70-dependent effects on alpha-synuclein degradation and toxicity. *Biochemical and Biophysical Research Communications, 325*, 367–373.

Klucken, J., Shin, Y., Masliah, E., Hyman, B. T., & McLean, P. J. (2004). Hsp70 reduces alpha-synuclein aggregation and toxicity. *The Journal of Biological Chemistry, 279*, 25497–25502.

Komander, D., & Rape, M. (2012). The ubiquitin code. *Annual Review of Biochemistry, 81*, 203–229.

Kraft, C., Peter, M., & Hofmann, K. (2010). Selective autophagy: Ubiquitin-mediated recognition and beyond. *Nature Cell Biology, 12*, 836–841.

Kurtishi, A., Rosen, B., Patil, K. S., Alves, G. W., & Moller, S. G. (2019). Cellular proteostasis in neurodegeneration. *Molecular Neurobiology, 56*, 3676–3689.

Kwon, Y. T., & Ciechanover, A. (2017). The ubiquitin code in the ubiquitin-proteasome system and autophagy. *Trends in Biochemical Sciences, 42*, 873–886.

Labbadia, J., & Morimoto, R. I. (2015). The biology of proteostasis in aging and disease. *Annual Review of Biochemistry, 84*, 435–464.

Lan, D. M., Liu, F. T., Zhao, J., Chen, Y., Wu, J. J., Ding, Z. T., et al. (2012). Effect of trehalose on PC12 cells overexpressing wild-type or A53T mutant alpha-synuclein. *Neurochemical Research, 37*, 2025–2032.

Lanzillotta, C., & Di Domenico, F. (2021). Stress responses in down syndrome neurodegeneration: State of the art and therapeutic molecules. *Biomolecules, 11*, 266.

Lanzillotta, C., Tramutola, A., Meier, S., Schmitt, F., Barone, E., Perluigi, M., et al. (2018). Early and selective activation and subsequent alterations to the unfolded protein response in down syndrome mouse models. *Journal of Alzheimer's Disease, 62*, 347–359.

Lanzillotta, C., Zuliani, I., Tramutola, A., Barone, E., Blarzino, C., Folgiero, V., et al. (2021). Chronic PERK induction promotes Alzheimer-like neuropathology in down syndrome: Insights for therapeutic intervention. *Progress in Neurobiology, 196*, 101892.

Lautenschlaeger, J., Prell, T., & Grosskreutz, J. (2012). Endoplasmic reticulum stress and the ER mitochondrial calcium cycle in amyotrophic lateral sclerosis. *Amyotrophic Lateral Sclerosis, 13*, 166–177.

Lee, H., Noh, J. Y., Oh, Y., Kim, Y., Chang, J. W., Chung, C. W., et al. (2012). IRE1 plays an essential role in ER stress-mediated aggregation of mutant huntingtin via the inhibition of autophagy flux. *Human Molecular Genetics, 21*, 101–114.

Leitman, J., Ron, E., Ogen-Shtern, N., & Lederkremer, G. Z. (2013). Compartmentalization of endoplasmic reticulum quality control and ER-associated degradation factors. *DNA and Cell Biology, 32*, 2–7.

Li, Q., Liu, Y., & Sun, M. (2017). Autophagy and Alzheimer's disease. *Cellular and Molecular Neurobiology, 37*, 377–388.

Lim, K. L., & Tan, J. M. (2007). Role of the ubiquitin proteasome system in Parkinson's disease. *BMC Biochemistry, 8*(Suppl 1), S13.

Lipton, J. O., & Sahin, M. (2014). The neurology of mTOR. *Neuron, 84*, 275–291.
Liu, J., Chen, M., Wang, X., Wang, Y., Duan, C., Gao, G., et al. (2016). Piperine induces autophagy by enhancing protein phosphotase 2A activity in a rotenone-induced Parkinson's disease model. *Oncotarget*, 7, 60823–60843.
Maraganore, D. M., Lesnick, T. G., Elbaz, A., Chartier-Harlin, M. C., Gasser, T., Kruger, R., et al. (2004). UCHL1 is a Parkinson's disease susceptibility gene. *Annals of Neurology, 55*, 512–521.
Mathis, S., Goizet, C., Soulages, A., Vallat, J. M., & Masson, G. L. (2019). Genetics of amyotrophic lateral sclerosis: A review. *Journal of the Neurological Sciences, 399*, 217–226.
Matts, R. L., Brandt, G. E., Lu, Y., Dixit, A., Mollapour, M., Wang, S., et al. (2011). A systematic protocol for the characterization of Hsp90 modulators. *Bioorganic & Medicinal Chemistry, 19*, 684–692.
Matus, S., Valenzuela, V., Medinas, D. B., & Hetz, C. (2013). ER dysfunction and protein folding stress in ALS. *International Journal of Cell Biology, 2013*, 674751.
Medinas, D. B., Gonzalez, J. V., Falcon, P., & Hetz, C. (2017). Fine-tuning ER stress signal transducers to treat amyotrophic lateral sclerosis. *Frontiers in Molecular Neuroscience, 10*, 216.
Mejzini, R., Flynn, L. L., Pitout, I. L., Fletcher, S., Wilton, S. D., & Akkari, P. A. (2019). ALS genetics, mechanisms, and therapeutics: Where are we now? *Frontiers in Neuroscience, 13*, 1310.
Mercado, G., Castillo, V., Soto, P., Lopez, N., Axten, J. M., Sardi, S. P., et al. (2018). Targeting PERK signaling with the small molecule GSK2606414 prevents neurodegeneration in a model of Parkinson's disease. *Neurobiology of Disease, 112*, 136–148.
Metaxakis, A., Ploumi, C., & Tavernarakis, N. (2018). Autophagy in age-associated neurodegeneration. *Cell*, 7(5), 37.
Milisav, I., Suput, D., & Ribaric, S. (2015). Unfolded protein response and macroautophagy in Alzheimer's, Parkinson's and Prion diseases. *Molecules, 20*, 22718–22756.
Mitchell, J. D., & Borasio, G. D. (2007). Amyotrophic lateral sclerosis. *Lancet, 369*, 2031–2041.
Montibeller, L., & de Belleroche, J. (2018). Amyotrophic lateral sclerosis (ALS) and Alzheimer's disease (AD) are characterised by differential activation of ER stress pathways: Focus on UPR target genes. *Cell Stress & Chaperones, 23*, 897–912.
Morimoto, N., Nagai, M., Ohta, Y., Miyazaki, K., Kurata, T., Morimoto, M., et al. (2007). Increased autophagy in transgenic mice with a G93A mutant SOD1 gene. *Brain Research, 1167*, 112–117.
Mueed, Z., Tandon, P., Maurya, S. K., Deval, R., Kamal, M. A., & Poddar, N. K. (2018). Tau and mTOR: The hotspots for multifarious diseases in Alzheimer's development. *Frontiers in Neuroscience, 12*, 1017.
Mulder, M. P. C., Witting, K. F., & Ovaa, H. (2020). Cracking the ubiquitin code: The ubiquitin toolbox. *Current Issues in Molecular Biology, 37*, 1–20.
Murrow, L., & Debnath, J. (2013). Autophagy as a stress-response and quality-control mechanism: Implications for cell injury and human disease. *Annual Review of Pathology, 8*, 105–137.
Myeku, N., & Duff, K. E. (2018). Targeting the 26S proteasome to protect against proteotoxic diseases. *Trends in Molecular Medicine, 24*, 18–29.
Nag, S., Yu, L., Capuano, A. W., Wilson, R. S., Leurgans, S. E., Bennett, D. A., et al. (2015). Hippocampal sclerosis and TDP-43 pathology in aging and Alzheimer disease. *Annals of Neurology*, 77, 942–952.
Nillegoda, N. B., & Bukau, B. (2015). Metazoan Hsp70-based protein disaggregases: Emergence and mechanisms. *Frontiers in Molecular Biosciences, 2*, 57.
Nishitoh, H., Kadowaki, H., Nagai, A., Maruyama, T., Yokota, T., Fukutomi, H., et al. (2008). ALS-linked mutant SOD1 induces ER stress- and ASK1-dependent motor neuron death by targeting Derlin-1. *Genes & Development, 22*, 1451–1464.

Nixon, R. A. (2007). Autophagy, amyloidogenesis and Alzheimer disease. *Journal of Cell Science, 120*, 4081–4091.

Nixon, R. A., & Yang, D. S. (2011). Autophagy failure in Alzheimer's disease—Locating the primary defect. *Neurobiology of Disease, 43*, 38–45.

Oddo, S. (2008). The ubiquitin-proteasome system in Alzheimer's disease. *Journal of Cellular and Molecular Medicine, 12*, 363–373.

Oddo, S. (2012). The role of mTOR signaling in Alzheimer disease. *Frontiers in Bioscience (Scholar Edition), 4*, 941–952.

Oddo, S., Vasilevko, V., Caccamo, A., Kitazawa, M., Cribbs, D. H., & LaFerla, F. M. (2006). Reduction of soluble Abeta and tau, but not soluble Abeta alone, ameliorates cognitive decline in transgenic mice with plaques and tangles. *The Journal of Biological Chemistry, 281*, 39413–39423.

Olzscha, H., Schermann, S. M., Woerner, A. C., Pinkert, S., Hecht, M. H., Tartaglia, G. G., et al. (2011). Amyloid-like aggregates sequester numerous metastable proteins with essential cellular functions. *Cell, 144*, 67–78.

Orr, M. E., & Oddo, S. (2013). Autophagic/lysosomal dysfunction in Alzheimer's disease. *Alzheimer's Research & Therapy, 5*, 53.

Pakos-Zebrucka, K., Koryga, I., Mnich, K., Ljujic, M., Samali, A., & Gorman, A. M. (2016). The integrated stress response. *EMBO Reports, 17*, 1374–1395.

Park, H., Kang, J. H., & Lee, S. (2020). Autophagy in neurodegenerative diseases: A hunter for aggregates. *International Journal of Molecular Sciences, 21*(9), 3369.

Pasinelli, P., & Brown, R. H. (2006). Molecular biology of amyotrophic lateral sclerosis: Insights from genetics. *Nature Reviews. Neuroscience, 7*, 710–723.

Pasinetti, G. M. (2001). Use of cDNA microarray in the search for molecular markers involved in the onset of Alzheimer's disease dementia. *Journal of Neuroscience Research, 65*, 471–476.

Perluigi, M., Di Domenico, F., & Butterfield, D. A. (2015). mTOR signaling in aging and neurodegeneration: At the crossroad between metabolism dysfunction and impairment of autophagy. *Neurobiology of Disease, 84*, 39–49.

Petrucelli, L., Dickson, D., Kehoe, K., Taylor, J., Snyder, H., Grover, A., et al. (2004). CHIP and Hsp70 regulate tau ubiquitination, degradation and aggregation. *Human Molecular Genetics, 13*, 703–714.

Picca, A., Calvani, R., Coelho-Junior, H. J., Landi, F., Bernabei, R., & Marzetti, E. (2020). Mitochondrial dysfunction, oxidative stress, and neuroinflammation: Intertwined roads to neurodegeneration. *Antioxidants (Basel), 9*(8), 647.

Pickrell, A. M., & Youle, R. J. (2015). The roles of PINK1, parkin, and mitochondrial fidelity in Parkinson's disease. *Neuron, 85*, 257–273.

Pringsheim, T., Jette, N., Frolkis, A., & Steeves, T. D. (2014). The prevalence of Parkinson's disease: A systematic review and meta-analysis. *Movement Disorders, 29*, 1583–1590.

Pritzkow, S., Morales, R., Lyon, A., Concha-Marambio, L., Urayama, A., & Soto, C. (2018). Efficient prion disease transmission through common environmental materials. *The Journal of Biological Chemistry, 293*, 3363–3373.

Putcha, P., Danzer, K. M., Kranich, L. R., Scott, A., Silinski, M., Mabbett, S., et al. (2010). Brain-permeable small-molecule inhibitors of Hsp90 prevent alpha-synuclein oligomer formation and rescue alpha-synuclein-induced toxicity. *The Journal of Pharmacology and Experimental Therapeutics, 332*, 849–857.

Querfurth, H. W., & LaFerla, F. M. (2010). Alzheimer's disease. *The New England Journal of Medicine, 362*, 329–344.

Rochet, J. C., & Lansbury, P. T., Jr. (2000). Amyloid fibrillogenesis: Themes and variations. *Current Opinion in Structural Biology, 10*, 60–68.

Romani-Aumedes, J., Canal, M., Martin-Flores, N., Sun, X., Perez-Fernandez, V., Wewering, S., et al. (2014). Parkin loss of function contributes to RTP801 elevation and neurodegeneration in Parkinson's disease. *Cell Death & Disease, 5*, e1364.

Ron, D., & Hubbard, S. R. (2008). How IRE1 reacts to ER stress. *Cell, 132*, 24–26.
Roussel, B. D., Kruppa, A. J., Miranda, E., Crowther, D. C., Lomas, D. A., & Marciniak, S. J. (2013). Endoplasmic reticulum dysfunction in neurological disease. *Lancet Neurology, 12*, 105–118.
Rozas, P., Bargsted, L., Martinez, F., Hetz, C., & Medinas, D. B. (2017). The ER proteostasis network in ALS: Determining the differential motoneuron vulnerability. *Neuroscience Letters, 636*, 9–15.
Ruegsegger, C., & Saxena, S. (2016). Proteostasis impairment in ALS. *Brain Research, 1648*, 571–579.
Ruggiano, A., Foresti, O., & Carvalho, P. (2014). Quality control: ER-associated degradation: Protein quality control and beyond. *The Journal of Cell Biology, 204*, 869–879.
Saha, S., Ash, P. E., Gowda, V., Liu, L., Shirihai, O., & Wolozin, B. (2015). Mutations in LRRK2 potentiate age-related impairment of autophagic flux. *Molecular Neurodegeneration, 10*, 26.
Saha, S., Panigrahi, D. P., Patil, S., & Bhutia, S. K. (2018). Autophagy in health and disease: A comprehensive review. *Biomedicine & Pharmacotherapy, 104*, 485–495.
Saibil, H. (2013). Chaperone machines for protein folding, unfolding and disaggregation. *Nature Reviews. Molecular Cell Biology, 14*, 630–642.
Sarkar, M., Kuret, J., & Lee, G. (2008). Two motifs within the tau microtubule-binding domain mediate its association with the hsc70 molecular chaperone. *Journal of Neuroscience Research, 86*, 2763–2773.
Sasaki, S. (2010). Endoplasmic reticulum stress in motor neurons of the spinal cord in sporadic amyotrophic lateral sclerosis. *Journal of Neuropathology and Experimental Neurology, 69*, 346–355.
Sasaki, S. (2011). Autophagy in spinal cord motor neurons in sporadic amyotrophic lateral sclerosis. *Journal of Neuropathology and Experimental Neurology, 70*, 349–359.
Scheper, W., Nijholt, D. A. T., & Hoozemans, J. J. M. (2011). The unfolded protein response and proteostasis in Alzheimer disease preferential activation of autophagy by endoplasmic reticulum stress. *Autophagy, 7*, 910–911.
Schmidt, M. F., Gan, Z. Y., Komander, D., & Dewson, G. (2021). Ubiquitin signalling in neurodegeneration: Mechanisms and therapeutic opportunities. *Cell Death and Differentiation, 28*, 570–590.
Shin, W. H., Park, J. H., & Chung, K. C. (2020). The central regulator p62 between ubiquitin proteasome system and autophagy and its role in the mitophagy and Parkinson's disease. *BMB Reports, 53*, 56–63.
Shulman, J. M., De Jager, P. L., & Feany, M. B. (2011). Parkinson's disease: Genetics and pathogenesis. *Annual Review of Pathology, 6*, 193–222.
Song, P., Li, S., Wu, H., Gao, R., Rao, G., Wang, D., et al. (2016). Parkin promotes proteasomal degradation of p62: Implication of selective vulnerability of neuronal cells in the pathogenesis of Parkinson's disease. *Protein & Cell, 7*, 114–129.
Sossin, W. S., & Costa-Mattioli, M. (2019). Translational control in the brain in health and disease. *Cold Spring Harbor Perspectives in Biology, 11*(8), a032912.
Soto, C., & Pritzkow, S. (2018). Protein misfolding, aggregation, and conformational strains in neurodegenerative diseases. *Nature Neuroscience, 21*, 1332–1340.
Spillantini, M. G., Schmidt, M. L., Lee, V. M., Trojanowski, J. Q., Jakes, R., & Goedert, M. (1997). Alpha-synuclein in Lewy bodies. *Nature, 388*, 839–840.
Spilman, P., Podlutskaya, N., Hart, M. J., Debnath, J., Gorostiza, O., Bredesen, D., et al. (2010). Inhibition of mTOR by rapamycin abolishes cognitive deficits and reduces amyloid-beta levels in a mouse model of Alzheimer's disease. *PLoS One, 5*, e9979.
Sulzer, D. (2010). Clues to how alpha-synuclein damages neurons in Parkinson's disease. *Movement Disorders, 25*(Suppl 1), S27–S31.
Switon, K., Kotulska, K., Janusz-Kaminska, A., Zmorzynska, J., & Jaworski, J. (2017). Molecular neurobiology of mTOR. *Neuroscience, 341*, 112–153.

Taldone, T., & Chiosis, G. (2009). Purine-scaffold Hsp90 inhibitors. *Current Topics in Medicinal Chemistry*, *9*, 1436–1446.

Taylor, R. C., & Dillin, A. (2011). Aging as an event of proteostasis collapse. *Cold Spring Harbor Perspectives in Biology*, *3*(5), a004440.

Thakur, P., & Nehru, B. (2014). Modulatory effects of sodium salicylate on the factors affecting protein aggregation during rotenone induced Parkinson's disease pathology. *Neurochemistry International*, *75*, 1–10.

Tramutola, A., Di Domenico, F., Barone, E., Perluigi, M., & Butterfield, D. A. (2016). It is all about (U)biquitin: Role of altered ubiquitin-proteasome system and UCHL1 in Alzheimer disease. *Oxidative Medicine and Cellular Longevity*, *2016*, 1–12.

Tramutola, A., Lanzillotta, C., Barone, E., Arena, A., Zuliani, I., Mosca, L., et al. (2018). Intranasal rapamycin ameliorates Alzheimer-like cognitive decline in a mouse model of Down syndrome. *Translational Neurodegeneration*, 7(28), 1–22.

Tramutola, A., Triplett, J. C., Di Domenico, F., Niedowicz, D. M., Murphy, M. P., Coccia, R., et al. (2015). Alteration of mTOR signaling occurs early in the progression of Alzheimer disease (AD): Analysis of brain from subjects with pre-clinical AD, amnestic mild cognitive impairment and late-stage AD. *Journal of Neurochemistry*, *133*, 739–749.

Tseng, B. P., Green, K. N., Chan, J. L., Blurton-Jones, M., & LaFerla, F. M. (2008). Abeta inhibits the proteasome and enhances amyloid and tau accumulation. *Neurobiology of Aging*, *29*, 1607–1618.

Uddin, M. S., Stachowiak, A., Mamun, A. A., Tzvetkov, N. T., Takeda, S., Atanasov, A. G., et al. (2018). Autophagy and Alzheimer's disease: From molecular mechanisms to therapeutic implications. *Frontiers in Aging Neuroscience*, *10*, 04.

Upadhya, S. C., & Hegde, A. N. (2007). Role of the ubiquitin proteasome system in Alzheimer's disease. *BMC Biochemistry*, *8*(Suppl 1), S12.

Urano, F., Wang, X., Bertolotti, A., Zhang, Y., Chung, P., Harding, H. P., et al. (2000). Coupling of stress in the ER to activation of JNK protein kinases by transmembrane protein kinase IRE1. *Science*, *287*, 664–666.

Van Skike, C. E., & Galvan, V. (2018). A perfect sTORm: The role of the mammalian target of rapamycin (mTOR) in cerebrovascular dysfunction of Alzheimer's disease: A mini-review. *Gerontology*, *64*, 205–211.

Walter, P., & Ron, D. (2011). The unfolded protein response: From stress pathway to homeostatic regulation. *Science*, *334*, 1081–1086.

Wang, S., & Kaufman, R. J. (2012). The impact of the unfolded protein response on human disease. *The Journal of Cell Biology*, *197*, 857–867.

Wate, R., Takahashi, S., Ito, H., Kusaka, H., Kubota, Y., Suetomi, K., et al. (2005). Radio-sensitivity of the cells from amyotrophic lateral sclerosis model mice transfected with human mutant SOD1. *Journal of Radiation Research*, *46*, 67–73.

Weng, F. L., & He, L. (2021). Disrupted ubiquitin proteasome system underlying tau accumulation in Alzheimer's disease. *Neurobiology of Aging*, *99*, 79–85.

Wilkinson, B., & Gilbert, H. F. (2004). Protein disulfide isomerase. *Biochimica et Biophysica Acta*, *1699*, 35–44.

Woerner, A. C., Frottin, F., Hornburg, D., Feng, L. R., Meissner, F., Patra, M., et al. (2016). Cytoplasmic protein aggregates interfere with nucleocytoplasmic transport of protein and RNA. *Science*, *351*, 173–176.

Wu, Y., Li, X., Zhu, J. X., Xie, W., Le, W., Fan, Z., et al. (2011). Resveratrol-activated AMPK/SIRT1/autophagy in cellular models of Parkinson's disease. *Neurosignals*, *19*, 163–174.

Yamamoto, K., Sato, T., Matsui, T., Sato, M., Okada, T., Yoshida, H., et al. (2007). Transcriptional induction of mammalian ER quality control proteins is mediated by single or combined action of ATF6alpha and XBP1. *Developmental Cell*, *13*, 365–376.

Yang, Y., Turner, R. S., & Gaut, J. R. (1998). The chaperone BiP/GRP78 binds to amyloid precursor protein and decreases Abeta40 and Abeta42 secretion. *The Journal of Biological Chemistry*, *273*, 25552–25555.

Yerbury, J. J., Ooi, L., Dillin, A., Saunders, D. N., Hatters, D. M., Beart, P. M., et al. (2016). Walking the tightrope: Proteostasis and neurodegenerative disease. *Journal of Neurochemistry*, *137*, 489–505.

Zavodszky, E., Seaman, M. N., Moreau, K., Jimenez-Sanchez, M., Breusegem, S. Y., Harbour, M. E., et al. (2014). Mutation in VPS35 associated with Parkinson's disease impairs WASH complex association and inhibits autophagy. *Nature Communications*, *5*, 3828.

Zhang, Y., Chen, X., Zhao, Y., Ponnusamy, M., & Liu, Y. (2017). The role of ubiquitin proteasomal system and autophagy-lysosome pathway in Alzheimer's disease. *Reviews in the Neurosciences*, *28*, 861–868.

Zheng, L., Terman, A., Hallbeck, M., Dehvari, N., Cowburn, R. F., Benedikz, E., et al. (2011). Macroautophagy-generated increase of lysosomal amyloid beta-protein mediates oxidant-induced apoptosis of cultured neuroblastoma cells. *Autophagy*, 7, 1528–1545.

Zhu, P. J., Huang, W., Kalikulov, D., Yoo, J. W., Placzek, A. N., Stoica, L., et al. (2011). Suppression of PKR promotes network excitability and enhanced cognition by interferon-gamma-mediated disinhibition. *Cell*, *147*, 1384–1396.

Zhu, P. J., Khatiwada, S., Cui, Y., Reineke, L. C., Dooling, S. W., Kim, J. J., et al. (2019). Activation of the ISR mediates the behavioral and neurophysiological abnormalities in Down syndrome. *Science*, *366*, 843–849.

CHAPTER THREE

Computational and structural investigation of Palmitoyl-Protein Thioesterase 1 (PPT1) protein causing Neuronal Ceroid Lipofuscinoses (NCL)

D. Thirumal Kumar[a,*], Nishaat Shaikh[b], S. Udhaya Kumar[b], and C. George Priya Doss[b,*]

[a]Faculty of Allied Health Sciences, Meenakshi Academy of Higher Education and Research, Chennai, Tamil Nadu, India

[b]Laboratory of Integrative Genomics, Department of Integrative Biology, School of Bio Sciences and Technology, Vellore Institute of Technology, Vellore, Tamil Nadu, India

*Corresponding authors: e-mail address: thirumalkumar.d@gmail.com; georgepriyadoss@vit.ac.in

Contents

Abstract

The Neuronal Ceroid Lipofuscinoses (NCL) are a group of progressive neurodegenerative disorders, associated with 14 Ceroid Lipofuscinosis Neuronal genes (CLN1–14). The mutations in the Palmitoyl-Protein Thioesterase 1 (PPT1) protein serve as one of the

Advances in Protein Chemistry and Structural Biology, Volume 132
ISSN 1876-1623
https://doi.org/10.1016/bs.apcsb.2022.07.002

major reasons for the causative of NCL. The PPT1 involves degrading and modifying cysteine residues in proteins or peptides by removing thioester-linked fatty acyl groups like palmitate prefers acyl chains of 14–18 carbons in length. In this study, we have analyzed the impact of PPT1 mutations on the deleteriousness, stability, conservative nature of amino acid, and impact of mutations on the protein structure. We have also used molecular dynamics simulations using GROMACS to perceive the alteration in the dynamic behavior of the PPT1 at the residual level. In this study, we have retrieved 23 PPT1 mutations from the UniProt database, and these were subjected to a series of analyses using varied computer algorithms. From these analyses, out of 23 mutations, 16 mutations were identified as deleterious. Among 16, eight mutations were identified to destabilize the protein structure, and finally, two mutations (W38C and L222P) were found to be positioned in the highly conserved region. The structural impact study observed that the mutant proline could disrupt the alpha helix formed by the leucine at position 222. Finally, from the molecular dynamics simulations, we observed that due to the mutations (W38C and L222P), the protein had experienced higher deviation, fluctuation, and lower compactness. These structural changes elucidate that these mutations can impact the structure and function of the PPT1 protein.

Abbreviations

CLN1 Ceroid lipofuscinosis, neuronal, 1
CSP Cysteine String Protein
MDS Molecular dynamics simulation
NCL Neuronal Ceroid Lipofuscinosis
PCA Principal Component Analysis
PPT1 Palmitoyl-Protein Thioesterase 1

1. Introduction

NCLs are characterized by visual loss, aberrant movement patterns, seizures, dementia, and early mortality, with symptoms typically beginning in infancy or adolescence (Cotman, Karaa, Staropoli, & Sims, 2013). Except for CLN4, NCLs are primarily brought on by a recessive loss of function mutations and are characterized by the accumulation of lipofuscin (Palmer, Barry, Tyynelä, & Cooper, 2013; Winklhofer, Tatzelt, & Haass, 2008). Most NCLs are primarily caused by dysregulated ER-lysosomal trafficking, lysosomal dysfunction, or atypical lipid changes following the known function of the mutant genes (Mole & Haltia, 2015; Warrier, Vieira, & Mole, 2013). The adult-onset of the CLN4, an autosomal dominantly inherited NCL, is between the ages of 25 and 46. The human *DNAJC5* gene produces the SV protein known as CSP, either contains the deletion of amino acid L116Δ or the amino acid substitution L115R that results in CLN4 (Benitez et al., 2011; Cadieux-Dion et al., 2013). As Cysteine String Protein (CSP) encodes an SV protein that is not known to have any lysosome-related activities, it stands out among NCL-associated genes. To sustain synaptic function and stop neurodegeneration, select clients are folded in the presynaptic terminal by CSP via its J and C-terminal domains, which interact with HSC70 to create a chaperone complex on synaptic vesicles (Fig. 1). In flies and mice, gene deletions result in paralysis, advanced locomotor abnormalities, and early death from neurodegeneration (Tobaben et al., 2001; Zhang et al., 2012).

In general, the removal of long-chain fatty acids from transformed cysteine residues in polypeptides is carried out by a thioesterase known as PPT1. Since palmitate is the primary lipid present at S-acylation sites, adding long-chain fatty acids, also known as S-acylation, is frequently referred to as palmitoylation (Ahearn et al., 2011). The transition period between the forms of the protein that have been palmitoylated and those that have not is thought to range between minutes and hours because this post-translational modification process is dynamic and reversible (El-Husseini & Bredt, 2002).

PPT1 is a soluble protein found in lysosomes and plays a significant role in other cellular localizations. The protein is also present in the cell medium, and a consensus signal sequence facilitates its secretion (Hellsten, Vesa, Olkkonen, Jalanko, & Peltonen, 1996; Lehtovirta et al., 2001). PPT1 was discovered to be directed to lysosomes with the help of mannose 6-phosphate receptor route, similar to many other soluble lysosomal enzymes. Before being endocytosed, mannose 6-phosphate receptors can adhere extracellular cargos to cell surfaces or in the trans-Golgi network (Verkruyse & Hofmann, 1996). For this pathway to function properly, PPT1's glycosylation is crucial. PPT1 can therefore be found in endocytosed vesicles, organelles, and vesicles of the secretory route (Lehtovirta et al., 2001). NCL's pathogenesis includes a malfunctioning Endoplasmic Reticulum (RA) as one of its components; PPT1 function in ER is also evident. PPT1 was found in synaptic vesicles, synaptosomes, and axons of neurons. Since the recycling of synaptic vesicles was impacted in the knockout mouse model, these localizations are crucial for understanding

Fig. 1 Schematic diagram of Cysteine String Protein (CSP) domain.

the disease mechanism (Salonen, Heinonen-Kopra, Vesa, & Jalanko, 2001; Zhang et al., 1999). Several therapeutic interventions have been tried in recent years, including gene therapy in combination with other approaches, small-molecule treatments, and enzyme replacement (Cooper et al., 2021; Hawkins-Salsbury, Cooper, & Sands, 2013; Sands, 2013). Reversible S-acylation controls PPT1, an essential depalmitoylation enzyme. The signal peptide's cysteine six contains the primary PPT1 palmitoylation site.

It is believed that PPT1 is essential for synaptic depalmitoylation and function because PPT1 function loss is known to affect the presynaptic function and leads to abnormal palmitoylation NCL type 1 (CLN1). Except for CSP, Go, and the mitochondrial synthase subunit O (F1 ATP), the majority of the metabolic substrates of PPT1 are unknown (Liu et al., 2019). The moderately selective enzyme PPT1 depalmitoylates around 10% of the synaptic palmitome, with the remaining substrates being depalmitoylated by subsequent depalmitoylating enzymes. It is possible that PPT1 and other depalmitoylating proteins contain related substrates. The severity and sudden onset of CLN1 illness illustrate PPT1's substrate-specificity (Gorenberg et al., 2022).

The examination of PPT1 substrates across the nine functional categories suggests that depalmitoylation appears to have a common role in the trafficking of synaptic proteins. Previous studies have shown that palmitoylation is necessary for the trafficking of synaptic proteins from the soma to the nerve terminal (Fukata & Fukata, 2010). While certain proteins are palmitoylated on the cytosolic surfaces of these organelles, resident PATs in the ER or Golgi lumen are responsible for palmitoylating synaptic proteins designated for the secretory pathway (Bechtel, Li, Kisty, Maurais, & Weerapana, 2020). Although the precise PATs executing these various tasks are unknown, palmitoylation facilitates the sorting and adherence of synaptic proteins to the membrane of the nascent synaptic vesicle (Gorenberg et al., 2022; Greaves, Prescott, Gorleku, & Chamberlain, 2009). Studies have revealed that PPT1 trafficking in neurons and non-neural cells is remarkably comparable (Hellsten et al., 1996). It is secreted in the soma, trafficked along the axons via the synaptic vesicles, and released into the synaptic cleft. This information explains how PPT1 functions as a key enzyme by controlling ion channels, synaptic adhesion molecules, and transporters (Gorenberg et al., 2022) (Fig. 2).

Although the precise mechanism is yet unknown, it is thought that some PPT1 escapes from the presynaptic terminal during endocytosis and accesses the cytosolic substrates. The PATs DHHC5 and DHHC17 reportedly carry

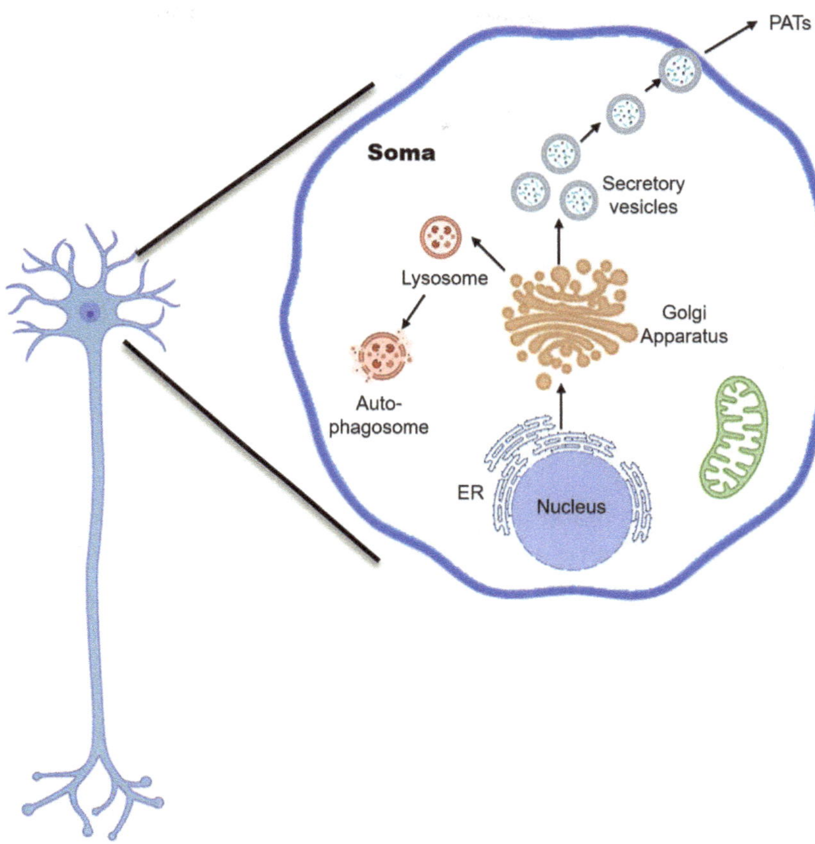

Fig. 2 Schematic diagram of the palmitoylation of synaptic cell adhesion molecules by PPT1.

out the palmitoylation of synaptic substrates continuously and locally. Lemonidis et al. have found nine proteins with the DHHC17 recognition motif in a study, indicating a potential partnership between DHHC17 and PPT1 for palmitoylation-depalmitoylation cycles in the synapse (Lemonidis et al., 2015). These comprise, among example, the synaptic vesicle recycling genes CLH1, DYN1, SYNJ1, and DNJC5. Previous studies discovered that PPT1 KO neurons showed poor synaptic vesicle endocytosis as seen by decreased presynaptic vesicle size, defects in triggered release of presynaptic vesicle, and the diligence of SNAP25 and VAMP2 on the presynaptic membrane (Kim et al., 2008). Thus, during synaptic vesicle endocytosis, the palmitoylation-depalmitoylation mechanisms controlled by PPT1 and DHHC17 enable effective membrane attachment and alienation of endocytic

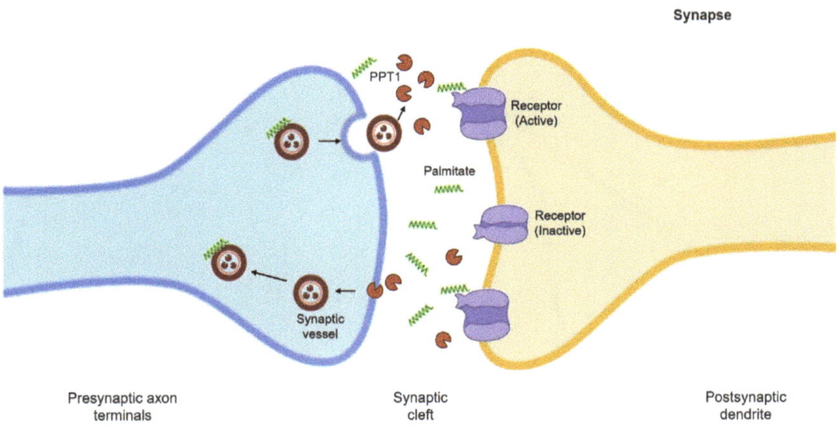

Fig. 3 Schematic diagram of the palmitoylation and depalmitoylation cycle.

proteins. The precise moment at which palmitoylation and depalmitoylation occur and the factors that set off this cycle are still unknown. Studies have documented that the palmitoylation and depalmitoylation cycles befall at the synapse, as shown in Fig. 3 (Gorenberg et al., 2022). In patients (CLN4) with DNJC5 mutations, increased level of PPT1 protein was seen. However, specific activity is reduced, leading to general alterations in palmitoylation, according to research by Henderson et al. Another study identified four proteins (NSF, STXB1, DYN1, and SYNPR) PPT1 substrates in CLN4 brains had aberrant palmitoylation, demonstrating the strong association between CLN1 and CLN4 illnesses. Unfortunately, the difficulties in collecting CLN brains and the damaged nature of end-stage brains make it impossible to corroborate the findings in human samples (Gorenberg et al., 2022; Henderson et al., 2015). The major mutations in the pool of mutations published in the public database were identified using a set of computational algorithms in this study, and we were able to anticipate their effects on the deleteriousness and stability of the PPT1 protein. To get a thorough grasp of the residual level fluctuations and structural impact on the protein due to the detected protein structures, molecular dynamics simulations were also carried out.

2. Materials and methods

2.1 Sequence and mutations retrieval

The amino acid sequence of the PPT1 protein in FASTA format was retrieved from the UniProt database with the UniProt ID P50897

(Poux et al., 2017). The mutations in the PPT1 protein for this study were retrieved from the Disease & Variants section of the UniProt with UniProt ID P50897.

2.2 Deleteriousness, stability, and conservation analysis

The deleterious effect of the mutations, changes in the stability of the proteins due to mutations, and conserved nature of amino acids prone to mutations were explored with the help of PredictSNP, iStable, and ConSurf algorithm, respectively (Ashkenazy et al., 2016; Bendl et al., 2014; Chen, Lin, & Chu, 2013). The FASTA sequence of the PPT1 protein and the mutations retrieved from the UniProt were provided as the input for these analyses. The analyses were performed in series using the default parameters. The entire pool of mutations was provided as the input for the deleteriousness prediction. Further, the mutations anticipated to be deleterious in all the embedded algorithms of PredictSNP were provided as the input for the stability prediction. Then, the positions of the mutations anticipated to be harmful and destabilizing the protein structure in all the algorithms of PredictSNP and iStable were analyzed using the ConSurf algorithm. Finally, the mutations identified as deleterious, destabilizing, and amino acids positioned at highly conserved positions were provided as the input for the molecular dynamics simulations analyses using GROMACS (Abraham et al., 2015). These mutations were considered the most significant mutations.

2.3 Structural effects prediction

The effects on protein structure due to mutation were analyzed using the online HOPE algorithm (Venselaar, te Beek, Kuipers, Hekkelman, & Vriend, 2010). The FASTA sequence of the PPT1 and the identified significant mutations (W38C and L222P) were provided as the input for prediction.

2.4 Molecular dynamics simulation (MDS) analysis

MDS were accomplished with significant mutations for the native PPT1 protein and the PPT1 protein. Utilizing the GROMOS force field and the most recent GROMACS program, molecular dynamics simulations were carried out (Abraham et al., 2015; Schmid et al., 2011). We used pdb2gmx to convert the PDBs of the native and mutant PPT1 proteins to gromacs files with the -ignh option to remove the hydrogen atoms from the protein structure. The protein was positioned at least 1.0 nm from the box's border in the middle of a fixed-volume cubical box. The solvents were replaced with ions to maintain neutrality. The system was minimized energy to ensure no steric conflicts and improper geometry. The equilibration procedure was conducted in two parts to prevent unconstrained dynamics. All bonds were bound with 50,000 steps in phase I's NVT ensemble (constant Number of Particles, Constant Volume, and Constant Temperature), which also used a Berendsen thermostat to keep the temperature at 300 K for a time constant of 1 ps (Berendsen, Postma, van Gunsteren, DiNola, & Haak, 1984). Phase II, NPT ensemble, is a continuation of NVT and is carried out using Parrinello Rahman barostat pressure coupling without producing velocity (Parrinello & Rahman, 1981). The well-adjusted system was subsequently put through a 100 ns molecular dynamics simulation. The SETTLE method limited water molecules, whereas non-water bonds were reserved by the LINCS procedure (Hess, Bekker, Berendsen, & Fraaije, 1997). The trajectories were collected as output after 100 ns of MD. The trajectories were further analyzed using the embedded packages in GROMACS and protocol was followed from our previous articles (Thirumal Kumar, Jain, Udhaya Kumar, George Priya Doss, & Zayed, 2021; Thirumal Kumar, Udhaya Kumar, Magesh, & George Priya Doss, 2021; Udhaya Kumar, Bithia, Thirumal Kumar, George Priya Doss, & Zayed, 2022; Udhaya Kumar, Thirumal Kumar, Siva, George Priya Doss, & Zayed, 2020).

3. Results

3.1 Sequence and mutations retrieval

The amino acid sequence of the PPT1 protein with a length of Length 306 amino acids and a mass of 34,193 Da was retrieved from the UniProt database. A total of 23 mutations reported in CLN1 deposited in the UniProt database of PPT1 protein was considered for further study.

3.2 Deleteriousness, stability, and conservation analysis

The deleteriousness analysis was performed using the PredictSNP algorithm, which embeds other algorithms such as MAPP, PhD-SNP, PhD-SNP,

PolyPhen – 1, PolyPhen-2, SIFT, SNAP, nsSNPAnalyzer, and PANTHER. From the analyses, 16 mutations out of the 23 mutations (W38C, H39Q, G42E, G108R, Y109D, R122W, S138L, C152Y, Q177E, V181M, P189R, L219Q, L222P, V228G, Y247H, and L305P) were anticipated to be harmful using all the algorithms (Table 1). The nsSNPAnalyzer algorithm predicted unknown for all the mutations; thus, the study ignored this algorithm.

These 16 mutations were provided as the input for the stability prediction using the iStable algorithm. The iStable algorithm, which embeds other algorithms such as i-Mutant2.0 and MUpro, was used to predict the stability changes of the protein due to mutations. From the analyses, we found that out of 16 deleterious mutations, eight deleterious mutations (W38C, V181M, P189R, L219Q, L222P, V228G, Y247H, and L305P) were predicted to be destabilized by all the algorithms (Table 2).

The amino acid positions of these eight mutations were further subjected to analyzing the conservation scale using the ConSurf algorithm. From the analysis, we identified the mutation positions 38, and 222 (W38C and L222P) were positioned in the highly conserved position with a Scale of 9 (Table 3). Both W38 and L222 amino acids were predicted to be in the buried position and predicted to be structural residues (Fig. 4).

3.3 Structural effects prediction

From the HOPE algorithm prediction, we observed that native tryptophan residue is located in its ideal β-strand secondary structure. The mutant cysteine residue prefers a different secondary structure. Therefore, the local conformation will be slightly destabilized. For the second mutation, the native Leucine is anticipated (using the Reprof software) to be placed in a α-helix and the mutant Proline disrupts a α-helix when the native Leucine not located at one of the first three sites of that helix. If the current mutation occurs, the helix will be disrupted, which can severely affect the protein's structure.

3.4 Molecular dynamics simulations

MDS were performed with the native PPT1 and PPT1 proteins with W38C and L222P mutations. GROMACS software with embedded tools was used for the simulations and analyses. The simulations were performed for 100 ns,

Table 1 Prediction of deleteriousness of PPT1 protein mutations using online PredictSNP.

S No	UniProt ID	Mutation	PredictSNP prediction	PredictSNP expected accuracy	MAPP prediction	MAPP expected accuracy	PhD-SNP prediction	PhD-SNP expected accuracy	PolyPhen-1 prediction	PolyPhen-1 expected accuracy
1	VAR_058434	W38C	D	0.869084	D	0.588897	D	0.858228	D	0.744912
2	VAR_005548	H39Q	D	0.869084	D	0.426537	D	0.858228	D	0.744912
3	VAR_005549	G42E	D	0.869084	D	0.806597	D	0.87524	D	0.744912
4	VAR_066874	C45Y	D	0.606973	N	0.710132	D	0.87524	D	0.744912
5	VAR_005550	T75P	N	0.631518	D	0.587706	N	0.552027	N	0.668841
6	VAR_005551	D79G	D	0.718713	D	0.766117	D	0.773385	N	0.668841
7	VAR_018511	G108R	D	0.869084	D	0.877061	D	0.87524	D	0.744912
8	VAR_005552	Y109D	D	0.869084	D	0.842579	D	0.817312	D	0.744912
9	VAR_005553	R122W	D	0.869084	D	0.771193	D	0.858228	D	0.744912
10	VAR_066875	S138L	D	0.869084	D	0.91979	D	0.858228	D	0.59445
11	VAR_066876	C152Y	D	0.869084	D	0.783358	D	0.893258	D	0.744912
12	VAR_005555	Q177E	D	0.869084	D	0.766117	D	0.87524	D	0.59445
13	VAR_005556	V181L	D	0.549464	N	0.63348	D	0.773385	N	0.668841
14	VAR_005557	V181M	D	0.869084	D	0.573143	D	0.588855	D	0.744912
15	VAR_066877	H187R	N	0.738345	N	0.795595	N	0.58231	N	0.668841
16	VAR_066878	P189R	D	0.869084	D	0.91979	D	0.88475	D	0.744912
17	VAR_005558	L219Q	D	0.869084	D	0.718141	D	0.858228	D	0.744912
18	VAR_066879	L222P	D	0.869084	D	0.774944	D	0.87524	D	0.744912
19	VAR_066880	V228G	D	0.869084	D	0.508996	D	0.858228	D	0.744912
20	VAR_005559	Y247H	D	0.869084	D	0.766117	D	0.817312	D	0.744912
21	VAR_005560	G250V	D	0.505959	N	0.642291	D	0.87524	N	0.668841
22	VAR_066881	W296R	D	0.869084	NA	NA	D	0.773385	D	0.59445
23	VAR_066882	L305P	D	0.869084	D	0.751124	D	0.87524	D	0.744912

D—deleterious; N—neutral; U—unknown.

PolyPhen-2 prediction	PolyPhen-2 expected accuracy	SIFT prediction	SIFT expected accuracy	SNAP prediction	SNAP expected accuracy	nsSNPAnalyzer prediction	nsSNPAnalyzer expected accuracy	PANTHER prediction	PANTHER expected accuracy
D	0.811429	D	0.792808	D	0.805103	U	0	D	0.744511
D	0.811429	D	0.792808	D	0.885196	U	0	D	0.76
D	0.811429	D	0.792808	D	0.848454	U	0	D	0.742515
D	0.474293	D	0.792808	N	0.500147	U	0	D	0.698603
N	0.724725	D	0.527785	N	0.500147	U	0	D	0.61
D	0.649743	D	0.792808	D	0.622082	U	0	U	0
D	0.811429	D	0.792808	D	0.805103	U	0	D	0.744511
D	0.811429	D	0.792808	D	0.622082	U	0	D	0.842315
D	0.811429	D	0.792808	D	0.848454	U	0	D	0.874252
D	0.649743	D	0.792808	D	0.622082	U	0	D	0.766467
D	0.811429	D	0.792808	D	0.805103	U	0	D	0.78
D	0.811429	D	0.792808	D	0.720388	U	0	D	0.698603
D	0.407455	D	0.527785	D	0.555519	U	0	D	0.61
D	0.811429	D	0.792808	D	0.555519	U	0	D	0.686627
D	0.502571	N	0.608192	N	0.708784	U	0	N	0.645423
D	0.811429	D	0.792808	D	0.720388	U	0	D	0.744511
D	0.811429	D	0.792808	D	0.555519	U	0	D	0.78
D	0.811429	D	0.792808	D	0.885196	U	0	D	0.842315
D	0.593188	D	0.45233	D	0.555519	U	0	D	0.61
D	0.811429	D	0.792808	D	0.805103	U	0	D	0.744511
D	0.398458	D	0.792808	N	0.554397	U	0	D	0.744511
D	0.431234	D	0.459498	D	0.555519	U	0	D	0.566866
D	0.811429	D	0.792808	D	0.555519	U	0	D	0.78

Table 2 Prediction of impact on stability by PPT1 protein mutations using online iStable.

S.·No.	UniProt ID	Mutation	i-Mutant2.0	DDG	MUpro	Conf. Score	iStable	Conf. Score
1	VAR_058434	W38C	D	−1	D	−1	D	0.709622
2	VAR_005548	H39Q	D	0.1	I	0.2592715	I	0.638928
3	VAR_005549	G42E	I	−0.16	D	−0.17966527	D	0.548674
4	VAR_018511	G108R	I	0.1	D	−0.13965328	D	0.626144
5	VAR_005552	Y109D	I	−0.17	D	−0.11701906	I	0.560578
6	VAR_005553	R122W	I	0.6	D	−1	D	0.537064
7	VAR_066875	S138L	I	0.51	I	0.36340729	I	0.703609
8	VAR_066876	C152Y	I	0.1	D	−0.36854408	D	0.550055
9	VAR_005555	Q177E	I	0.15	D	−0.0050788317	D	0.52387
10	VAR_005557	V181M	D	−0.49	D	−0.32716939	D	0.793218
11	VAR_066878	P189R	D	−0.49	D	−0.63878536	D	0.824543
12	VAR_005558	L219Q	D	−1.65	D	−1	D	0.848051
13	VAR_066879	L222P	D	−1.3	D	−1	D	0.842802
14	VAR_066880	V228G	D	−2.22	D	−1	D	0.765941
15	VAR_005559	Y247H	D	−1.03	D	−0.44347368	D	0.767396
16	VAR_066882	L305P	D	−1.34	D	−1	D	0.707966

D – decrease; I – increase.

Table 3 Prediction of conservative Scale of PPT1 protein mutations using online ConSurf.

S. No.	UniProt ID	Mutation	Score
1	VAR_058434	W38C	9
2	VAR_005557	V181M	8
3	VAR_066878	P189R	8
4	VAR_005558	L219Q	8
5	VAR_066879	L222P	9
6	VAR_066880	V228G	7
7	VAR_005559	Y247H	7
8	VAR_066882	L305P	7

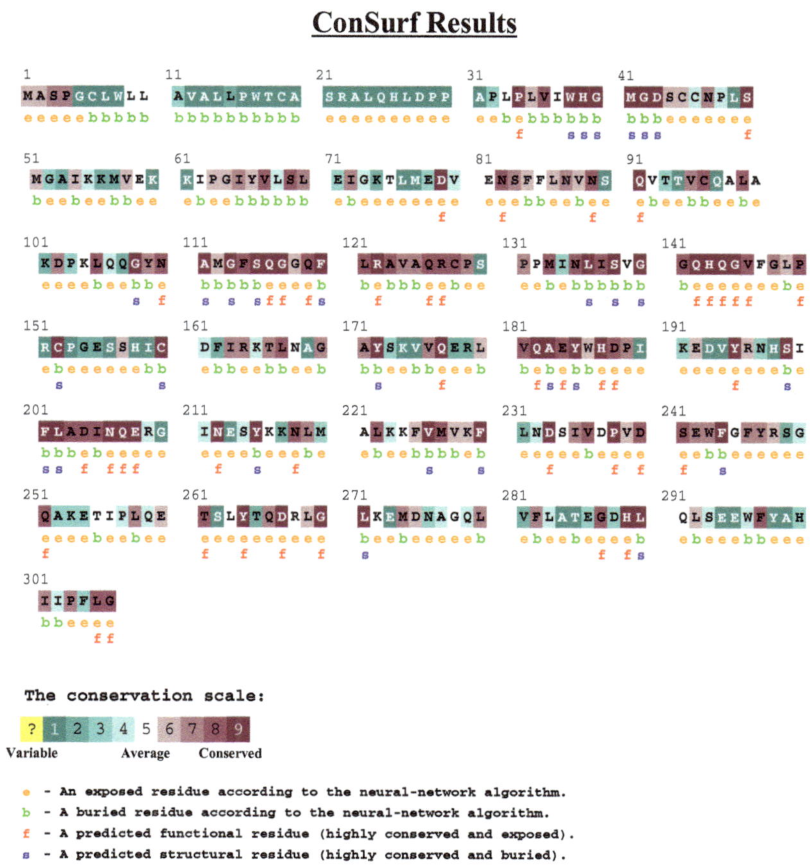

Fig. 4 Prediction of conservation Scale of each amino acid in PPT1 protein using ConSurf.

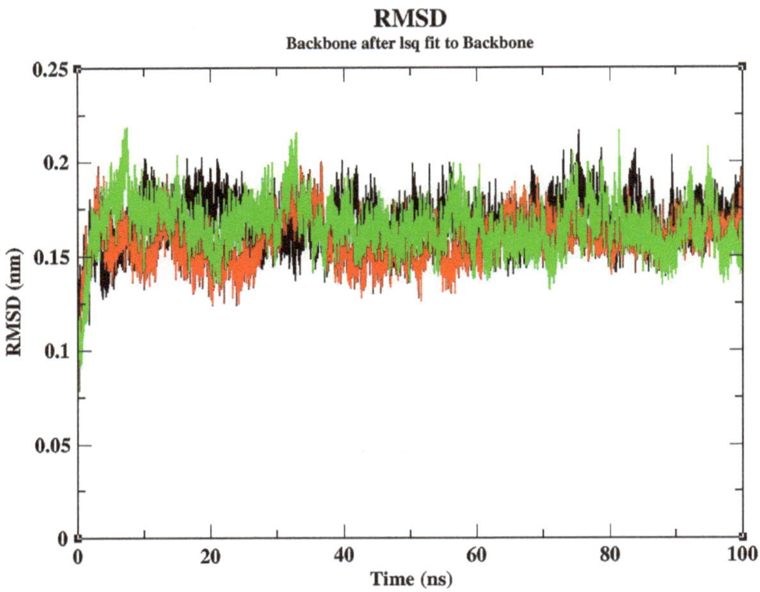

Fig. 5 Root Mean Square Deviation (RMSD) plot showing the deviation between native PPT1 protein and proteins with mutations (W38C and L222). Color scheme: Native—Green, L222P—Black, W18C—Red.

and the trajectories were analyzed. The RMSD was analyzed using the gmx rms to find if the molecules had attained convergence during the simulation period. From our analysis, we observed that the native PPT1 and the mutants (W38C and L222P) converged at the end of 100 ns (Fig. 5). This elucidated that molecules are suitable for further analyses. The RMSF was analyzed using the gmx rmsf plugin of the GROMACS to understand the protein's residual level changes due to the mutations. We perceived that the L222P mutation in PPT1 had fluctuated more than the native PPT1 and W38C in PPT1 (Fig. 6). The radius of gyration (Rg) was analyzed using the gmx gyrate plugin of the GROMACS. From the Rg analysis, we observed that the native PPT1 molecule had a lower deviation when compared to the proteins with mutations W38C and L222P (Fig. 7). The Principal Component Analysis (PCA) was analyzed using the gmx covar and gmx anaeig plugins of the GROMACS. From the analysis, we observed that the native PPT1 has lesser molecular convergence when compared to the proteins with mutations W38C and L222P (Fig. 8).

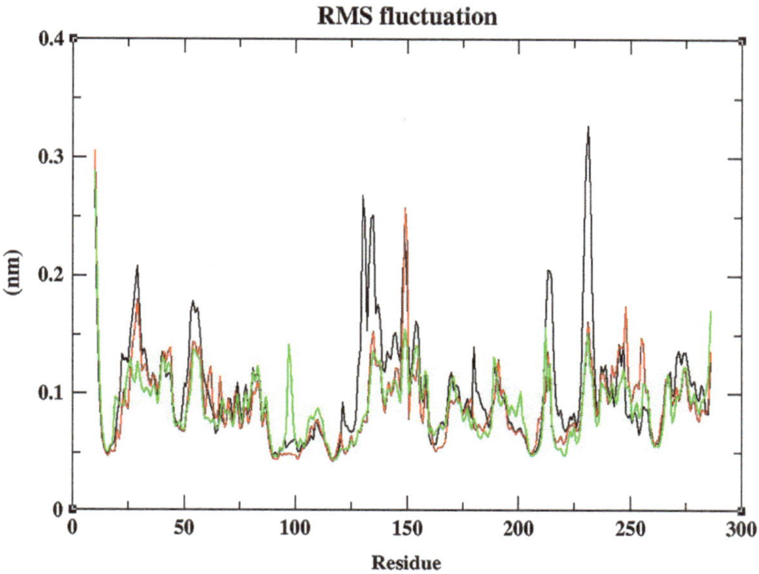

Fig. 6 Root Mean Square Fluctuation (RMSF) plot showing the fluctuation between native PPT1 protein and proteins with mutations (W38C and L222). Color scheme: Native—Green, L222P—Black, W18C—Red.

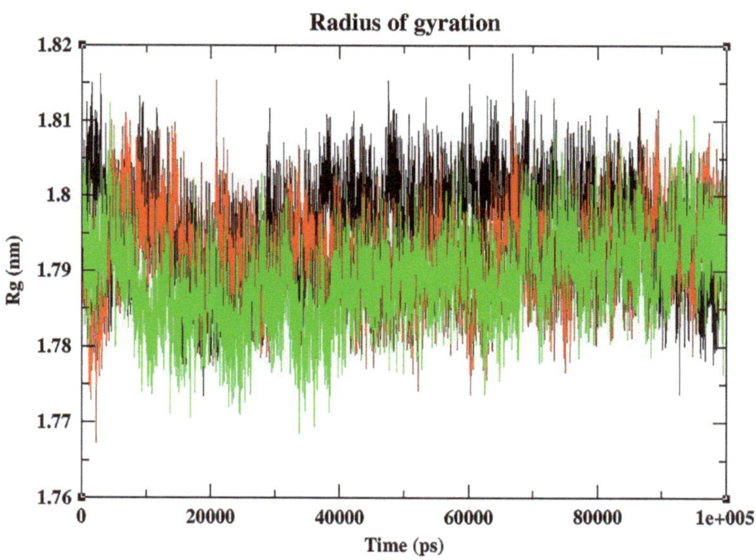

Fig. 7 Radius of Gyration plot (Rg) showing the deviation between native PPT1 protein and proteins with mutations (W38C and L222). Color scheme: Native—Green, L222P—Black, W18C—Red.

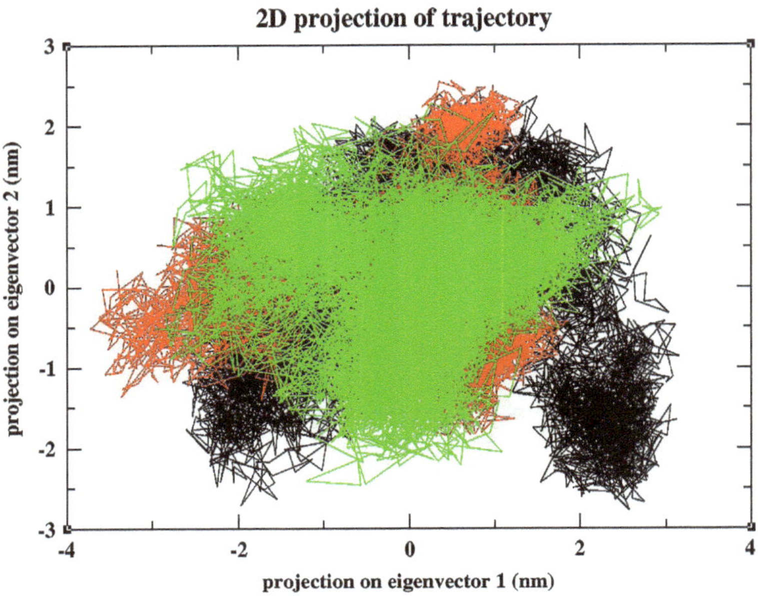

Fig. 8 Principal Component Analysis (PCA) plot showing the structural deviation between native PPT1 protein and proteins with mutations (W38C and L222). Color scheme: Native—Green, L222P—Black, W18C—Red.

4. Discussion

During axonal trafficking, palmitoylation is essential for maintaining the inactive state of synaptic proteins, it suppresses enzymatic activity or ectopic binding connections by preventing the creation of disulfide bonds. Additionally, this maintains the vesicle-bound nature of synaptically targeted proteins. The membrane-bound synaptic adhesion molecules must be introduced into the synaptic membrane once at the synapse and must act together in both the cis and trans directions. The precursor synaptic vesicle must release the proteins into the cytoplasm. According to studies, PPT1 carries out these functions by locally depalmitoylating synapses. It is thought that disulfide bonds forming on their own in an oxidative extracellular space cause functional maturation by the depalmitoylated cysteines in membrane proteins with extracellular depalmitoylation (Bechtel et al., 2020; Cijsouw et al., 2018; Gorenberg et al., 2022). This study aimed to evaluate the physical and functional influence of the mutations in the PPT1 protein. We have retrieved 23 mutations from the UniProt database, and they were subjected to a series of computational analyses to identify the most significant mutations and study their impact on the structure of a protein using molecular dynamics simulations. Initially, we used the PredictSNP online algorithm to classify the mutations as deleterious and neutral. The collective algorithms embedded within the PredictSNP classified 16 mutations as deleterious by all the tools (Table 1).

Furthermore, out of these 16 mutations, eight negatively affected the PPT1 protein structure (Table 2). The mutations occurring in the most conserved region of the protein tend to highly disrupt the protein structure contrasted with the mutations occurring in the variable region of the protein (Ashkenazy et al., 2016; Guharoy & Chakrabarti, 2005). Considering the importance of the conserved nature of amino acids, we have studied the Scale of conservation of the eight mutations responsible for deleteriousness and destabilization. From the conservation analysis using the ConSurf server, we identified the two mutations W38C and L222P to be positioned in the most conserved site with the Scale of 9 (Table 3 and Fig. 4). Both W38 and L222 amino acids were predicted to be in the buried position and predicted to be structural residues thus adding evidence of their role in disrupting the protein structure. Furthermore, the MDS were performed using GROMACS to know the residual level changes of the protein due to the mutations W38C and L222P. From the RMSD analysis, the molecules were well converged at the end of 100 ns and suitable for further analyses. The protein molecules with mutations W38C and L222P were found to possess higher deviation in comparison to the native PPT1 molecule (Fig. 5). The RMSF was analyzed to comprehend the behavior of each amino acid on the protein with and without mutations. From the RMSF analysis, we observed that the proteins with mutations W38C and L222P showed comparatively higher fluctuation in comparison to the PPT1 (native) molecule (Fig. 6). This finding coincides with the results obtained in RMSD. The Rg was analyzed to study the alteration in compactness of the PPT1 due to identified mutations. The deviation of the Rg is inversely related to the compactness of the protein (Lobanov, Bogatyreva, & Galzitskaia, 2008). From the Rg analysis, we observed that the PPT1 (native) molecule influenced the least deviation compared to the proteins with W38C and L222P mutations. This elucidates that the compactness of the proteins with mutations was disrupted (Fig. 7). Finally, the PCA was performed to confirm the structural changes in the PPT1 protein due to mutations. We observed from PCA analysis that both the proteins with mutation W38C and L222P have been disrupted compared to the native PPT1 protein (Fig. 8).

5. Conclusion

One of the main causes of NCL is a protein called Palmitoyl-Protein Thioesterase (PPT1) that has undergone mutations. In this study, we examined the deleteriousness, stability, conservatism, and effect on the protein structure of PPT1 mutations. Additionally, we have employed GROMACS molecular dynamics simulations to comprehend how the protein's dynamic behavior has changed at the residual level. In this study, 23 PPT1 mutations were retrieved from the UniProt database and subjected to a number of analyses with various computational methods. Out of the 23 mutations identified for these analyses, 16 mutations were deemed deleterious, 8 mutations were determined to destabilize the protein structure, and 2 mutations (W38C and L222P) were shown to be located in the highly conserved area. According to the structural impact analysis, the leucine at position 222 of the alpha helix undergoes the disruption of an alpha helix with the mutant proline. Finally, we found from the molecular dynamics simulations that the protein has experienced increased deviation, fluctuation, and decreased compactness due to the mutations (W38C and L222P). These structural alterations show that these mutations can affect the PPT1 protein's structure and function.

Acknowledgment

The authors take this opportunity to thank the management of Meenakshi Academy of Higher Education and Research and Vellore Institute of Technology for providing facilities to carry out this research work.

Role of funding source

No funding agency was involved in the present study.

Author contributions

TKD, NS, and UKS contributed substantially to the conception and design, acquisition of data, or analysis and interpretation of data. TKD, NS, UKS, and GPDC involved critically drafting or revising the article for important intellectual content. TKD and GPDC were involved in finalizing the version to be published.

Conflict of interest

The authors have declared that no conflicts of interest exist.

References

Abraham, M. J., Murtola, T., Schulz, R., Páll, S., Smith, J. C., Hess, B., et al. (2015). GROMACS: High performance molecular simulations through multi-level parallelism from laptops to supercomputers. *SoftwareX*, *1–2*, 19–25. https://doi.org/10.1016/j.softx.2015.06.001.

Ahearn, I. M., Tsai, F. D., Court, H., Zhou, M., Jennings, B. C., Ahmed, M., et al. (2011). FKBP12 binds to acylated H-ras and promotes depalmitoylation. *Molecular Cell*, *41*(2), 173–185. https://doi.org/10.1016/J.MOLCEL.2011.01.001.

Ashkenazy, H., Abadi, S., Martz, E., Chay, O., Mayrose, I., Pupko, T., et al. (2016). ConSurf 2016: An improved methodology to estimate and visualize evolutionary conservation in macromolecules. *Nucleic Acids Research*, *44*(W1), W344–W350. https://doi.org/10.1093/nar/gkw408.

Bechtel, T. J., Li, C., Kisty, E. A., Maurais, A. J., & Weerapana, E. (2020). Profiling cysteine reactivity and oxidation in the endoplasmic reticulum. *ACS Chemical Biology*, *15*(2), 543–553. https://doi.org/10.1021/ACSCHEMBIO.9B01014/SUPPL_FILE/CB9B01014_SI_002.XLSX.

Bendl, J., Stourac, J., Salanda, O., Pavelka, A., Wieben, E. D., Zendulka, J., et al. (2014). PredictSNP: Robust and accurate consensus classifier for prediction of disease-related mutations. *PLoS Computational Biology*, *10*(1), e1003440. https://doi.org/10.1371/journal.pcbi.1003440.

Benitez, B. A., Alvarado, D., Cai, Y., Mayo, K., Chakraverty, S., Norton, J., et al. (2011). Exome-sequencing confirms DNAJC5 mutations as cause of adult neuronal ceroid-Lipofuscinosis. *PLoS One*, *6*(11), e26741. https://doi.org/10.1371/JOURNAL.PONE.0026741.

Berendsen, H. J. C., Postma, J. P. M., van Gunsteren, W. F., DiNola, A., & Haak, J. R. (1984). Molecular dynamics with coupling to an external bath. *The Journal of Chemical Physics*, *81*(8), 3684. https://doi.org/10.1063/1.448118.

Cadieux-Dion, M., Andermann, E., Lachance-Touchette, P., Ansorge, O., Meloche, C., Barnabé, A., et al. (2013). Recurrent mutations in DNAJC5 cause autosomal dominant Kufs disease. *Clinical Genetics*, *83*(6), 571–575. https://doi.org/10.1111/CGE.12020.

Chen, C. W., Lin, J., & Chu, Y. W. (2013). iStable: Off-the-shelf predictor integration for predicting protein stability changes. *BMC Bioinformatics*, *14*(Suppl 2), S5. https://doi.org/10.1186/1471-2105-14-S2-S5.

Cijsouw, T., Ramsey, A. M., Lam, T. K. T., Carbone, B. E., Blanpied, T. A., & Biederer, T. (2018). Mapping the proteome of the synaptic cleft through proximity labeling reveals new cleft proteins. *Proteomes*, *6*(4). https://doi.org/10.3390/PROTEOMES6040048.

Cooper, J. D., Puhl, A. C., Wang, S. H., Eultgen, E. M., Takahashi, K., Le, S. Q., et al. (2021). Devising effective enzyme replacement therapy for infantile onset neuronal ceroid lipofuscinosis (CLN1 disease). *Molecular Genetics and Metabolism*, *132*(2), S28. https://doi.org/10.1016/J.YMGME.2020.12.048.

Cotman, S. L., Karaa, A., Staropoli, J. F., & Sims, K. B. (2013). Neuronal ceroid Lipofuscinosis: Impact of recent genetic advances and expansion of the clinicopathologic spectrum. *Current Neurology and Neuroscience Reports*, *13*(8), 366. https://doi.org/10.1007/S11910-013-0366-Z.

El-Husseini, A. E. D., & Bredt, D. S. (2002). Protein palmitoylation: A regulator of neuronal development and function. *Nature Reviews. Neuroscience*, *3*(10), 791–802. https://doi.org/10.1038/NRN940.

Fukata, Y., & Fukata, M. (2010). Protein palmitoylation in neuronal development and synaptic plasticity. *Nature Reviews Neuroscience*, *11*(3), 161–175. https://doi.org/10.1038/nrn2788.

Gorenberg, E. L., Tieze, S. M., Yücel, B., Zhao, H. R., Chou, V., Wirak, G. S., et al. (2022). Identification of substrates of palmitoyl protein thioesterase 1 highlights roles of depalmitoylation in disulfide bond formation and synaptic function. *PLoS Biology*, *20*(3), e3001590. https://doi.org/10.1371/JOURNAL.PBIO.3001590.

Greaves, J., Prescott, G. R., Gorleku, O. A., & Chamberlain, L. H. (2009). The fat controller: Roles of palmitoylation in intracellular protein trafficking and targeting to membrane microdomains (review). *Molecular Membrane Biology*, *26*(1), 67–79. https://doi.org/10.1080/09687680802620351.

Guharoy, M., & Chakrabarti, P. (2005). Conservation and relative importance of residues across protein-protein interfaces. *Proceedings of the National Academy of Sciences of the United States of America*, *102*(43), 15447–15452. https://doi.org/10.1073/pnas.0505425102.

Hawkins-Salsbury, J. A., Cooper, J. D., & Sands, M. S. (2013). Pathogenesis and therapies for infantile neuronal ceroid lipofuscinosis (infantile CLN1 disease). *Biochimica et Biophysica Acta*, *1832*(11), 1906–1909. https://doi.org/10.1016/J.BBADIS.2013.05.026.

Hellsten, E., Vesa, J., Olkkonen, V. M., Jalanko, A., & Peltonen, L. (1996). Human palmitoyl protein thioesterase: Evidence for lysosomal targeting of the enzyme and disturbed cellular routing in infantile neuronal ceroid lipofuscinosis. *The EMBO Journal*, *15*(19), 5240. https://doi.org/10.1002/j.1460-2075.1996.tb00909.x.

Henderson, M. X., Wirak, G. S., Zhang, Y. Q., Dai, F., Ginsberg, S. D., Dolzhanskaya, N., et al. (2015). Neuronal ceroid lipofuscinosis with DNAJC5/CSPα mutation has PPT1 pathology and exhibit aberrant protein palmitoylation. *Acta Neuropathologica*, *131*(4), 621–637. https://doi.org/10.1007/S00401-015-1512-2.

Hess, B., Bekker, H., Berendsen, H. J. C., & Fraaije, J. G. E. M. (1997). LINCS: A linear constraint solver for molecular simulations. *Journal of Computational Chemistry*, *18*(12), 1463–1472. https://doi.org/10.1002/(SICI)1096-987X(199709)18:12<1463::AID-JCC4>3.0.CO;2-H.

Kim, S. J., Zhang, Z., Sarkar, C., Tsai, P. C., Lee, Y. C., Dye, L., et al. (2008). Palmitoyl protein thioesterase-1 deficiency impairs synaptic vesicle recycling at nerve terminals, contributing to neuropathology in humans and mice. *The Journal of Clinical Investigation*, *118*(9), 3075–3086. https://doi.org/10.1172/JCI33482.

Lehtovirta, M., Kyttälä, A., Eskelinen, E. L., Hess, M., Heinonen, O., & Jalanko, A. (2001). Palmitoyl protein thioesterase (PPT) localizes into synaptosomes and synaptic vesicles in neurons: Implications for infantile neuronal ceroid lipofuscinosis (INCL). *Human Molecular Genetics*, *10*(1), 69–75. https://doi.org/10.1093/HMG/10.1.69.

Lemonidis, K., Werno, M. W., Greaves, J., Diez-Ardanuy, C., Sanchez-Perez, M. C., Salaun, C., et al. (2015). The zDHHC family of S-acyltransferases. *Biochemical Society Transactions*, *43*(2), 217–221. https://doi.org/10.1042/BST20140270.

Lobanov, M. I., Bogatyreva, N. S., & Galzitskaia, O.v. (2008). Radius of gyration is indicator of compactness of protein structure. *Molekuliarnaia Biologiia*, *42*(4), 701–706. http://www.ncbi.nlm.nih.gov/pubmed/18856071.

Mole, S. E., & Haltia, M. (2015). *The neuronal ceroid-lipofuscinoses (batten disease)* (5th ed., pp. 793–808). Rosenberg's Molecular and Genetic Basis of Neurological and Psychiatric Disease. https://doi.org/10.1016/B978-0-12-410529-4.00070-X.

Palmer, D. N., Barry, L. A., Tyynelä, J., & Cooper, J. D. (2013). NCL disease mechanisms. *Biochimica et Biophysica Acta*, *1832*(11), 1882–1893. https://doi.org/10.1016/J.BBADIS.2013.05.014.

Parrinello, M., & Rahman, A. (1981). Polymorphic transitions in single crystals: A new molecular dynamics method. *Journal of Applied Physics*, *52*(12), 7182–7190. https://doi.org/10.1063/1.328693.

Poux, S., Arighi, C. N., Magrane, M., Bateman, A., Wei, C.-H., Lu, Z., et al. (2017). On expert curation and scalability: UniProtKB/Swiss-Prot as a case study. *Bioinformatics*, *33*(21), 3454–3460. https://doi.org/10.1093/bioinformatics/btx439.

Salonen, T., Heinonen-Kopra, O., Vesa, J., & Jalanko, A. (2001). Neuronal trafficking of palmitoyl protein thioesterase provides an excellent model to study the effects of different mutations which cause infantile neuronal ceroid lipofuscinocis. *Molecular and Cellular Neuroscience, 18*(2), 131–140. https://doi.org/10.1006/MCNE.2001.1010.

Sands, M. S. (2013). Considerations for the treatment of infantile neuronal ceroid lipofuscinosis (infantile batten disease). *Journal of Child Neurology, 28*(9), 1151–1158. https://doi.org/10.1177/0883073813495960.

Schmid, N., Eichenberger, A. P., Choutko, A., Riniker, S., Winger, M., Mark, A. E., et al. (2011). Definition and testing of the GROMOS force-field versions 54A7 and 54B7. *European Biophysics Journal, 40*(7), 843–856. https://doi.org/10.1007/s00249-011-0700-9.

Thirumal Kumar, D., Jain, N., Udhaya Kumar, S., George Priya Doss, C., & Zayed, H. (2021). Identification of potential inhibitors against pathogenic missense mutations of PMM2 using a structure-based virtual screening approach. *Journal of Biomolecular Structure and Dynamics, 39*(1), 171–187. https://doi.org/10.1080/07391102.2019.1708797.

Thirumal Kumar, D., Udhaya Kumar, S., Magesh, R., & George Priya Doss, C. (2021). Investigating mutations at the hotspot position of the ERBB2 and screening for the novel lead compound to treat breast cancer—A computational approach. *Advances in Protein Chemistry and Structural Biology, 123*, 49–71. https://doi.org/10.1016/bs.apcsb.2020.10.001.

Tobaben, S., Thakur, P., Fernández-Chacón, R., Südhof, T. C., Rettig, J., & Stahl, B. (2001). A trimeric protein complex functions as a synaptic chaperone machine. *Neuron, 31*(6), 987–999. https://doi.org/10.1016/S0896-6273(01)00427-5.

Udhaya Kumar, S., Bithia, R., Thirumal Kumar, D., George Priya Doss, C., & Zayed, H. (2022). Mutational landscape of K-Ras substitutions at 12th position—A systematic molecular dynamics approach. *Journal of Biomolecular Structure and Dynamics, 40*(4), 1571–1585. https://doi.org/10.1080/07391102.2020.1830177.

Udhaya Kumar, S., Thirumal Kumar, D., Siva, R., George Priya Doss, C., & Zayed, H. (2020). An extensive computational approach to analyze and characterize the functional mutations in the galactose-1-phosphate uridyl transferase (GALT) protein responsible for classical galactosemia. *Computers in Biology and Medicine, 117*, 103583. https://doi.org/10.1016/j.compbiomed.2019.103583.

Venselaar, H., te Beek, T. A. H., Kuipers, R. K. P., Hekkelman, M. L., & Vriend, G. (2010). Protein structure analysis of mutations causing inheritable diseases. An e-Science approach with life scientist friendly interfaces. *BMC Bioinformatics, 11*, 548. https://doi.org/10.1186/1471-2105-11-548.

Verkruyse, L. A., & Hofmann, S. L. (1996). Lysosomal targeting of palmitoyl-protein thioesterase. *Journal of Biological Chemistry, 271*(26), 15831–15836. https://doi.org/10.1074/JBC.271.26.15831.

Warrier, V., Vieira, M., & Mole, S. E. (2013). Genetic basis and phenotypic correlations of the neuronal ceroid lipofusinoses. *Biochimica et Biophysica Acta, 1832*(11), 1827–1830. https://doi.org/10.1016/J.BBADIS.2013.03.017.

Winklhofer, K. F., Tatzelt, J., & Haass, C. (2008). The two faces of protein misfolding: Gain- and loss-of-function in neurodegenerative diseases. *The EMBO Journal, 27*(2), 336. https://doi.org/10.1038/SJ.EMBOJ.7601930.

Zhang, Y. Q., Henderson, M. X., Colangelo, C. M., Ginsberg, S. D., Bruce, C., Wu, T., et al. (2012). Identification of CSPα clients reveals a role in dynamin 1 regulation. *Neuron, 74*(1), 136–150. https://doi.org/10.1016/J.NEURON.2012.01.029.

Zhang, Z., Mandal, A. K., Wang, N., Keck, C. L., Zimonjic, D. B., Popescu, N. C., et al. (1999). Palmitoyl-protein thioesterase gene expression in the developing mouse brain and retina: Implications for early loss of vision in infantile neuronal ceroid lipofuscinosis. *Gene, 231*(1–2), 203–211. https://doi.org/10.1016/S0378-1119(99)00050-5.

CHAPTER FOUR

Translation initiation and its relationship with metabolic mechanisms in cancer development, progression and chemoresistance

Andrea Muñoz-Ayala[a,b], Brenda Chimal-Vega[a,b], and Victor García-González[a,b,*]

[a]Departamento de Bioquímica, Facultad de Medicina Mexicali, Universidad Autónoma de Baja California, Mexicali, México
[b]Laboratorio Multidisciplinario de Estudios Metabólicos y Cáncer, Universidad Autónoma de Baja California, Mexicali, México
*Corresponding author: e-mail address: vgarcia62@uabc.edu.mx

Contents

Advances in Protein Chemistry and Structural Biology, Volume 132
ISSN 1876-1623
https://doi.org/10.1016/bs.apcsb.2022.05.011

Abstract

Pathways that regulate protein homeostasis (proteostasis) in cells range from mRNA processing to protein degradation; perturbations in regulatory mechanisms of these pathways can lead to oncogenic cellular processes. Protein synthesis modulation failures are common phenomena in cancer cells, wherein specific conditions that promote the translation of protein factors promoting carcinogenesis are present. These specific conditions may be favored by metabolic lipid alterations like those found in metabolic syndrome and obesity. Protein translation modifications have been described in obesity, favoring the translation of protein targets that benefit lipid accumulation; a determining factor is the activity of the cap-binding eukaryotic translation initiation factor 4E (eIF4E), a crosstalk in protein translation and lipogenesis. Besides, alterations of protein translation initiation steps are critical participants for the development of both pathogenic conditions, cancer, and obesity. This chapter is focused on the regulation of recognition and processing of carcinogenic-mRNA and the connections among lipid metabolism and cell signaling pathways that promote oncogenesis, tumoral microenvironment generation and potentially the development of chemoresistance. We performed an in-depth analysis of events, such as those occurring in obesity and dyslipidemias, that may influence protein translation, driving the recognition of certain mRNAs and favoring cancer development and chemoresistance.

1. Introduction

Several diseases have been described as having a molecular etiology alterations of protein synthesis mechanisms, folding, and shielding of the native structure, phenomena that could lead to an accumulation of misfolded proteins (Díaz-Villanueva, Díaz-Molina, & García-González, 2015).

Therefore, mechanisms that regulate critical steps in the protein lifecycle such as transcription, translation, folding, and degradation are susceptible targets for pharmacologic interventions. Moreover, the human proteome contains approximately 20,000 proteins, and estimations indicate that more than 600 of them could be functionally important for several types of cancers (Dale et al., 2021).

Global translational rates in cancer cells are generally enhanced, but aren't necessarily associated with augmented transcription, suggesting that alterations occur in protein-translation steps. Regulation at the protein translational level allows a rapid response to internal and external signals in the life cycle of cells (Somers, Pöyry, & Willis, 2013). mRNA translation is performed in three steps: initiation, elongation, and termination (Siddiqui & Sonenberg, 2015). In this chapter, we will focus on the initiation step. Evidence suggests that metabolic overload and mainly the lipids-effect could be a critical factor in deregulation of protein translation; for instance, the modulation on mTOR/pS6K signaling is broadly documented and could regulate lipid metabolism(Rui, 2007), however, other mechanism are also important. Therefore, the connection between protein initiation and the influence of lipid overload could allow the elucidation of critical mechanisms in cancer development and progression, tumoral microenvironment generation and chemoresistance.

2. Translation initiation and its regulation

The initiation of translation is described as the limiting step in protein synthesis (Jackson, Hellen, & Pestova, 2010; Silvera, Formenti, & Schneider, 2010; Truitt & Ruggero, 2016; Yang, Zhong, & Cao, 2020). This process needs the assembly of the 80S ribosome from the 40S and 60S subunits, an mRNA, and a transfer RNA (Met-tRNA$_i$). In this initiation step, the Met-tRNA$_i$ and the start codon in the mRNA are positioned in the ribosomal P site (Passmore et al., 2007). Almost every eukaryotic mRNA possesses a 5′methylguanosine cap at the seventh position of the first Guanosine (m7G) and a poly adenine (poly(A)) tail at the 3′ end (Bentley, 2014). These modifications allow mRNA interaction with some of the eukaryotic initiation factors (eIFs). There are around 12 known eIFs proteins (Merrick & Pavitt, 2018; Pestova et al., 2001) whose function favors a subtle regulation of protein synthesis, representing a high energy cost for cells. However, their deregulation can lead to the development of different pathological conditions, such as cancer (Biffo, Manfrini, & Ricciardi, 2018; Merrick & Pavitt, 2018; Silvera et al., 2010; Truitt & Ruggero, 2016). The translation initiation process can be described in four major steps (Pestova et al., 2001).

2.1 The formation of the preinitiation complex (PIC)

Initially, the ternary complex (TC) is formed by the eukaryotic initiation factor 2 (eIF2) bound to GTP and the Met-tRNA$_i$. Later, TC binds to the 40S subunit of the ribosome along with other eIFs; eIF5, eIF3, eIF1, and eIF1A giving place to 43S PIC (Pestova et al., 2001), whose main function is to allow or prevent mRNA entry and scanning in the latter steps changing into an open or close conformation. Specifically, eIF1 and eIF1A stabilize the 43S complex in an open conformation (Passmore et al., 2007).

Regulation at this stage by several kinases is focused on the phosphorylation of serine 51 of the eIF2 (p-eIF2α) α subunit, in response to cellular stress conditions (Wek, 2018). Critically, stress conditions can be generated by oncogenic pathways. eIF2 is a GDP/GTP binding protein that has a high affinity to Met-tRNA$_i$ only when bound to GTP. eIF2B is a nucleotide exchange factor (GEF) that switches eIF2 to its active form bound to GTP (Merrick & Pavitt, 2018). Phosphorylation of eIF2α induces a high affinity to GDP preventing activation by eIF2B, therefore lowering TC concentrations and PIC formation (Weiss, Allen, Kloehn, Abid, & Curran, 2021; Wek, 2018).

2.2 mRNA activation by assembly with eIF4F complex and binding to the 43S PIC

The heterotrimeric complex eIF4F is formed by eIF4A (with helicase activity), eIF4E (which binds to the mRNA 5′ 7-methylguanosine cap) (m^7G cap), and eIF4G (a scaffold protein) (Gross et al., 2003; Hinnebusch, 2014). eIF4F is required to directly recruit the 40S ribosomal subunit in the PIC to mRNAs through a set of protein-protein interactions and to unfold RNA secondary structures located at the 5′ untranslated region (5'UTR) of mRNAs, which is the non-coding segment upstream of the main protein coding sequence (Malka-Mahieu, Newman, Désaubry, Robert, & Vagner, 2017). mRNA binds to eIF4E, present in eIF4F complex, and poly (A) binding protein (PABP), thus, forming a circular mRNA. eIF4A, whose activity is enhanced by eIF4B, is responsible to unfold mRNAs secondary structures. eIF4G also activates eIF4A by an allosteric mechanism (Hinnebusch, 2014). eIF3 is crucial for the interaction among eIF4F, containing the activated mRNA and the PIC complex (Jackson et al., 2010; Merrick & Pavitt, 2018).

Regulation in this step primarily depends on eIF4E, which is the limiting factor on the formation of the eIF4F complex. Considered as a proto-oncogene, eIF4E can be overexpressed in around 30% of all human tumors (Weiss et al., 2021). eIF4E is susceptible to regulation through the interaction with eIF4E binding proteins (4E-BPs). When 4E-BPs bind to eIF4E, the interaction with eIF4G to form the eIF4F complex is prevented. In turn, 4E-BPs are regulated by a mechanism of phosphorylation. Interaction with eIF4E is diminished in hyperphosphorylated 4E-BPs. 4E-BPs are mTOR complex 1 (mTORc1) substrates (Richter & Sonenberg, 2005), and mTORc1 is downstream of the PI3K/AKT signaling pathway.

In non-tumoral cells, nutrient-rich conditions generate mTORc1 activation and phosphorylation of 4E-BPs, leading to a protein translation increase. Moreover, PI3K/AKT signaling is commonly overactivated in cancer cells (Biffo et al., 2018). In addition, eIF4E can be phosphorylated at Ser-209 by the mitogen-activated protein kinase (MAPK)-interacting kinases (MNKs), which first bind eIF4G (Pyronnet et al., 1999; Yang et al., 2020). eIF4G phosphorylation by PKCα at Ser 1186 has been described to modulate its interaction with MNKs (Dobrikov, Dobrikova, Shveygert, & Gromeier, 2011). eIF4G can also be phosphorylated in multiple serine residues by several serine/threonine kinases (Martínez-Alonso et al., 2022).

eIF4A is also susceptible to regulation by programmed cell death 4 (Pdcd4) protein, which regulates its activity in a concentration-dependent manner. It has recently been found that in diverse types of tumors Pdcd4 is downregulated (Gonzalez-Ortiz, Galindo-Hernandez, Hernandez-Acevedo, Hurtado-Ureta, & Garcia-Gonzalez, 2021; Lankat-Buttgereit & Göke, 2009), a condition that could potentiate a high eIF4A activity, and ultimately, translation initiation.

2.2.1 Formation of mRNA structures in 5′ untranslated region (5′UTR)

mRNAs compete for the association to the eIF4F complex (Piserà, Campo, & Campo, 2018). mRNAs have varied 5′and 3′-UTR that directly affect their interaction with translational machinery. Emphasizing on 5′UTR, the human median length is about 218 nucleotides but varies hugely among individual genes (Leppek, Das, & Barna, 2018). Depending on 5′UTR length and composition, it can adopt different secondary structures that can influence translation efficiency (Korneeva, Song, Gram, Edens, & Rhoads, 2016; Pelletier & Sonenberg, 1987). mRNAs with high GC content in 5′UTRs and high negative folding free energy have a lower rate of initiation (Leppek et al., 2018). These mRNAs require high levels of eIF4F complex and therefore of the limiting factor eIF4E to unfold their secondary structure to be effectively translated (Piserà et al., 2018).

Considering their dependence on eIF4E mRNAs can be classified as strong or weak mRNAs. Strong mRNAs have short and unstructured 5′-UTRs, which are typically transcripts of housekeeping genes, and are barely affected by eIF4E levels. In comparison, weak mRNAs have long 5′UTR with high content of GC capable to form stable secondary structures such as RNA-G quadruplex involved in cell cycle progression and cancer (Piserà et al., 2018; Silvera et al., 2010).

2.3 Scanning of the mRNA-bound ribosomal complex along with the 5′UTR mRNA

For traditional CAP dependent translation, in this step, the 5′UTR of the mRNA is scanned in a $5' \rightarrow 3'$ direction base by base until complementarity to the Met-tRNA$_i$ anticodon is found. Commonly, the first AUG codon functions as the start codon. The sequence context is important for its recognition (Hinnebusch, 2014) since the initiation complex makes contact with 17 nucleotides upstream and 11 nucleotides downstream the AUG (Elfakess et al., 2011). AUG codons are more effectible recognized when surrounded by the Kozak consensus sequence (GCCACCAUGG) (Sonneveld, Verhagen, & Tanenbaum, 2020).

The 40S ribosomal subunit occasionally skips the first AUG and initiates translation at a downstream AUG, a phenomenon known as leaky scanning (Elfakess et al., 2011; Sonneveld et al., 2020). The scanning and the unwinding of the secondary structure of the RNAs, primarily by eIF4A, and then the eIF4F, requires the expenditure of energy in form of ATP (Hinnebusch, 2014).

2.4 Formation of the 80S ribosome

The AUG recognition generates a change in PIC from an open to a closed conformation to prevent further scanning (Hinnebusch, 2014; Merrick & Pavitt, 2018; Passmore et al., 2007). This conformational change causes eIF2-GTP hydrolysis and the release of eIFs associated with the 40S subunit eIF1, eIF1A, eIF2-GDP, and eIF5. Those events facilitate the recruitment of the 60S subunit to give place to the functional 80S ribosome (Weiss et al., 2021), at the initiation codon poised to begin the elongation stage of translation (Elfakess et al., 2011; Merrick & Pavitt, 2018).

A regulator mechanism is associated with eIF-6, which has been described as an anti-association factor that binds to the 60S ribosomal subunit and prevents its interaction with the 40S subunit (Russell & Spremulli, 1980). In addition, eIF-6 is involved in control of mRNA specific translation. eIF-6 levels are regulated by conditions of demand of ribosome synthesis and translational rate (Brina, Miluzio, Ricciardi, & Biffo, 2015), such as cancer.

3. The impact of translation initiation in cancer

Cancer is a complex disease; tumorigenesis involves a progressive acquisition of changes in the initially normal cells. The traditional hallmarks of cancer are evading apoptosis, sustaining proliferative signaling, evading growth suppressors, sustained angiogenesis, limitless replicative potential, and activation of invasion and metastasis (Hanahan & Weinberg, 2000). Although these are critical for cancer development, phenomena such as the misbalance of protein translation and lipid metabolism may be critical triggering factors.

Considering of high rate of proliferation, an elevated rate of protein synthesis is required. Cancerous cells generally have enlarged nucleoli, where it takes place ribosomal RNA synthesis and ribosome assembly(Silvera et al., 2010).

Almost every major oncogenic signaling pathways found to be altered in human cancers (Truitt & Ruggero, 2016).

Oncogenes have effects on the cellular proteome, many of them act in the initiation step of translation (Kuzuoglu-Ozturk et al., 2021). Several exert their effects over eIF4E, which is the limiting factor in the formation of the eIF4F complex. Among them, we can mention phosphatidylinositol 3-kinase (PI3K)/Akt and Myc which increase transcription, Ras activates phosphorylation by regulating mitogen-activated protein kinase (MAPK) which in turn interacts with MNK1, regulating the expression and hyperactivity of eIF4E. eIF4E is also regulated through the mTOR pathway by suppressors of eIF4E-binding protein (4EBPs), which inhibits its activity (Kuzuoglu-Ozturk et al., 2021). Critically, p-eIF4E favors translation and promotes metastasis-associated pathways in cancer cells (Siddiqui & Sonenberg, 2015; Xie, Merrett, Jensen, & Proud, 2019; Yang et al., 2020).

3.1 eIF4E structure and function

eIF4E is a single 25-kDa polypeptide in mammals, its primary structure is highly conserved in all eukaryotes. The three-dimensional structure consists of eight antiparallel β strands and three α-helices on the dorsal surface, which shows an invariant hydrophobic-acid portion responsible for the binding of eIF4E to proteins partners such as eIF4G and 4EBPs (Fig. 1). The cap binding site is determined by the formation of cation-π bond stacking between the 7-methylguanine, and the two conserved tryptophan residues located within the cap-binding pocket Trp-56 and Trp-102 (Fig. 1). Critically, the interaction of eIF4E with eIF4G involves Trp-43 and Trp-73 in human eIF4E-1. The eIF4E dorsal surface is responsible for the binding of eIF4E-associated proteins, including eIF4G and 4E-BPs. Therefore, the interaction of eIF4G and 4E-BPs is mutually exclusive because they bind to the same site of eIF4E (Piserà et al., 2018).

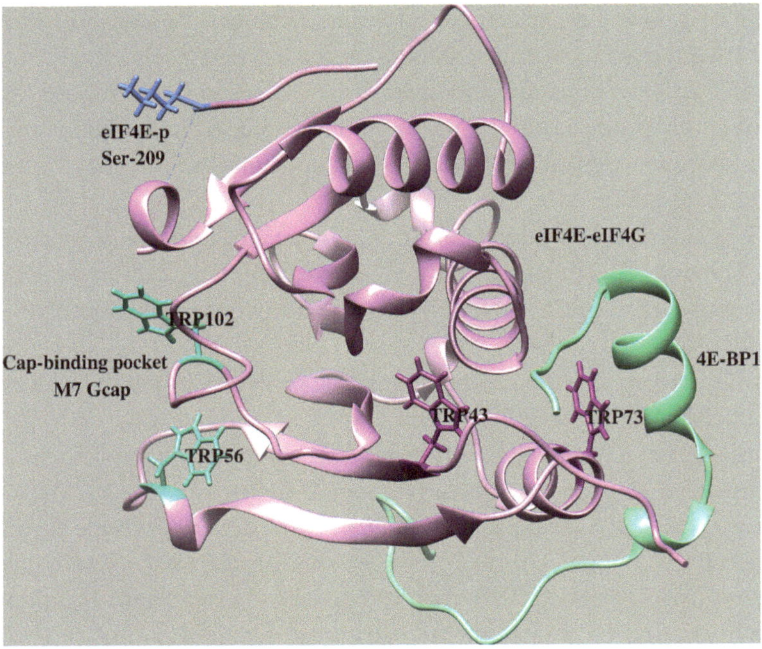

Fig. 1 Human eIF4E/4E-BP1 complex, identifying the critical residues that control the function. The crystallographic structure of eIF4E is characterized by the binding with the protein partners eIF4G and 4E-BPs occurring between the Trp 43 and Trp 73 (green-cyan and purple Trp). Phosphorylation at Ser-209 (blue) and the interaction site in the cap-binding pocket of eIF4E residues located Trp-56 and Trp-102 (Cyan) are identified. Image adapted from Piserà et al., 2018. PDB structure: ID 4UED.

Although eIF4E is mainly localized in the cytoplasm, it has also been found in the nucleus. eIF4E function is involved in the export of specific mRNAs from the nucleus to the cytoplasm (Culjkovic-Kraljacic & Borden, 2013) On the other hand, it has been shown that it may control translation with the TOP mRNAs (mRNAs with 50 terminal oligopyrimidine tracts), these mRNAs encode elongation factors, ribosomal proteins, and fused proteins. Likewise, eIF4E promotes the translation of histone H4 mRNA without the need for cap structure, which has a paired-stem loop element within its open reading frame similar to the 4E-SE element of several mRNAs (Piserà et al., 2018).

3.2 eIF4E abundance and regulation by 4E-BP1 and its implications in cancer

Depending on eIF4E abundance, it can favor the translation of proteins related to transformation and tumorigenesis (Batool, Aashaq, & Andrabi, 2019), specifically those with long and complex 5′UTR. Decreases in eIF4E expression levels do not affect normal development in eIF4E$^{+/-}$ mice models, whose translational levels remains unchanged. However, normal or increased expression levels of eIF4E are essential for cancer cells to maintain a selective translatome that promotes oncogenic transformation (Truitt & Ruggero, 2016).

4E-BP1 hyperphosphorylation by several pathways in cancer cells increases the amount of available eIF4E to form the eIF4F complex and promote translation. Hyper-phosphorylated 4E-BP1 has been associated with metastasis in different types of cancer, such as breast, ovarian, and prostate tumors with poor prognosis and malignant progression independent of other previous oncogenic factors (Castellvi et al., 2006). In breast cancer, it was found that phosphorylated 4E-BP1 was expressed in poorly differentiated tumors showing an increase compared to total 4E-BP1. In addition, tumors express higher levels of eIF4E than 4E-BP1. In ovarian cancer the same effect was observed, a correlation with high levels of p-4E-BP1, regardless of HER2/neu, EGFR, PTEN, or PI3K status, suggesting that it could be a therapeutic target for breast and ovarian cancer screening (Armengol et al., 2007).

Likewise, in high-grade prostatic intraepithelial neoplasia, activation of the Akt/mTOR cascade was detected, wherein the expression of phosphorylated Akt and phosphorylated p70S6K was similar, as well as 4E-BP-p, suggesting that this factor could help identify patients at high risk. On the other hand, in colon carcinoma, high expression of 4EBP-p did not correlate with poorly differentiated tumors, but an association with the presence of lymph node metastases was registered. Although the evidence is not robust, p-4EBP1 is proposed as a biomarker, considering its overexpression is highly related to different types of cancer as it is associated with eIF4E, which plays a key role in the control of protein synthesis, cell survival, and growth (Armengol et al., 2007; Piserà et al., 2018).

3.3 eIF4E phosphorylation and its implication in cancer

It has been demonstrated that inhibition of eIF4E phosphorylation can prevent tumorigenesis in various cancer models (Furic et al., 2010; Wendel et al., 2007; Wheater, Johnson, & Blaydes, 2010). Determined events upstream ERK/MAPK and p38 cause Thr^{209} and Thr^{214} phosphorylation in MNK1 and Ser^{209}-eIF4E phosphorylation, under the activation of mitogens, osmotic stress, heat shock, and proinflammatory cytokines

(Wang et al., 2020; Yang et al., 2020). Likewise, certain oncogenes can cause MNK1 phosphorylation, such as in leukemia Friend leukemia integration 1 (FLI1) transcription factor and AML fusion genes (Wang et al., 2020; Yang et al., 2020). It has been proved that substitution of serine 209 residue to alanine in eIF4E (eIF4E^{S209A}), generating a non-phosphorylatable variant, remarkably reduces tumorous processes in multiple cell lines and murine models (Ruan et al., 2020; Xie et al., 2019). p-eIF4E favors the translation of mRNAs involved in tumorigenesis (Yang et al., 2020), mainly those with the characteristic weak mRNAs 5′UTR (Table 1). However, the mechanism in which this phosphorylation promotes specific mRNA translation continues under study (Yang et al., 2020).

p-eIF4E contributes in sustaining proliferative signaling by increasing expression of cyclin D1, c-Myc and β-catenin, key molecules for cell division (Yang et al., 2020). Contributing to the hallmark of cell death resistance, p-eIF4E favors the translation of anti-apoptotic proteins, Mcl-1 (Tailler et al., 2019; Wendel et al., 2007), B-cell lymphoma 2 (Bcl-2), and survivin (Wang et al., 2020).

In breast cancer cell line MDA-MB-435 expressing $\alpha_6\beta_4$ integrin, p-eIF4E favored vascular endothelial growth factor (VEGF) translation (Korneeva et al., 2010). VEGF functions as an autocrine signaling molecule, promoting survival and metastasis phenomena (Goel & Mercurio, 2014). Robichaud et al. showed that eIF4E phosphorylation in response to TGFβ promotes translation of SNAIL and MMP-3 stimulating epithelial to mesenchymal transitions (EMTs) in normal murine mammary gland cell model (NMuMG). They also proved in a xerograph model, mice with eIF4E^{S209A} mutation displayed a twofold reduction in lung metastases (Robichaud et al., 2014).

Finally, it has been observed that p-eIF4E can favor transcription of molecules involved in chemoresistance in different cancer types (Geter et al., 2017; Khalil & De Benedetti, 2022; Sunavala-Dossabhoy et al., 2005). This point will be addressed later. (Fig. 2).

Inhibition of p-eIF4E for cancer treatment is a strategy that has been proved in several cancer models. For example, Njar workgroup has evaluated in breast and castration resistant prostate cancer (CRPC) cell lines and xerographs compounds called retinoic acid metabolism blocking agents (RAMBAs) retinamides (RRs), which induce MNKs degradation through the ubiquitin proteasome pathway obtaining promising results(Kwegyir-Affu, Ramalingam, Purushottamachar, Ramamurthy, & Njar, 2015; Ramalingam et al., 2014, 2019; Ramamurthy et al., 2018).

Table 1 Protein targets modulated by eIF4E phosphorylation and/or overexpression.

	Protein	Example of evaluated model	Advantage that confers to tumor cells *in vivo*	Effects of protein enhanced translation	References
Cell proliferation	Cyclin D1	Castration-resistant prostate cancer cells	Promotes cell division	Allows progression of the cell cycle through the G1 to S phase.	Ramamurthy, Ramalingam, Gediya, & Njar, 2018; Yang et al., 2020
	c-Myc	Colorectal cancer (HCT116, DLD1, RKO, SW480, HT29)	Promotes cell growth, proliferation, and loss of differentiation	Activates transcription of genes involved in those processes by binding to response elements or remodeling chromatin.	van Riggen, Yetil, and Felsher (2010); Ruan et al., 2020; Yang et al., 2020
	β-catenin	Nasopharyngeal Carcinoma (NP69, CNE1, HNE1, HNE2, 5-8F, and 6-10B)	Promotes proliferation and metastasis	Increases nuclear localization and transcriptional upregulation of Ccnd1, c-Myc, MMP-7 and AXIN2 genes as in Wnt pathway	Ruan et al., 2020; Xu et al., 2017)
Apoptosis evasion	MCL1	Leukemia (NALM-6, HL-60, REH, RS4, K562)	Overall resistance to programed cell death	Prevent activation of BAX and BAK proapoptotic members of the BCL-2 family	Tailler, Lindqvist, Gibson, & Adams, 2019
	B-cell lymphoma 2 (Bcl-2)	Castration-resistant prostate cancer cells	Overall resistance to programed cell death	Prevent activation of BAX and BAK proapoptotic members of the BCL-2 family	Ramamurthy et al., 2018; Yang et al., 2020
	Survivin	Erythro- leukemia (HEL and K562)	Resistance to programmed cell death	Inhibits caspases	Jaiswal, Goel, and Mittal (2015); Wang et al., 2020

Continued

Table 1 Protein targets modulated by eIF4E phosphorylation and/or overexpression.—cont'd

	Protein	Example of evaluated model	Advantage that confers to tumor cells *in vivo*	Effects of protein enhanced translation	References
EMT and metastasis	SNAIL	Normal murine mammary gland	Promotes metastasis increasing steam cell like features	Regulates signaling pathways leading to EMT.	Ramamurthy et al., 2018; Robichaud et al., 2014
	MMP-3	Normal murine mammary gland	Promotes tissue invasion and metastasis.	Extracellular matrix remodeling	Ramamurthy et al., 2018; Robichaud et al., 2014; Robichaud et al. (2018)
	Vascular endothelial growth factor (VEGF)	Triple negative breast cancer (MDAMB-435)	Promotes dedifferentiation and an epithelial-mesenchymal transition phenotype	Activates receptor tyrosine kinases (RTKs) and neuropilins in an autocrine way	Goel & Mercurio, 2014; Korneeva et al., 2010
Chemoresistance	Tousled-like kinase 1 isoform B (TLK1B)[a]	Castration-resistant prostate cancer cells	Resistance over DNA damaging chemotherapeutics like anthracyclines (doxorubicin)	Phosphorylates NEK1, activating ATR > Chk1 > DNA damage response	Khalil & De Benedetti, 2022; Sunavala-Dossabhoy, Balakrishnan, Sen, Nuthalapaty, & De Benedetti, 2005
	Runx2	Estrogen receptor positive breast cancer (MCF-7)	Tamoxifen resistance	Functions as a molecular scaffold to recruit chromatin remodeling enzymes and stablish a molecular program that opposes Erα signaling.	Geter et al., 2017

[a]Overexpression by p-eIF4E is not yet described.

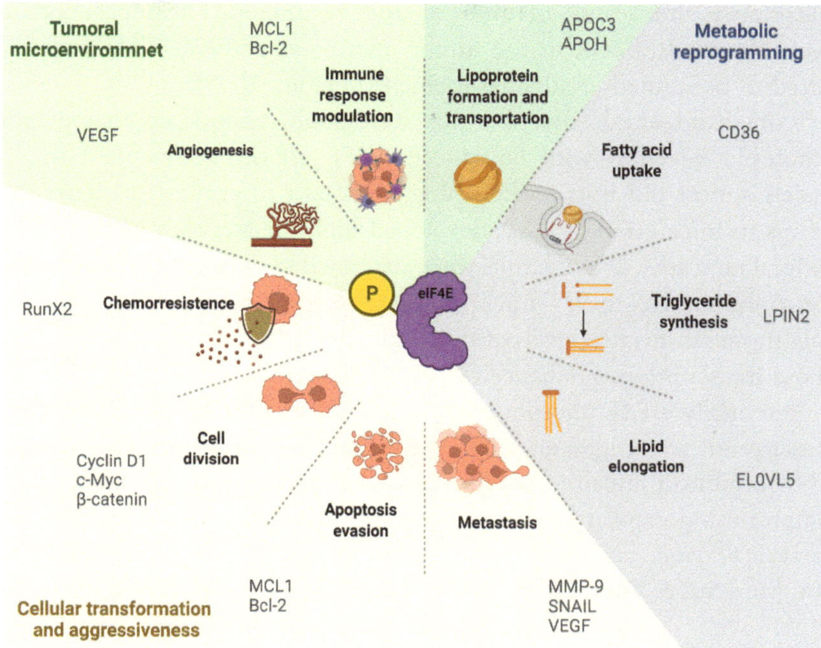

Fig. 2 Relationship between eIF4E phosphorylation cancer development and progression. p-eIF4E favors cellular transformation processes by increasing transcription of proteins involved in metastasis (MMP-9, SNAIL, VEGF, β-catenin), apoptosis evasion (MCL1, Bcl2), cell division (Cyclin D1, c-Myc), and chemoresistance (RunX2). p-eIF4E also can contribute to metabolic reprogramming increasing fatty acid uptake (CD36), triglyceride synthesis (LPIN2), and lipid elongation (ELOVL5) favoring energy accumulation in tumor cells. Moreover, p-eIF4E can increase lipoprotein formation and transportation (APOC3, APOH) by the liver, increasing lipid availability in the tumor microenvironment, in addition, to promote angiogenesis (VEGF) and promoting survival of prometastatic neutrophils (MCL1, Bcl2). Created with BioRender.com

3.4 Impact of eIF4E phosphorylation in tumor microenvironment

Cancer cells are exposed to highly fluctuating metabolic conditions, signaling metabolites, stromal biomolecules, extracellular vesicles and, chemotherapeutic agents, which together generate a microenvironment that contributes to cancer development or changes in phenotype (Garcia-Gonzalez, Galindo-Hernandez, & Gonzalez-Ortiz, 2020).

In a solid tumor, cancerous cells are surrounded by different resident or recruited non-cancerous cell types that establish collaborative interactions (Hanahan & Coussens, 2012). Importantly is common to find endothelial cells, fibroblast, and adaptative and innate immune cells that together contribute to the creation of tumor microenvironment. The impact of translational control in cells in the tumor microenvironment (TME) recently started to be studied (Nathaniel Robichaud et al., 2018).

Robichaud et al. describe that xerograph tumors originated from orthotopic injection with 66cl4 cells (WT eIF4E) in mice $eIF4E^{S209A}$ models almost did not develop lung metastasis in contrast to xerograph tumors in mice $eIF4E^{WT}$ that developed high metastases of up to 50% of the total lung area. These results were attributed to enhanced pro-metastatic neutrophil survival in WT mice. These neutrophils turned out to be apparently more sensitive to levels of BCL2 and MCL1 (Robichaud et al., 2018) whose transcription is enhanced by p-eIF4E.

Potentially eIF4E phosphorylation in endothelial cells can favor VEGF transcription and angiogenesis in consequence. Cercospormide, a potent MNK inhibitor, impaired hepato-carcinoma endothelial cells angiogenesis compromising capillary network formation, migration proliferation, and survival *in vitro,* and suppressed tumor growth in a xenograft model (Liu, Sun, Su, & Guo, 2016).

3.5 Therapeutic strategies on eIF4E

Therefore, several eIF4E inhibition strategies have been developed (Hsieh & Ruggero, 2010) like the use of antisense oligonucleotide (ASO) therapy. The effect of second-generation ASO against eIF4E was evaluated, in combination with irinotecan in solid tumors and irinotecan-refractory colorectal cancer in phase I/II translational clinical trials (Duffy et al., 2016). On the other hand, evidence suggests that ribavirin, an eIF4E inhibitor, might be successfully used as a cancer treatment. Ribavirin treatment in a phase II clinical trial led to clinical benefit in patients with poor-prognosis of acute myeloid leukemia (Assouline et al., 2009). Critically, a preclinical study reported an effect of ribavirin treatment on breast cancers overexpressing eIF4E (Pettersson et al., 2012).

Evidence of the impact of the mTOR/eIF4E axis is represented by the drug everolimus, an orally administrated inhibitor of mTOR. In a pioneer study, a pharmacodynamic phase I report in patients with advanced solid tumors was performed (Mauthe et al., 2018) A dose- and schedule-dependent inhibition of the mTOR pathway was reported in 55 patients, a correlation was found between everolimus plasma levels and inhibition of peIF4G and p4E-BP1 (Tabernero et al., 2008). Moreover, clinical evidence suggests everolimus used in combination with endocrine therapy is an adequate first-line treatment option for postmenopausal women with ER+, HER2- advanced breast cancer (Royce et al., 2018; Safra et al., 2017).

3.6 eIF4E and the association with metabolic alterations

Metabolic syndrome (MetS) favors the presence of higher levels of lipoproteins, and adipokines in the tumor microenvironment that promote neoplastic progression processes like anchor-independent growth and epithelial-mesenchymal transitions (EMT) (Garcia-Gonzalez et al., 2020). EMT are a determinant hallmark of cellular invasion and metastasis (Fedele, Sgarra, Battista, & Cerchia, 2022). In this sense, metastasis is the process in which cancer cells are dispersed from the site of tumorigenesis and disseminated to specific tissues; metastasis remains the major cause of morbidity and mortality in cancer patients (Chaffer & Weinberg, 2011). During these transitions, metabolic adaptations allow cells to withstand a new tumor microenvironment in the metastatic site.

It is important to denote that eIF4E phosphorylation can favor obesity and hypertriglyceridemia as well metabolic adaptation in cancer cells, promoting cancer development or progression, as will be analyzed later.

4. Obesity and cancer

Obesity is considered a chronic disease that induces functional changes, altering systemic metabolic homeostasis. Specifically, abdominal obesity is associated with insulin resistance and the development of dyslipidemias such as high LDL, low HDL, hypertriglyceridemia, and hyperglycemia (Huang, 2009). Obesity has been related to pathologies such as hypertension, type 2 diabetes, cardiovascular disease, osteoarthritis, kidney failure, liver disease, and the development of different types of cancer including hepatocellular carcinoma, pancreatic cancer, and breast cancer (Chen, Wen, Li, Luo, & Zhang, 2016).

Altered adipokine-concentration, high levels of circulating and local estrogens, disrupted insulin signaling, modifications within the microbiome, and local and systemic effects of inflammation could contribute to the obesity and cancer relationship (Seiler, Chen, Brown, & Fagundes, 2018). Obesity and dyslipidemias have been considered triggering conditions of neoplastic pathology, contributing to the increase of mortality rate. BC has been classified luminal type with estrogen receptor (ER+) expression, HER2-positive, and triple negative (TNBC) (Kolb, Sutterwala, & Zhang, 2016). Diagnosis is performed by immunohistochemical criteria.

It has been shown that postmenopausal obese women have a higher risk of developing ER+ luminal BC than women with normal weight (Healy et al., 2010; Kolb et al., 2016). Likewise, reports suggested an increased risk of recurrence in obese *versus* normal weight women with diagnosed BC (Chan et al., 2014). Moreover, several oncogenes drive metabolic and translational changes in cancer cells (Biffo et al., 2018).

In the Apolipoprotein Mortality Risk (AMORIS) study, comprised of 812,073 Swedish individuals, compared lipid profiles from BC patients, measured for 8 years previous to BC diagnosis, results showed weak evidence associated with impaired triglycerides concentrations (Melvin et al., 2012). Likewise, in a report of 208 BC patients *versus* a control group of 176 healthy women, patients with lymph node metastasis showed plasmatic altered metabolites such as glucose, triglycerides, elevated LDL, and low HDL (Raza, Asif, Rehman, & Sheikh, 2018). The presence of obesity before diagnosis, insulin resistance during diagnosis, and hyperinsulinemia after treatment have been associated with an increased risk of mortality and recurrence of BC (Caan et al., 2008; Duggan et al., 2011).

In another report, components of MetS were associated with a modest risk of BC-recurrence and BC-specific mortality (Calip et al., 2014). This study included 4216 women between the period of 1990 and 2008, the presence of $\geq$3 MetS components in 26% of women, was associated with increased risk of a second breast cancer event (SCBE) when compared (Calip et al., 2014). Although. data of the NIH-American Association of Retired Persons Diet and Health Study cohort (NIH-AARP), suggested individual MetS components, such as waist circumference, high cholesterol, and hypertension, were independently associated with an increased risk of BC mortality (Dibaba, Braithwaite, & Akinyemiju, 2018).

In this sense, a role of lipid metabolism alterations as biomarkers of cancer prognosis and survival has been proposed (Garcia-Gonzalez et al., 2018). Changes of expression of lipid-metabolism genes are found in several tumors, considering lipid metabolism modulates cellular processes ranging from plasmatic and organelle membrane organization and plasticity (Zhao et al., 2016), ATP production (Carracedo, Cantley, & Pandolfi, 2013), to intracellular signaling alterations (Röhrig & Schulze, 2016).

4.1 The role of lipid misbalance

In breast cancer malignant transformation lipid metabolism unbalance, especially in fatty acid metabolism represents a key factor (Simeone et al., 2021), since it is associated with abnormal expression of targets that could dysregulate metabolic pathways. For instance, *de novo* lipogenesis provides the substrates to maintain membrane composition and energetic functions in tumor cells. Critically, genes involved in the metabolism of exogenous-fatty acids, triglyceride, and lipid droplets synthesis, as well as fatty acid oxidation are overexpressed in aggressive basal-like negative receptor subtype (Giudetti et al., 2019; Monaco, 2017).

EMTs could involucrate rearrangements on lipid metabolism, therefore, lipid profiles could aid distinguishing epithelial cells before and after their transition to mesenchymal cells. Excess fat triggers extensive remodeling of the adipose tissue microenvironment in terms of its size, vascularity, and cellular and matrix composition is accompanied by an exacerbated secretion of adipokines (Ouchi, Parker, Lugus, & Walsh, 2011), which is correlated with a systemic chronic low-grade inflammatory condition and metabolism alterations (Cox, West, & Cripps, 2015). Indeed, inflammation and oxidative stress as central processes of breast cancer development and chemoresistance have been proposed (Pires et al., 2019).

Clinically important apolipoproteins related to lipid metabolism and atherogenesis include apolipoprotein B-100, the major structural component of VLDL, IDL, LDL, and lipoprotein(a) (Mehta et al., 2022). VLDL and LDL modulate intracellular metabolism and have been associated with the development of neoplastic phenomena, such as proliferation, anchorage-independent growth, development of EMTs, angiogenesis, and metastasis in cellular models (Gonzalez-Ortiz et al., 2021; Lu et al., 2017). Specifically, the treatment of L1 and L5 subfractions of LDL and VLDL in BC cells increased the levels of mesenchymal markers Slug, vimentin, and β-catenin, promoting migration and invasion, distinctive events of cells under the EMT process (Lu et al., 2017). Furthermore, secretion of angiogenic factors in BC-cells and formation of new blood vessels were also registered (Lu et al., 2017). Therefore, VLDL and LDL could be critical in several mechanisms of tumor progression.

In addition, a positive correlation of higher VLDL concentrations and lymph node metastasis has been reported in BC patients. It has been proposed that these lipoproteins could transport triglycerides, cholesterol, and metabolites such as oxysterols from the liver to the tumor cells (Flote et al., 2016). Elevated concentrations of VLDL have been associated with enhanced VEGF-signaling, which promotes lymphatic angiogenesis, inflammatory cascades by tumor associated macrophages, are phenomena that contribute to nodal metastasis disease (Flote et al., 2016).

In an analysis of 1054 BC patients compared with 2483 healthy women stratified by age, Li et al. reported that lipid profiles become worsened after chemotherapy, increasing concentrations of total cholesterol, triglycerides, LDL, and apolipoprotein B; while levels of HDL and Apo–A1 decrease compared with pre-chemotherapeutic status (Li et al., 2018). Likewise, in The Malmo Diet and Cancer Study (MDCS), which enrolled 17,035 women (1991–1996), BC risk was inversely associated with Apo-B concentrations (Borgquist et al., 2016), the main protein of VLDL and LDL.

Importantly, in another report, MDCS participants with incident invasive breast cancer diagnosed between 1991 and 2014 were identified, and during the 5 years of follow-up, 90 invasive breast cancer recurrences were diagnosed over 3807 person-years. Harborg et al. reported that high pre-diagnostic levels of Apo-B were associated with an increased risk of recurrence among breast cancer patients (Borgquist et al., 2016; Harborg et al., 2022).

Increased ApoB levels and hypertriglyceridemia reflect a metabolic misbalance originated by obesity, a critical risk factor for cancer. In this sense phosphorylation by mTOR inactivates members of the 4E-BP family, a mechanism of translational repressors of 5′ cap-dependent protein synthesis, involving eIF4E and the complex eIF4F. Deletion of 4E-BP1 and 4E-BP2 in mice results in conditions of insulin resistance, energy expenditure decrease, and increased adipogenesis (Bacquer et al., 2007). Therefore, a link between the misbalance of lipid metabolism, the role of cancer progression, and the control of translation initiation could be present. Moreover, in the ob/ob obese mice model, attenuation of obesity induced VLDL hyperlipidemia relates to reduction of tumor volume by 50% (Hussain et al., 2022), this was reached by inhibition of microsomal triglyceride transfer protein (MTP), key in triglyceride synthesis, then VLDL (Hussain et al., 2022).

4.2 Connection with protein translation

Unexpectedly, protein translation-initiation represents a point of regulation on lipid metabolic targets. For instance, the lipid overload could promote the deregulation of the activity of the eIF4F complex, and specifically, the effect could be on eIF4E-related mechanisms. Anabolic and catabolic hormonal and nutritional signals can control protein synthesis and mRNA translation regulated by mTOR, which plays an important role in body weight regulation through regulation of eIF4-E and S6K-1 (Drake, Alway, Hollander, & Williamson, 2010). Indeed, eIF4E activity could be a regulatory factor for obesity, since studies in mice with lipid-enriched nutrients induced obesity showed an increase in eIF4E phosphorylation at serine 209. The effect of eIF4E was analyzed in eIF4E^{S209A} mice, where it was observed no lipid-rich diet induced weight gain (Conn et al., 2021). eIF4E phosphorylation causes an increased expression of critical features on lipid metabolism, transport and storage, just like CD36 in fatty acid uptake, Elongation of very long chain fatty acids protein 5 (ELOVL5) in lipid elongation, phosphatidate phosphatase(LPIN2) in triglyceride synthesis, Apolipoprotein C-III (APOC3) in lipoprotein formation, Beta-2-glycoprotein 1 (APOH) in lipoprotein transportation and perilipin 2 (PLIN2) in lipid droplets formation (Conn et al., 2021) (Fig. 2). Potentially, these adaptations could impact liver lipoprotein formation and accumulation of neutral lipids on lipid droplets, contributing to steatosis. On the other hand, they could influence cellular transformation.

Moreover, in specific conditions under protein deprivation in a rat model, a low-protein diet increases the amount of 4E-BP1, leading to triglyceride accumulation in the rat liver. In addition, knockdown of hepatic 4E-BP1 attenuated the TG accumulation concomitantly with an increase of fatty acid oxidation through carnitine palmitoyltransferase 1A (CPT1A) (Toyoshima et al., 2020).

Other translation factors play a critical role in obesity/metabolic syndrome; for instance, eIF6 is associated with translational control of adipogenic transcription factors such as C/EBPβ, C/EBPδ, and ATF4 that have G/C rich regions or uORF sequences in their 5′UTR (Brina et al., 2015; Brina, Miluzio, Ricciardi, & Biffo, 2015) A hub target is the regulation of fatty acid synthase (Fasn) (Brina, Miluzio, Ricciardi, Clarke, et al., 2015). Moreover, through the transcription factor C/EBPβ, eIF6 promotes the expression of biomarkers connected with the progression from Non-Alcoholic Fatty Liver Disease (NAFLD) to hepatocellular carcinoma (Scagliola et al., 2021); specifically, eIF6 depletion is associated with lipid accumulation blunting and fatty acid oxidation stimulation (Scagliola et al., 2021).

This is consistent with hepatocellular carcinoma, which is highly associated with obesity and type II Diabetes (Baidal & Lavine, 2016), due to excess caloric intake promoting the production of triglycerides in the form of lipid droplets, where lipid accumulation leads to increased levels of free fatty acids and other lipid metabolites causing hepatocyte injury, endoplasmic reticulum (ER) stress and inflammation (Kolb et al., 2016).

Therefore, a circuit for homeostatic lipid accumulation mediated by protein translation could be present. Moreover, the discussed evidence has shown the fundamental role of lipid metabolism in the development of hallmarks of cancer; describing the mechanisms involved will allow the identification of drug-targets and the design of novel therapies in BC patients. However, the lack of sensitivity of cancer cells to therapy is one of the key factors in the failure of chemotherapy during breast cancer treatment. Chemoresistance is also implicated in cancer recurrence and is one of the main causes of death in patients.

4.3 eIF4E-lipid overload and their impact on chemoresistance

Resistance to therapeutic treatments can be intrinsic, with a lack of response from the beginning of treatment, or acquired, due to the loss of therapeutic efficacy after a time of treatment. Emphasizing acquired resistance, the mechanisms include changes in the drug absorption; transport and efflux of drugs on the cell membrane; enzymatic metabolism of drugs leading to their inactivation; alteration in the quantity and affinity of receptors or drug targets; the expression of genes related to cancer and DNA repair (Ji et al., 2019).

eIF4E and the complex eIF4F are critical effectors and rate limiting elements of protein translation and in several cases selective to mRNA targets implicated in cancer development and lipid metabolism misbalances. Moreover, scopes may extend to other cellular phenomena such as chemoresistance. Indeed, chemoresistance development is closely related to the tumor microenvironment, wherein highly fluctuating metabolic conditions are present, including signaling elements, stromal biomolecules, and chemotherapeutic drugs. Together, these conditions are capable of inducing changes in the phenotype of tumor cells (Galindo-Hernandez, Gonzales-Vazquez, & Cortes-Reynosa, 2015) and potentially lead to the development of chemoresistance (Zheng, 2017).

Several reports suggest a connection of overexpression with chemoresistance. For instance, in a prospective study of 103 TNBC patients the eIF4E levels in cancer specimens were quantified as x-fold over benign samples from non-cancer patients. Results suggest that patients with triple-negative neoplasms with a high eIF4E expression ($\geq$15-fold) showed greater rates of cancer recurrence ($P=0.04$) and cancer-related death ($P=0.02$) than the low eIF4E group (Flowers et al., 2009).

For instance, in a metastatic castration-resistant prostate cancer (mCRPC) model, eIF4E overexpression enhances translation of Tousled-like kinase 1 isoform B (TLK1B), a threonine kinase with a long 5' UTR mRNA. Increased expression confers cancer cells resistance over DNA damaging chemotherapeutics like anthracyclines (doxorubicin) by activating NEK1 > ATR > Chk1 > DNA damage response (Khalil & De Benedetti, 2022). It was previously known that TLK1 upregulation confers cancer cells resistance to radiotherapy *in vitro* (Sunavala-Dossabhoy et al., 2005) and predicts cancer recurrence in irradiated breast cancer (Wolfort et al., 2006). In a prospective study of 152 stage I to III BC patients treated with a doxorubicin-based chemotherapy adjuvant, tumor expression of eIF4E and TLK1Bwere characterized. Results indicated TLK1B overexpression was correlated with eIF4E elevation; and high TLK1B was associated with a higher risk for cancer recurrence (Byrnes et al., 2007). In this sense, in a prospective study to detect cancer recurrence in patients who had been treated with adjuvant radiation therapy, 158 patients with stage I to III breast cancer were accrued. Data suggested that patients in the highest TLK1B group had a 3.0-fold increase in relative risk for cancer recurrence after adjuvant radiation therapy ($P = 0.036$; 95% CI, 1.0–5.0), compared with the low TLK1B group (Wolfort et al., 2006) Moreover, seventeen locally advanced breast cancer patients were accrued, and tumor specimens were obtained before and after neoadjuvant therapy. Breast cancer patients whose tumors had low eIF4E overexpression after neoadjuvant chemotherapy had lower cancer recurrence compared with those who did not (Hiller et al., 2009).

In stage IV breast cancer patients ($N = 200$), eIF4E was highly phosphorylated at serine 209 in response to chemotherapy, which significantly correlated with poorer clinical responses and outcomes (Li et al., 2017). In a complementary way, in breast cancer cells depletion of eIF4E enhanced the anti-proliferative and pro-apoptotic effects of chemotherapeutic drugs (doxorubicin, cyclophosphamide, and fluorouracil). And the inhibition of events such as eIF4E-p inhibition and β-catenin activation by MNK-inhibitors enhanced the BC cell response to chemotherapy *in vitro* and *in vivo* (Li et al., 2017). Therefore, it is possible that the MNK-eIF4E-β-catenin axis could be involucrate in chemoresistance. This proposal has been strengthened by carcinoma squamous cells characterization, wherein the eIF4E/β-catenin axis plays a positive regulatory role in the resistance of cervical cancer cells to chemotherapy (Xu et al., 2017). Therefore, overexpression of eIF4E and downstream targets could represent biomarkers for breast cancer chemoresistance, however, the cellular biology of tumors is complex, and is necessary more studies to validate this proposal.

The phenomenon of resistance to docetaxel, a chemotherapy drug for BC treatment, could occur in ~50% of patients. In this sense, a cellular chemoresistant BC model, gene regulation through miR-141 has been proposed to play an important role in cancer drug resistance, evidence suggest that downregulation of eIF4E may be a molecular mechanism by which miR-141 exerted its functions of inducer of chemoresistance on human BC cells (Yao et al., 2015). Being more complex the phenomenon and the regulation by the translation initiation complex can be a trigger.

Additionally, eIF4E expression levels and phosphorylation has been implicated in chemoresistance. MCF7 Tamoxifen resistant cells (TamR) appeared to be more sensitive to eIF4E depletion by eIF4E shRNAs, also, silencing eIF4E or inhibiting eIF4E phosphorylation restored tamoxifen sensitivity. p-eIF4E promotes Runx2 transcription which is a transcription factor involved in cell reprogramming known for being an important inhibitor of estrogen signaling, favoring a poorly differentiated state. Association between p-eIF4E and tamoxifen resistance was also evaluated *in vivo*, biopsy specimens from ER+ tamoxifen-resistant tumors showed significantly increased p-eIF4E compared with tamoxifen-sensitive tumors (Geter et al., 2017) (Fig. 2).

Therefore, overcoming intrinsic and acquired drug resistance is a major issue in treating cancer patients considering the effects of chemoresistance. Moreover, the effect of lipid metabolism on chemoresistance phenomena has been little explored; however, considering recent evidence about the impact of eIF4E on control of lipid metabolism, and the broad evidence about dyslipidemias and the risk of breast cancer, it is very possible that dyslipidemias and intracellular metabolism in tumor cells represent a triggering factor in the development of chemoresistance. Even more, the metabolic reprogramming of tumors and the effect in chemoresistance could be controlled by mechanisms such as epigenetic factors that regulate gene expression (Gonzalez-Ortiz et al., 2021), and impact on the oncogenic drivers and potentially in hallmarks of cancers processes.

5. Final considerations

The use of novel strategies for drug development have allow identification of genes whose modulation could inhibit growth of cancer cells, representing suitable therapeutic targets (Tsherniak et al., 2017). Knowing the impact of lipid metabolism on the critical step of translation initiation is required for a better understanding of cancer hallmarks; the analysis of this interactome with cutting-edge techniques can guide the selection of better therapeutic targets. For instance, since a broad-spectrum perspective, 627 cancer protein targets have been identified, of which 232 are enzymes (37%), being the focus of characterization (Behan et al., 2019). Notwithstanding, 395 non-enzyme targets, including transcription factors and signaling molecules considered oncogenic driver-proteins are described as undruggable targets (Behan et al., 2019), with three-dimensional structures that show restrictions to occupancy-driven small-molecule inhibitors. However, new opportunities could appear; molecular inhibitors on specific translation factors have been on the rise. For instance, inhibitors that block the ability of eIF4E to recruit PIC such as 4E1RCat and 4E2RCat, have shown anti-tumoral effects (L. Chen et al., 2012).

Likewise, several effective treatments are often not durable considering that tumor cells develop chemoresistance (Salami & Crews, 2017). In this sense, the eIF4G- eIF4E interaction represents a major drug target for repressing specific translation. eIF4E is described as an oncogenic driver (Kuzuoglu-Ozturk et al., 2021); and through a genome-wide CRISPRi screen, more than 600 genetic interactions that support eIF4E oncogenic activity has been identified, among these, Bcl-xL and exon-junction complex (EJC) (Kuzuoglu-Ozturk et al., 2021). In addition to the involvement of eIF4E in proteostasis, it plays an important role in the modulation of cellular metabolism by favoring the effective translation of certain lipid metabolic targets, and consequently its deregulation can lead to dyslipidemias. Therefore, a crosstalk mechanism among cancer development and dyslipidemias may be a reality, triggering phenomena such as proliferation, migration, and invasion of cancer cells. Moreover, therapeutic strategies must consider these connecting points to improve their effectiveness.

In conclusion, the regulation of translation initiation, and specifically eIF4F represents a hub that connects specific cellular processes in cancer and obesity which could explain the association between these conditions; moreover, the identification and characterization of new targets with a pharmacological potential are critical.

Acknowledgments

The authors recognize the helpful contribution of Luis Pablo Ávila Barrientos and Roberto León-Gutiérrez, who critically reviewed and edited the manuscript. This study was supported by Fondo Sectorial de Investigación para la Educación CB 2017-2018 (A1-S-28653/SEP/CONACYT) and Programa Presupuestario F003 (CONACYT) "Programas Nacionales Estratégicos de Ciencia, Tecnología y Vinculación con los Sectores Social, Público y Privado".

References

Armengol, G., Rojo, F., Castellví, J., Iglesias, C., Cuatrecasas, M., Pons, B., et al. (2007). 4E-binding protein 1: A key molecular 'funnel factor' in human cancer with clinical implications. *Cancer Research*, *67*(16), 7551–7555. https://doi.org/10.1158/0008-5472.CAN-07-0881.

Assouline, S., Culjkovic, B., Cocolakis, E., Rousseau, C., Beslu, N., Amri, A., et al. (2009). Brief report molecular targeting of the oncogene eIF4E in acute myeloid leukemia (AML): A proof-of-principle clinical trial with ribavirin. *Blood Advances*, *114*(2), 257–260. https://doi.org/10.1182/blood-2009-02-205153.

Baidal, J. A. W., & Lavine, J. E. (2016). The intersection of nonalcoholic fatty liver disease and obesity. *Science Translational Medicine*, *8*(323). https://doi.org/10.1126/scitranslmed.aad8390.

Batool, A., Aashaq, S., & Andrabi, K. I. (2019). Eukaryotic initiation factor 4E (eIF4E): A recap of the cap-binding protein. *Journal of Cellular Biochemistry*, *120*(9), 14201–14212. https://doi.org/10.1002/jcb.28851.

Behan, F. M., Iorio, F., Picco, G., Gonçalves, E., Beaver, C. M., Migliardi, G., et al. (2019). Prioritization of cancer therapeutic targets using CRISPR – Cas9 screens. *Nature*, *568*. https://doi.org/10.1038/s41586-019-1103-9.

Bentley, D. L. (2014). Coupling mRNA processing with transcription in time and space. *Nature Reviews Genetics*, *15*(3), 163–175. https://doi.org/10.1038/nrg3662.

Biffo, S., Manfrini, N., & Ricciardi, S. (2018). Crosstalks between translation and metabolism in cancer. *Current Opinion in Genetics and Development*, *48*, 75–81. https://doi.org/10.1016/j.gde.2017.10.011.

Borgquist, S., Butt, T., Almgren, P., Shiffman, D., Stocks, T., Orho-Melander, M., et al. (2016). Apolipoproteins, lipids and risk of cancer. *International Journal of Cancer*, *138*(11), 2648–2656. https://doi.org/10.1002/ijc.30013.

Brina, D., Miluzio, A., Ricciardi, S., & Biffo, S. (2015). eIF6 anti-association activity is required for ribosome biogenesis, translational control and tumor progression. *Biochimica et Biophysica Acta - Gene Regulatory Mechanisms*, *1849*(7), 830–835. https://doi.org/10.1016/j.bbagrm.2014.09.010.

Brina, D., Miluzio, A., Ricciardi, S., Clarke, K., Davidsen, P. K., Viero, G., et al. (2015). eIF6 coordinates insulin sensitivity and lipid metabolism by coupling translation to transcription. *Nature Communications*, *6*(8261), 1–15. https://doi.org/10.1038/ncomms9261.

Byrnes, K. W., DeBenedetti, A., Holm, N. T., Luke, J., Nunez, J., Chu, Q. D., et al. (2007). Correlation of TLK1B in elevation and recurrence in doxorubicin-treated breast Cancer patients with high eIF4E overexpression. *Journal of the American College of Surgeons*, *204*(5), 925–933. https://doi.org/10.1016/j.jamcollsurg.2007.02.027.

Caan, B. J., Kwan, M. L., Hartzell, G., Castillo, A., Slattery, M. L., Sternfeld, B., et al. (2008). Pre-diagnosis body mass index, post-diagnosis weight change, and prognosis among women with early stage breast cancer. *Cancer Causes & Control*, *19*(10), 1319–1328. https://doi.org/10.1007/s10552-008-9203-0.

Calip, G. S., Malone, K. E., Gralow, J. R., Stergachis, A., Hubbard, R. A., & Boudreau, D. M. (2014). Metabolic syndrome and outcomes following early-stage breast cancer. *Breast Cancer Research and Treatment*, *148*(2), 363–377. https://doi.org/10.1007/s10549-014-3157-6.

Carracedo, A., Cantley, L. C., & Pandolfi, P. P. (2013). Cancer metabolism: Fatty acid oxidation in the limelight. *Nature Reviews Cancer*, *13*(4), 227–232. https://doi.org/10.1038/nrc3483.

Castellvi, J., Garcia, A., Rojo, F., Ruiz-Marcellan, C., Gil, A., Baselga, J., et al. (2006). Phosphorylated 4E binding protein 1: A hallmark of cell signaling that correlates with survival in ovarian cancer. *Cancer*, *107*(8), 1801–1811. https://doi.org/10.1002/cncr.22195.

Chaffer, C. L., & Weinberg, R. A. (2011). A perspective on cancer cell metastasis. *Science, 331*(2011), 1559–1564. https://doi.org/10.1126/science.1203543.

Chan, D. S. M., Vieira, A. R., Aune, D., Bandera, E. V., Greenwood, D. C., McTiernan, A., et al. (2014). Body mass index and survival in women with breast cancer—Systematic literature review and meta-analysis of 82 follow-up studies. *Annals of Oncology, 25*(10), 1901–1914. https://doi.org/10.1093/annonc/mdu042.

Chen, L., Aktas, B. H., Wang, Y., He, X., Sahoo, R., Denoyelle, S., et al. (2012). Tumor suppression by small molecule inhibitors of translation initiation. *Oncotarget, 3*(8), 869–881.

Chen, Y., Wen, Y. Y., Li, Z. R., Luo, D. L., & Zhang, X. H. (2016). The molecular mechanisms between metabolic syndrome and breast cancer. *Biochemical and Biophysical Research Communications, 471*(4), 391–395. https://doi.org/10.1016/j.bbrc.2016.02.034.

Conn, C. S., Yang, H., Tom, H. J., Ikeda, K., Oses-Prieto, J. A., Vu, H., et al. (2021). The major cap-binding protein eIF4E regulates lipid homeostasis and diet-induced obesity. *Nature Metabolism, 3*(2), 244–257. https://doi.org/10.1038/s42255-021-00349-z.

Cox, A. J., West, N. P., & Cripps, A. W. (2015). Obesity, inflammation, and the gut microbiota. *The Lancet Diabetes and Endocrinology, 3*(3), 207–215. https://doi.org/10.1016/S2213-8587(14)70134-2.

Culjkovic-Kraljacic, B., & Borden, K. L. B. (2013). Aiding and abetting cancer: MRNA export and the nuclear pore. *Trends in Cell Biology, 23*(7), 328–335. https://doi.org/10.1016/j.tcb.2013.03.004.

Dale, B., Cheng, M., Park, K. S., Kaniskan, H.Ü., Xiong, Y., & Jin, J. (2021). Advancing targeted protein degradation for cancer therapy. *Nature Reviews Cancer, 21*(10), 638–654. https://doi.org/10.1038/s41568-021-00365-x.

Díaz-Villanueva, J. F., Díaz-Molina, R., & García-González, V. (2015). Protein folding and mechanisms of proteostasis. *International Journal of Molecular Sciences, 16*(8), 17193–17230. https://doi.org/10.3390/ijms160817193.

Dibaba, D., Braithwaite, D., & Akinyemiju, T. (2018). Metabolic syndrome and the risk of breast Cancer and subtypes by race, Menopause and BMI. *Cancers, 10*(9), 299. https://doi.org/10.3390/cancers10090299.

Dobrikov, M., Dobrikova, E., Shveygert, M., & Gromeier, M. (2011). Phosphorylation of eukaryotic translation initiation factor 4G1 (eIF4G1) by protein kinase C regulates eIF4G1 binding to Mnk1. *Molecular and Cellular Biology, 31*(14), 2947–2959. https://doi.org/10.1128/mcb.05589-11.

Drake, J. C., Alway, S. E., Hollander, J. M., & Williamson, D. L. (2010). AICAR treatment for 14 days normalizes obesity-induced dysregulation of TORC1 signaling and translational capacity in fasted skeletal muscle. *American Journal of Physiology - Regulatory Integrative and Comparative Physiology, 299*(6), 1546–1555. https://doi.org/10.1152/ajpregu.00337.2010.

Duffy, A., Makarova-Rusher, O., Ulahannan, V., Rahma, O., Fioravanti, S., Walker, M., et al. (2016). Modulation of tumor eIF4E by antisense inhibition: A phase I/II translational clinical trial of ISIS 183750—An antisense oligonucleotide against eIF4E—In combination with irinotecan in solid tumors and irinotecan-refractory colorectal cancer. *International Journal of Cancer, 139*, 1648–1657. https://doi.org/10.1002/ijc.30199.

Duggan, C., Irwin, M. L., Xiao, L., Henderson, K. D., Smith, A. W., Baumgartner, R. N., et al. (2011). Associations of insulin resistance and adiponectin with mortality in women with breast cancer. *Journal of Clinical Oncology, 29*(1), 32–39. https://doi.org/10.1200/JCO.2009.26.4473.

Elfakess, R., Sinvani, H., Haimov, O., Svitkin, Y., Sonenberg, N., & Dikstein, R. (2011). Unique translation initiation of mRNAs-containing TISU element. *Nucleic Acids Research, 39*(17), 7598–7609. https://doi.org/10.1093/nar/gkr484.

Fedele, M., Sgarra, R., Battista, S., & Cerchia, L. (2022). The epithelial – Mesenchymal transition at the crossroads between metabolism and tumor progression. *International Journal of Molecular Sciences*, *23*(800), 2–34. https://doi.org/10.3390/ijms23020800.

Flote, V. G., Vettukattil, R., Bathen, T. F., Egeland, T., McTiernan, A., Frydenberg, H., et al. (2016). Lipoprotein subfractions by nuclear magnetic resonance are associated with tumor characteristics in breast cancer. *Lipids in Health and Disease*, *15*(1), 1–12. https://doi.org/10.1186/s12944-016-0225-4.

Flowers, A., Chu, Q. D., Panu, L., Meschonat, C., Caldito, G., Lowery-Nordberg, M., et al. (2009). Eukaryotic initiation factor 4E overexpression in triple-negative breast cancer predicts a worse outcome. *Surgery*, *146*(2), 220–226. https://doi.org/10.1016/j.surg.2009.05.010.

Furic, L., Rong, L., Larsson, O., Koumakpayi, I. H., Yoshida, K., Brueschke, A., et al. (2010). EIF4E phosphorylation promotes tumorigenesis and is associated with prostate cancer progression. *Proceedings of the National Academy of Sciences of the United States of America*, *107*(32), 14134–14139. https://doi.org/10.1073/pnas.1005320107.

Galindo-Hernandez, O., Gonzales-Vazquez, C., & Cortes-Reynosa, P. (2015). Extracellular vesicles from women with breast cancer promote an epithelial-mesenchymal transition-like process in mammary epithelial cells MCF10A. *Tumor Biology*. https://doi.org/10.1007/s13277-015-3711-9.

Garcia-Gonzalez, V., Diaz-Villanueva, J. F., Galindo-Hernandez, O., Martinez-Navarro, I., Hurtado-Ureta, G., & Perez-Arias, A. A. (2018). Ceramide metabolism balance, a multifaceted factor in critical steps of breast cancer development. *International Journal of Molecular Sciences*, *19*(9). https://doi.org/10.3390/ijms19092527.

Garcia-Gonzalez, V., Galindo-Hernandez, O., & Gonzalez-Ortiz, A. (2020). Impact of cholesterol-pathways on breast cancer development, a metabolic landscape. *J Cancer*, *2*, 1–24.

Geter, P. A., Ernlund, A. W., Bakogianni, S., Alard, A., Arju, R., Giashuddin, S., et al. (2017). Hyperactive mTOR and MNK1 phosphorylation of eIF4E confer tamoxifen resistance and estrogen independence through selective mRNA translation reprogramming. *Genes and Development*, *31*(22), 2235–2249. https://doi.org/10.1101/gad.305631.117.

Giudetti, A. M., Domenico, S. D., Ragusa, A., Lunetti, P., Gaballo, A., Franck, J., et al. (2019). A specific lipid metabolic profile is associated with the epithelial mesenchymal transition program. *BBA - Molecular and Cell Biology of Lipids*, *1864*(3), 344–357. https://doi.org/10.1016/j.bbalip.2018.12.011.

Goel, H. L., & Mercurio, A. (2014). VEGF targets the tumour cell. *Nature Reviews Cancer*, *23*(1), 1–7. https://doi.org/10.1038/nrc3627.VEGF.

Gonzalez-Ortiz, A., Galindo-Hernandez, O., Hernandez-Acevedo, G. N., Hurtado-Ureta, G., & Garcia-Gonzalez, V. (2021). Impact of cholesterol-pathways on breast cancer development, a metabolic landscape. *Journal of Cancer*, *12*(14), 4307–4321. https://doi.org/10.7150/jca.54637.

Gross, J. D., Moerke, N. J., Von Der Haar, T., Lugovskoy, A. A., Sachs, A. B., McCarthy, J. E. G., et al. (2003). Ribosome loading onto the mRNA cap is driven by conformational coupling between eIF4G and eIF4E. *Cell*, *115*(6), 739–750. https://doi.org/10.1016/S0092-8674(03)00975-9.

Hanahan, D., & Coussens, L. M. (2012). Accessories to the crime : Functions of cells recruited to the tumor microenvironment. *Cancer Cell*, *21*(3), 309–322. https://doi.org/10.1016/j.ccr.2012.02.022.

Hanahan, D., & Weinberg, R. A. (2000). The Hallmarks of Cancer. *Cell Press*, *100*, 57–70. https://doi.org/10.1016/S0092-8674(00)81683-9.

Harborg, S., Ahern, T. P., Feldt, M., Rosendahl, A. H., Cronin-Fenton, D., Melander, O., et al. (2022). Circulating lipids and breast cancer prognosis in the Malmö diet and cancer study. *Breast Cancer Research and Treatment*, *191*(3), 611–621. https://doi.org/10.1007/s10549-021-06462-7.

Healy, L. A., Ryan, A. M., Rowley, S., Boyle, T., Connolly, E., Kennedy, M. J., et al. (2010). Obesity increases the risk of postmenopausal breast cancer and is associated with more advanced stage at presentation but no impact on survival. *Breast Journal, 16*(1), 95–97. https://doi.org/10.1111/j.1524-4741.2009.00861.x.

Hiller, D. J., Chu, Q., Meschonat, C., Panu, L., Burton, G., & Li, B. D. L. (2009). Predictive value of eIF4E reduction after neoadjuvant therapy in breast Cancer. *Journal of Surgical Research, 156*(2), 265–269. https://doi.org/10.1016/j.jss.2009.03.060.

Hinnebusch, A. G. (2014). The scanning mechanism of eukaryotic translation initiation. *Annual Review of Biochemistry, 83*(1), 779–812. https://doi.org/10.1146/annurev-biochem-060713-035802.

Hsieh, A. C., & Ruggero, D. (2010). Targeting eukaryotic translation initiation factor 4E (eIF4E) in Cancer. *Clinical Cancer Research, 16*(20), 4914–4920. https://doi.org/10.1158/1078-0432.CCR-10-0433.

Huang, P. L. (2009). A comprehensive definition for metabolic syndrome. *DMM Disease Models and Mechanisms, 2*(5–6), 231–237. https://doi.org/10.1242/dmm.001180.

Hussain, A., Lian, J., Watts, R., Gutiérrez, T., Nelson, R., Goping, I. S., et al. (2022). Attenuation of obesity-induced hyperlipidemia reduces tumor growth. *Biochimica et Biophysica Acta (BBA) - Molecular and Cell Biology of Lipids, 1867*(4), 159124.

Jackson, R. J., Hellen, C. U. T., & Pestova, T. V. (2010). The mechanism of eukaryotic translation initiation and principles of its regulation. *Nature Reviews Molecular Cell Biology, 11*(2), 113–127. https://doi.org/10.1038/nrm2838.

Jaiswal, P. K., Goel, A., & Mittal, R. D. (2015). Survivin: A molecular biomarker in cancer. *Indian Journal of Medical Research, 141*, 389–397.

Ji, X., Lu, Y., Tian, H., Meng, X., Wei, M., & Cho, W. C. (2019). Chemoresistance mechanisms of breast cancer and their countermeasures. *Biomedicine and Pharmacotherapy, 114*(March), 108800. https://doi.org/10.1016/j.biopha.2019.108800.

Khalil, M. I., & De Benedetti, A. (2022). Tousled-like kinase 1: A novel factor with multifaceted role in mCRPC progression and development of therapy resistance. *Cancer Drug Resistance*, 93–101. https://doi.org/10.20517/cdr.2021.109.

Kolb, R., Sutterwala, F. S., & Zhang, W. (2016). Obesity and cancer: Inflammation bridges the two. *Current Opinion in Pharmacology, 29*(3), 77–89. https://doi.org/10.1016/j.coph.2016.07.005.

Korneeva, N. L., Song, A., Gram, H., Edens, M. A., & Rhoads, R. E. (2016). Inhibition of mitogen-activated protein kinase (MAPK)-interacting kinase (MNK) preferentially affects translation of mRNAs containing both a 5′-terminal cap and hairpin. *Journal of Biological Chemistry, 291*(7), 3455–3467. https://doi.org/10.1074/jbc.M115.694190.

Korneeva, N. L., Soung, Y. H., Kim, H. I., Giordano, A., Rhoads, R. E., Gram, H., et al. (2010). Mnk mediates integrin α6β4-dependent eIF4E phosphorylation and translation of VEGF mRNA. *Molecular Cancer Research, 8*(12), 1571–1578. https://doi.org/10.1158/1541-7786.MCR-10-0091.

Kuzuoglu-Ozturk, D., Hu, Z., Rama, M., Devericks, E., Weiss, J., Chiang, G. G., et al. (2021). Revealing molecular pathways for cancer cell fitness through a genetic screen of the cancer translatome. *Cell Reports, 35*(13), 109321. https://doi.org/10.1016/j.celrep.2021.109321.

Kwegyir-Affu, A. K., Ramalingam, S., Purushottamachar, P., Ramamurthy, V. P., & Njar, V. C. O. (2015). Galeterone and VNPT55 induce proteasomal degradation of AR/AR-V7, induce significant apoptosis via cytochrome c release and suppress growth of castration resistant prostate cancer xenografts in vivo. *Oncotarget, 6*(29), 27440–27460. https://doi.org/10.18632/oncotarget.4578.

Lankat-Buttgereit, B., & Göke, R. (2009). The tumour suppressor Pdcd4: Recent advances in the elucidation of function and regulation. *Biology of the Cell, 101*(6), 309–317. https://doi.org/10.1042/bc20080191.

Bacquer, O. L., Petroulakis, E., Paglialunga, S., Poulin, F., Richard, D., Cianflone, K., et al. (2007). Elevated sensitivity to diet-induced obesity and insulin resistance in mice lacking 4E-BP1 and 4E-BP2. *The Journal of Clinical Investigation, 117*(2). https://doi.org/10.1172/JCI29528DS1.

Leppek, K., Das, R., & Barna, M. (2018). Functional 5′ UTR mRNA structures in eukaryotic translation regulation and how to find them. *Nature Reviews Molecular Cell Biology, 19*(3), 158–174. https://doi.org/10.1038/nrm.2017.103.

Li, X., Lui, Z.-l., Wu, Y.-T., Wu, H., Dai, W., Arshad, B., et al. (2018). Status of lipid and lipoprotein in female breast cancer patients at initial diagnosis and during chemotherapy. *Lipids in Health and Disease, 17*(91), 1–6.

Li, Z., Sun, Y., Qu, M., Wan, H., Cai, F., & Zhang, P. (2017). Inhibiting the MNK-eIF4E-β-catenin axis increases the responsiveness of aggressive breast cancer cells to chemotherapy. *Oncotarget, 8*(2), 2906–2915. https://doi.org/10.18632/oncotarget.13772.

Liu, Y., Sun, L., Su, X., & Guo, S. (2016). Inhibition of eukaryotic initiation factor 4E phosphorylation by cercosporamide selectively suppresses angiogenesis, growth and survival of human hepatocellular carcinoma. *Biomedicine and Pharmacotherapy, 84*, 237–243. https://doi.org/10.1016/j.biopha.2016.09.038.

Lu, C. W., Lo, Y. H., Chen, C. H., Lin, C. Y., Tsai, C. H., Chen, P. J., et al. (2017). VLDL and LDL, but not HDL, promote breast cancer cell proliferation, metastasis and angiogenesis. *Cancer Letters, 388*(December), 130–138. https://doi.org/10.1016/j.canlet.2016.11.033.

Malka-Mahieu, H., Newman, M., Désaubry, L., Robert, C., & Vagner, S. (2017). Molecular pathways: The eIF4F translation initiation complex-newopportunities for cancer treatment. *Clinical Cancer Research, 23*(1), 21–25. https://doi.org/10.1158/1078-0432.CCR-14-2362.

Martínez-Alonso, E., Guerra-Pérez, N., Escobar-Peso, A., Peracho, L., Vera-Lechuga, R., Cruz-Culebras, A., et al. (2022). Phosphorylation of eukaryotic initiation factor 4G1 (eIF4G1) at Ser1147 is specific for eIF4G1 bound to eIF4E in delayed neuronal death after ischemia. *International Journal of Molecular Sciences, 23*(3), 1830. https://doi.org/10.3390/ijms23031830.

Mauthe, M., Orhon, I., Rocchi, C., Zhou, X., Luhr, M., & Hijlkema, K. (2018). Chloroquine inhibits autophagic flux by decreasing autophagosome-lysosome fusion. *Autophagy, 14*(8), 1435–1455. https://doi.org/10.1080/15548627.2018.1474314.

Mehta, A., Vasquez, N., Ayers, C. R., Patel, J., Hooda, A., Khera, A., et al. (2022). Independent association of lipoprotein(a) and coronary artery calcification with atherosclerotic cardiovascular risk. *Journal of the American College of Cardiology, 79*(8), 757–768.

Melvin, J. C., Seth, D., Holmberg, L., Garmo, H., Hammar, N., Jungner, I., et al. (2012). Lipid profiles and risk of breast and ovarian cancer in the swedish AMORIS study. *Cancer Epidemiology Biomarkers and Prevention, 21*(8), 1381–1384. https://doi.org/10.1158/1055-9965.EPI-12-0188.

Merrick, W. C., & Pavitt, G. D. (2018). Protein synthesis initiation in eukaryotic cells. *Cold Spring Harbor Perspectives in Biology, 10*(12), 1–22. https://doi.org/10.1101/cshperspect.a033092.

Monaco, M. E. (2017). Fatty acid metabolism in breast cancer subtypes. *Oncotarget, 8*(17), 29487–29500.

Ouchi, N., Parker, J. L., Lugus, J. J., & Walsh, K. (2011). Adipokines in inflammation and metabolic disease. *Nature Reviews Immunology, 11*(2), 85–97. https://doi.org/10.1038/nri2921.

Passmore, L. A., Schmeing, T. M., Maag, D., Applefield, D. J., Acker, M. G., Algire, M. A. A., et al. (2007). The eukaryotic translation initiation factors eIF1 and eIF1A induce an open conformation of the 40S ribosome. *Molecular Cell, 26*(1), 41–50. https://doi.org/10.1016/j.molcel.2007.03.018.

Pelletier, J., & Sonenberg, N. (1987). The involvement of mRNA secondary structure in protein synthesis. *Biochemistry and Cell Biology = Biochimie et Biologie Cellulaire, 65*(6), 576–581. https://doi.org/10.1139/o87-074.

Pestova, T. V., Kolupaeva, V. G., Lomakin, I. B., Pilipenko, E. V., Shatsky, I. N., Agol, V. I., et al. (2001). Molecular mechanisms of translation initiation in eukaryotes. *Proceedings of the National Academy of Sciences of the United States of America, 98*(13), 7029–7036. https://doi.org/10.1073/pnas.111145798.

Pettersson, F., Yau, C., Dobocan, M. C., Culjkovic-Kraljacic, B., Retrouvay, H., Puckett, R., et al. (2012). Ribavirin treatment effects on breast cancers overexpressing eIF4E, a biomarker with prognostic specificity for luminal B-type breast cancer. *Clinical Cancer Research, 17*(March). https://doi.org/10.1158/1078-0432.CCR-10-2334.

Pires, B. R. B., Panis, C., Alves, V. D., Herrera, A. C. S. A., Binato, R., Pizzatti, L., et al. (2019). Label-free proteomics revealed oxidative stress and inflammation as factors that enhance Chemoresistance in luminal breast Cancer. *Oxidative Medicine and Cellular Longevity, 2019*, 1–15. https://doi.org/10.1155/2019/5357649.

Piserà, A., Campo, A., & Campo, S. (2018). Structure and functions of the translation initiation factor eIF4E and its role in cancer development and treatment. *Journal of Genetics and Genomics, 45*(1), 13–24. https://doi.org/10.1016/j.jgg.2018.01.003.

Pyronnet, S., Imataka, H., Gingras, A. C., Fukunaga, R., Hunter, T., & Sonenberg, N. (1999). Human eukaryotic translation initiation factor 4G (eIF4G) recruits Mnk1 to phosphorylate eIF4E. *EMBO Journal, 18*(1), 270–279. https://doi.org/10.1093/emboj/18.1.270.

Ramalingam, S., Gediya, L., Kwegyir-Afful, A. K., Ramamurthy, V. P., Purushottamachar, P., Mbatia, H., et al. (2014). First Mnks degrading agents block phosphorylation of eIF4E, induce apoptosis, inhibit cell growth, migration and invasion in triple negative and Her2-overexpressing breast cancer cell lines. *Oncotarget, 5*(2), 530–543. https://doi.org/10.18632/oncotarget.1528.

Ramalingam, S., Ramamurthy, V. P., Gediya, L. K., Murigi, F. N., Purushottamachar, P., Huang, W., et al. (2019). The novel Mnk1/2 degrader and apoptosis inducer VNLG-152 potently inhibits TNBC tumor growth and metastasis. *Cancers, 11*(3), 1–25. https://doi.org/10.3390/cancers11030299.

Ramamurthy, V. P., Ramalingam, S., Gediya, L. K., & Njar, V. C. O. (2018). The retinamide VNLG-152 inhibits f-AR/AR-V7 and MNK–eIF4E signaling pathways to suppress EMT and castration-resistant prostate cancer xenograft growth. *FEBS Journal, 285*(6), 1051–1063. https://doi.org/10.1111/febs.14383.

Raza, U., Asif, M. R., Rehman, A. B., & Sheikh, A. (2018). Hyperlipidemia and hyper glycaemia in breast Cancer patients is related to disease stage. *Pakistan Journal of Medical Sciences, 34*(1), 209–214.

Richter, J. D., & Sonenberg, N. (2005). Regulation of cap-dependent translation by eIF4E inhibitory proteins. *Nature, 433*(7025), 477–480. https://doi.org/10.1038/nature03205.

Robichaud, N., Del Rincon, S. V., Huor, B., Alain, T., Petruccelli, L. A., Hearnden, J., et al. (2014). Phosphorylation of eIF4E promotes EMT and metastasis via translational control of SNAIL and MMP-3. *Oncogene, 34*(16), 2032–2042. https://doi.org/10.1038/onc.2014.146.

Robichaud, N., Hsu, B. E., Istomine, R., Alvarez, F., Blagih, J., Ma, E. H., et al. (2018). Translational control in the tumor microenvironment promotes lung metastasis: Phosphorylation of eIF4E in neutrophils. *Proceedings of the National Academy of Sciences of the United States of America, 115*(10), E2202–E2209. https://doi.org/10.1073/pnas.1717439115.

Röhrig, F., & Schulze, A. (2016). The multifaceted roles of fatty acid synthesis in cancer. *Nature Reviews Cancer, 16*(11), 732–749. https://doi.org/10.1038/nrc.2016.89.

Royce, M., Bachelot, T., Villanueva, C., Özgüroğlu, M., Azebedo, S., Melo Cruz, F., et al. (2018). everolimus plus endocrine therapy for postmenopausal women with estrogen receptor–positive, human epidermal growth factor receptor 2–Negative Advanced Breast Cancer A Clinical Trial. *JAMA Oncology*, *4*(7), 977–984. https://doi.org/10.1001/jamaoncol.2018.0060.

Ruan, H., Li, X., Xu, X., Leibowitz, B. J., Tong, J., Chen, L., et al. (2020). eIF4E S209 phosphorylation licenses myc- and stress-driven oncogenesis. *eLife*, *9*, 1–25. https://doi.org/10.7554/eLife.60151.

Rui, L. (2007). A link between protein translation and body weight. *Journal of Clinical Investigation*, *117*(2), 310–313. https://doi.org/10.1172/JCI31289.

Russell, D. W., & Spremulli, L. L. (1980). Mechanism of action of the wheat germ ribosome dissociation factor: Interaction with the 60 S subunit. *Archives of Biochemistry and Biophysics*, *201*(2), 518–526. https://doi.org/10.1016/0003-9861(80)90540-8.

Safra, T., Kaufman, B., Kadouri, L., Ben-baruch, N. E., Ryvo, L., Nisenbaum, B., et al. (2017). Everolimus plus letrozole for treatment of patients with HR+, HER2– Advanced breast cancer progressing on endocrine therapy: An open-label, phase 2 trial. *Clinical Breast Cancer*, *18*(2), e197–e203. https://doi.org/10.1016/j.clbc.2017.09.004.

Salami, J., & Crews, C. M. (2017). Waste disposal—An attractive strategy for cancer therapy. *Frontiers in Cancer Therapy*, *1167*(March), 1163–1167.

Scagliola, A., Miluzio, A., Ventura, G., Oliveto, S., Cordiglieri, C., Manfrini, N., et al. (2021). Targeting of eIF6-driven translation induces a metabolic rewiring that reduces NAFLD and the consequent evolution to hepatocellular carcinoma. *Nature Communications*, *12*, 1–16. https://doi.org/10.1038/s41467-021-25195-1.

Seiler, A., Chen, M. A., Brown, R. L., & Fagundes, C. P. (2018). Obesity, dietary factors, nutrition, and breast Cancer risk. *Current Breast Cancer Reports*, *10*(1), 14–27. https://doi.org/10.1007/s12609-018-0264-0.

Siddiqui, N., & Sonenberg, N. (2015). Signalling to eIF4E in cancer. *Biochemical Society Transactions*, *43*, 763–772. https://doi.org/10.1042/BST20150126.

Silvera, D., Formenti, S. C., & Schneider, R. J. (2010). Translational control in cancer. *Nature Reviews Cancer*, *10*(4), 254–266. https://doi.org/10.1038/nrc2824.

Simeone, P., Tacconi, S., Longo, S., Lanuti, P., Bravaccini, S., Pirini, F., et al. (2021). Expanding roles of De novo lipogenesis in breast Cancer. *International Journal of Environmental Research and Public Health*, *18*(3375), 1–16.

Somers, J., Pöyry, T., & Willis, A. E. (2013). A perspective on mammalian upstream open reading frame function. *International Journal of Biochemistry and Cell Biology*, *45*(8), 1690–1700. https://doi.org/10.1016/j.biocel.2013.04.020.

Sonneveld, S., Verhagen, B. M. P., & Tanenbaum, M. E. (2020). Heterogeneity in mRNA translation. *Trends in Cell Biology*, *30*(8), 606–618. https://doi.org/10.1016/j.tcb.2020.04.008.

Sunavala-Dossabhoy, G., Balakrishnan, S. K., Sen, S., Nuthalapaty, S., & De Benedetti, A. (2005). The radioresistance kinase TLK1B protects the cells by promoting repair of double strand breaks. *BMC Molecular Biology*, *6*, 1–10. https://doi.org/10.1186/1471-2199-6-19.

Tabernero, J., Rojo, F., Calvo, E., Burris, H., Judson, I., Hazell, K., et al. (2008). Dose- and schedule-dependent inhibition of the mammalian target of rapamycin pathway with everolimus: A phase I tumor pharmacodynamic study in patients with advanced solid tumors. *Journal of Clinical Oncology*, *26*(10), 1603–1610. https://doi.org/10.1200/JCO.2007.14.5482.

Tailler, M., Lindqvist, L. M., Gibson, L., & Adams, J. M. (2019). By reducing global mRNA translation in several ways, 2-deoxyglucose lowers MCL-1 protein and sensitizes hemopoietic tumor cells to BH3 mimetic ABT737. *Cell Death and Differentiation*, *26*(9), 1766–1781. https://doi.org/10.1038/s41418-018-0244-y.

Toyoshima, Y., Yoshizawa, F., Tokita, R., Taguchi, Y., Takahashi, S. I., Kato, H., et al. (2020). A translation repressor, 4E-BP1, regulates the triglyceride level in rat liver during protein deprivation. *American Journal of Physiology - Endocrinology and Metabolism, 318*(5), E636–E645. https://doi.org/10.1152/ajpendo.00464.2019.

Truitt, M. L., & Ruggero, D. (2016). New frontiers in translational control of the cancer genome. *Nature Reviews Cancer, 16*(5), 288–304. https://doi.org/10.1038/nrc.2016.27.

Tsherniak, A., Vazquez, F., Montgomery, P. G., Golub, T. R., Boehm, J. S., Hahn, W. C., et al. (2017). Defining a cancer dependency map. *Cell Press, 170*(3), 564–576. https://doi.org/10.1016/j.cell.2017.06.010.

van Riggen, J., Yetil, A., & Felsher, D. W. (2010). MYC as a regulator of ribosome biogenesis and protein synthesis. *Nature Reviews Cancer, 10*(4), 301–309. https://doi.org/10.1038/nrc2819.

Wang, C., Song, J., Liu, W., Yao, Y., Kapranov, P., Sample, K. M., et al. (2020). FLI1 promotes protein translation via the transcriptional regulation of MKNK1 expression. *International Journal of Oncology, 56*(2), 430–439. https://doi.org/10.3892/ijo.2019.4943.

Weiss, B., Allen, G. E., Kloehn, J., Abid, K., & Curran, J. A. (2021). eIF4E3 forms an active eIF4F complex during stresses (eIF4F S) targeting mTOR and re-programs the translatome. *Nucleic Acid Research*. https://doi.org/10.1093/nar/gkab267.

Wek, R. C. (2018). Role of eIF2α kinases in translational control and adaptation to cellular stress. *Cold Spring Harbor Perspectives in Biology, 10*(7). https://doi.org/10.1101/cshperspect.a032870.

Wendel, H. G., Silva, R. L. A., Malina, A., Mills, J. R., Zhu, H., Ueda, T., et al. (2007). Dissecting eIF4E action in tumorigenesis. *Genes and Development, 21*(24), 3232–3237. https://doi.org/10.1101/gad.1604407.

Wheater, M. J., Johnson, P. W. M., & Blaydes, J. P. (2010). The role of MNK proteins and eIF4E phosphorylation in breast cancer cell proliferation and survival. *Cancer Biology and Therapy, 10*(7), 728–735. https://doi.org/10.4161/cbt.10.7.12965.

Wolfort, R., de Benedetti, A., Nuthalapaty, S., Yu, H., Chu, Q. D., & Li, B. D. (2006). Up-regulation of TLK1B by eIF4E overexpression predicts cancer recurrence in irradiated patients with breast cancer. *Surgery, 140*(2), 161–169. https://doi.org/10.1016/j.surg.2006.05.001.

Xie, J., Merrett, J. E., Jensen, K. B., & Proud, C. G. (2019). The MAP kinase-interacting kinases (MNKs) as targets in oncology. *Expert Opinion on Therapeutic Targets, 23*(3), 187–199. https://doi.org/10.1080/14728222.2019.1571043.

Xu, H., Wang, Z., Xu, L., Mo, G., Duan, G., Wang, Y., et al. (2017). Targeting the eIF4E/β-catenin axis sensitizes cervical carcinoma squamous cells to chemotherapy. *American Journal of Translational Research, 9*(3), 1203–1212.

Yang, X., Zhong, W., & Cao, R. (2020). Phosphorylation of the mRNA cap-binding protein eIF4E and cancer. *Cellular Signalling, 73*. https://doi.org/10.1016/J.CELLSIG.2020.109689.

Yao, Y. S., Qiu, W. S., Yao, R. Y., Zhang, Q., Zhuang, L. K., Zhou, F., et al. (2015). MiR-141 confers docetaxel chemoresistance of breast cancer cells via regulation of EIF4E expression. *Oncology Reports, 33*(5), 2504–2512. https://doi.org/10.3892/or.2015.3866.

Zhao, W., Prijic, S., Urban, B. C., Tisza, M. J., Zuo, Y., Li, L., et al. (2016). Candidate antimetastasis drugs suppress the metastatic capacity of breast cancer cells by reducing membrane fluidity. *Cancer Research, 76*(7), 2037–2049. https://doi.org/10.1158/0008-5472.CAN-15-1970.

Zheng, H. C. (2017). The molecular mechanisms of chemoresistance in cancers. *Oncotarget, 8*(35), 59950–59964. https://doi.org/10.18632/oncotarget.19048.

CHAPTER FIVE

Cytoplasmic fatty acid-binding proteins in metabolic diseases and cancers

Sanjay Basak[a], Rahul Mallick[b], Antara Banerjee[c], Surajit Pathak[c], and Asim K. Duttaroy[d,*]

[a]Molecular Biology Division, ICMR-National Institute of Nutrition, Indian Council of Medical Research, Hyderabad, India
[b]A.I. Virtanen Institute for Molecular Sciences, University of Eastern Finland, Finland
[c]Department of Medical Biotechnology, Faculty of Allied Health Sciences, Chettinad Academy of Research and Education (CARE), Chettinad Hospital and Research Institute (CHRI), Chennai, India
[d]Department of Nutrition, Institute of Basic Medical Sciences, Faculty of Medicine, University of Oslo, Oslo, Norway
*Corresponding author: e-mail address: a.k.duttaroy@medisin.uio.no

Contents

Abstract

Cytoplasmic fatty acid-binding proteins (FABPs) are multipurpose proteins that can modulate lipid fluxes, trafficking, signaling, and metabolism. FABPs regulate metabolic and inflammatory pathways, its inhibition can improve type 2 diabetes mellitus and atherosclerosis. In addition, FABPs are involved in obesity, metabolic disease, cardiac dysfunction, and cancers. FABPs are promising tissue biomarkers in solid tumors for diagnostic and/or prognostic targets for novel therapeutic strategies. The signaling responsive elements of FABPs and determinants of FABP-mediated functions may be exploited in preventing or treating these diseases.

Advances in Protein Chemistry and Structural Biology, Volume 132
ISSN 1876-1623
https://doi.org/10.1016/bs.apcsb.2022.05.009

1. Introduction

Fatty acids have wide-ranging roles in many aspects of biology, like cellular building blocks, fuel sources, regulating cell-specific expression in lipid metabolism, cellular signaling, cell growth and proliferation, inflammation, and gene expression(Ellinghaus, Wolfrum, Assmann, Spener, & Seedorf, 1999; Jump & Clarke, 1999; Mallick, Basak, & Duttaroy, 2021; Mallick & Duttaroy, 2022). Free fatty acids (FFAs) can modify the action or location of proteins signal via surface receptors and serve as ligands for transcription factors. Intracellular transport of long-chain fatty acids (LCFAs) requires the involvement of small molecular weights proteins (15 kDa) are, known as cytoplasmic fatty acid-binding proteins (FABPs) (Duttaroy, 2009). FABPs are suggested to facilitate fatty acid transport into the cellular compartments. The FABP was first detected in the rat jejunum (Ockner, Manning, Poppenhausen, & Ho, 1972). Subsequently, several other types of FABPs were identified in the liver, myocardium, adipose tissue, placenta, skin, brain, kidney, and prostate (Dutta-Roy, 2000).

Feeding a high-fat diet facilitates obesity, diabetes, dyslipidemia, atherosclerosis, and inappropriate immune responses (Mallick et al., 2021). However, the contribution of FABPs and their dysregulation in cardiovascular diseases (CVDs) and cancers are not yet fully understood (Hardman, 2004; Leckie et al., 2014). Therefore, targeting FABPs or therefore, modulators associated with their metabolism may offer therapeutic opportunities for a broad range of pathologies in metabolic diseases and cancers. This chapter discusses the roles of FABPs and their mechanisms in modulating the cardiovascular system and cancers.

2. FABPs: Their types, tissue distribution, and their putative function

FABPs are named after the first tissue from which they are identified (Esteves & Ehrlich, 2006). In mammals, nine different FABPs with a tissue-specific distribution were identified as: liver FABP (L-FABP or FABP-1), intestinal FABP (I-FABP or FABP-2), heart FABP (H-FABP or FABP-3), adipocyte FABP (A-FABP or FABP-4), epidermal FABP (E-FABP or FABP-5), ileal FABP (Il-FABP or FABP-6), brain B-FABP or FABP-7, myelin FABP (M-FABP or FABP-8), and testis FABP (T-FABP or FABP-9) (Chmurzynska, 2006). In addition, FABP-12 was identified in

Table 1 Tissue distribution of FABPs.

Family members of FABPs	Tissue distribution
FABP-1	Hepatocytes, enterocytes; renal tubular cells, the alveolar epithelium, Placenta(Campbell, Bush, Veerkamp, & Dutta-Roy, 1998; Cheng, Geng, Cheng, & Guo, 2018; Duttaroy, 2009; Hertzel & Bernlohr, 2000; Levi, Gatmaitan, & Arias, 1969)
FABP-2	Intestine (Khan & Sorof, 1990)
FABP-3	The myocardium, skeletal muscle, mammary gland, brain, stomach, placenta, lung (Kato, Land, Majumdar, Barnholtz-Sloan, & Severson, 2010; Leroy et al., 2017; Lopes-Coelho, Andre, Felix, & Serpa, 2018)
FABP-4	Adipocytes, monocytes, macrophages, placenta (Duttaroy, 2009; Thumser & Wilton, 1996)
FABP-5	Epidermis, brain, lung, liver; kidney; mammary gland, brown adipocytes (Furuhashi et al., 2008)
FABP-6	Ileal enterocytes, adrenal gland, ovary, placenta (Chen et al., 2012; Li et al., 2018)
FABP-7	Brain development- post-natal cerebellum (Payne, Bernstein, Dvorak, & Bernstein, 2008)
FABP-8	Schwann cell (Liu, Monckton, & Godbout, 2012)
FABP-9	Testis (Slipicevic et al., 2008)
FABP-12	Human retinoblastoma cell lines, rodent retina, and testis (Liu et al., 2008)

humans and animals (Liu, Li, & Godbout, 2008). Table 1 describes the distribution of FABPs in various tissues. Despite the variable sequence identity, FABPs' amino acid sequence similarity varies from 20% to 70%. FABPs share a common tertiary structure composed of 10 anti-parallel β-sheets and two α-helixes. The orthogonal β-sheets enclose a large solvent-accessible ligand-binding cavity centered at one end of the barrel, where the helix–turn–helix motif facilitates the entry of FFAs (Storch & McDermott, 2009). FABPs ligand-binding pocket, capped by an N-terminal helix-turn-helix motif which regulates binding portal. The helical N-terminus is involved in transferring fatty acid from membranes via electrostatic interactions (Chmurzynska, 2006; Corsico, Cistola, Frieden, & Storch, 1998;

Liou, Kahn, & Storch, 2002). The fatty acid ligand-binding occurs in a solvent-accessible area of the FABP through a dynamic region made of α-helix II and the turns between βC-βD and βE-βF loops before binding in the cavity of the protein (Sacchettini, Gordon, & Banaszak, 1989).

FABPs bind LCFAs (c14-20) with different selectivity and affinity (Chmurzynska, 2006; Furuhashi & Hotamisligil, 2008). Except for FABP-1, which binds two molecules of fatty acids, all other FABPs bind only one fatty acid per protein molecule. Apart from fatty acids, FABPs bind bile salts, eicosanoids, heme, and peroxisome proliferators (Duttaroy, 2009; Dutta-Roy, Gopalswamy, & Trulzsch, 1987). Ligand affinities are generally increased with the hydrophobicity of the ligands, and their Kd values are in the nanomolar to micromolar range (Furuhashi & Hotamisligil, 2008). For example, FABP-3 preferably binds n-6 fatty acids, whereas FABP-7 binds n-3 fatty acids. In addition to LCFAs, FABP-5 binds peroxisome proliferator-activated receptor (PPAR) δ ligands such as retinoic acid, thus regulating cell growth and proliferation.

FABPs have tissue-specific functions despite their ligand affinity and tertiary structures (Dutta-Roy, 2000). The divergent functions of FABPs may be due to, in part, tissue-specific protein-protein and protein-membrane interactions. FABP-4 is highly expressed in microvascular endothelial cells in tissues (Cataltepe et al., 2012; Elmasri et al., 2012, 2009). Both FABP-4 and FABP-5 are expressed in the microvessels of the liver and adipose tissues (Antohe et al., 1998; Yu et al., 2016). FABP-1, FABP-3, and FABP-4 are expressed in the human placental trophoblasts (Basak, Sarkar, Mathapati, & Duttaroy, 2018; Duttaroy, 2009; Leroy et al., 2017). In addition, FABP-3 and FABP-5 are present in the microvascular heart and skeletal muscles. FABP9 is expressed in Leydig cells in the human testis (Moradi, Forootan, Hosseini, & Pouresmaeili, 2019).

The uptake of exogenous fats can modulate metabolic reprogramming in various cells, including cancer cells. Metabolic disorders such as abnormal lipid metabolism, high lipids storage, trafficking, and abnormal signaling of macrophages and adipocytes contribute to cardiovascular diseases (CVDs). FABPs are highly expressed in metabolically active tissues, including adipose, endothelial cells (Furuhashi et al., 2008). Consequently, FABPs are responsible for lipid metabolic homeostasis and lipid-mediated biological processes. Pathologically metabolic disorders are associated with altered expression of FABPs in metabolic syndrome, CVD, obesity, cancers, and inflammation.

3. FABPs in metabolic diseases

Triacylglycerols are stored as an energy source in white adipose tissue. Adipose tissue is also a vital endocrine organ that secretes many factors such as resistin, leptin, FABP-4, and inflammatory cytokines (Sitar-Taut et al., 2021). Therefore, the metabolism of adipose tissue has essential effects on inflammatory processes, insulin resistance, obesity, and metabolic diseases. FABP-4 is expressed in high levels in adipose tissue and very low levels of FABP-5 (Queipo-Ortuno et al., 2012; Sitar-Taut et al., 2021).

FABP-4 in adipocytes and macrophages plays a crucial role in insulin resistance and vascular pathology (Furuhashi & Hotamisligil, 2008; Furuhashi, Ishimura, Ota, & Miura, 2011; Furuhashi, Saitoh, Shimamoto, & Miura, 2014; Hotamisligil & Bernlohr, 2015). FABP-4 deficient mice with high-fat diets exhibited insulin resistance; however, there is no effect of FABP-4 on insulin sensitivity in lean mice (Hotamisligil et al., 1996; Uysal, Scheja, Wiesbrock, Bonner-Weir, & Hotamisligil, 2000). FABP-4 is involved in differentiation into functional macrophages and adipocytes mediated via a cellular signaling system (Furuhashi et al., 2014; Makowski & Hotamisligil, 2005; Shum et al., 2006). Adipocytes contribute to blood FABP-4 levels more than macrophages or endothelial cells (Cao et al., 2013; Furuhashi et al., 2008). An elevated plasma FABP-4 level is associated with obesity, insulin resistance, and atherosclerosis (Xu et al., 2006). Plasma FABP-4 is modestly associated with CVD risk even after adjustment for standard CVD risk factors (Egbuche et al., 2020). The plasma FABP-4 levels are associated with hyperlipidemia, hyperglycemia, and hypertension (Hsu et al., 2010; Mohlig et al., 2007; Stejskal & Karpisek, 2006). In addition, plasma FABP-4 levels are positively correlated with endothelial dysfunction (Xiao et al., 2010), atherosclerosis (Miyoshi et al., 2010), and CVD risk factors (Rhee et al., 2009; Yeung et al., 2009). The plasma levels of FABP-4 also promote increased numbers of stenotic vessels, suggesting an etiological role in the progression of coronary artery disorders (Rhee et al., 2009). Plasma FABP-4 levels were significantly high in diabetic patients with macrovascular complications (Yeung et al., 2009). In addition, the level of urinary FABP-1 is involved in diabetic nephropathy (Kamijo-Ikemori, Sugaya, & Kimura, 2014). FABP-1 and FABP-2 may be used as novel biomarkers of diabetic nephropathy (Tsai et al., 2020).

FABP-4 level in adipose tissue is regulated by factors such as fatty acids, PPARγ agonists glucocorticoid, and insulin (Amri, Bertrand,

Ailhaud, & Grimaldi, 1991; Baxa et al., 1989; Cook et al., 1988; Melki & Abumrad, 1993).

Atypical chronic low-grade inflammation/metaflammation from adipose tissue is triggered by excess energy and metabolites in canonical and non-canonical inflammatory signaling pathways (Gregor & Hotamisligil, 2011; Hotamisligil, 2006; Mallick & Duttaroy, 2022). During obesity-induced stress condition, FABP-4 binds palmitic acid and produce dependent inflammatory response such as chemokine- and TNFα-NF-κB signaling pathways (Furuhashi et al., 2016). Palmitic acid (16:0)-mediated inflammatory responses by FABP-4 were reported in macrophages, endothelial cells, and vascular smooth muscle cells (VSMC) (Furuhashi et al., 2016). The cross-talk between adipocytes and macrophages leads to metabolic dysregulation in adipose tissue (Hotamisligil, 2017). FABP-4 also regulates endothelial cells' inflammatory activity via expressing genes such as eNOS and intercellular adhesion molecule 1 (ICAM-1) involved in the activation of endothelial cells. The unstable atherosclerotic plaques trigger acute cardiovascular events via inflammatory and extrinsic coagulation pathway. The plasma FABP-4 level was shown to be an independent risk factor for vascular inflammation in a case control study (Yoo et al., 2011). The plasma FABP-4 levels correlate with several pro-inflammatory markers positively such as lipocalin-2, hsCRP, TNF-α receptors, and IL-6 but inversely with adiponectin level (Xu et al., 2007). All these data support that the FABP-4 acts as a critical pro-inflammatory mediator that connects obesity with CVD in humans.

FABP-4 involves the proliferation and migration of VSMC and hepatic glucose production (Cao et al., 2013; Girona et al., 2013). FABP-4 decreases the contraction of cardiomyocytes and activation of insulin-sensitive eNOS (Aragones et al., 2012; Lamounier-Zepter et al., 2009). Deletion of the *FABP-4* gene decreases lipolysis and improves insulin sensitivity in mice (Coe, Simpson, & Bernlohr, 1999; Uysal et al., 2000). FABP-4 increases cholesterol levels in macrophages and induces inflammation, which is crucial for foam cell formation (Llaverias et al., 2004; Makowski, Brittingham, Reynolds, Suttles, & Hotamisligil, 2005). Fig. 1 describes the putative physiological and pathological functions of FABPs.

In contrast to FABP-4 (aP2), FABP-5 (mal1) is least expressed in adipose tissue (Makowski & Hotamisligil, 2005; Simpson, LiCata, Ribarik Coe, & Bernlohr, 1999). But FABP-4 deficiency upregulates FABP-5 expression in adipose tissue, but not vice versa (Maeda et al., 2003; Simpson et al., 1999), although FABP-4 and FABP-5 expression are identical in macrophages

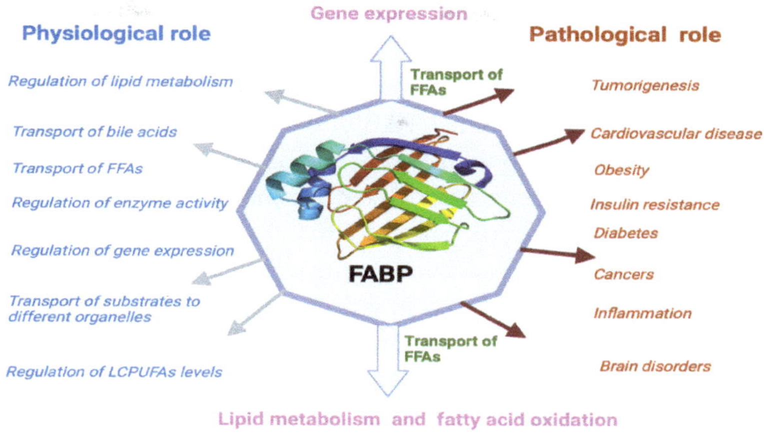

Fig. 1 FABPs and their physiological and pathological roles.

without any compensatory upregulation (Makowski et al., 2001). Endothelial cells can regulate the expression of FABP-5 depending on tissue types and microvasculature (Masouye et al., 1997). Despite insufficient information about a specific function, FABP-5 is also known to induce lipolytic activities and insulin resistance (Hertzel, Bennaars-Eiden, & Bernlohr, 2002; Maeda et al., 2003). Cross-talk between adipocyte and macrophage to contribute to metabolic disorientation is dependent on FABP-4 and FABP-5 (Furuhashi et al., 2008). Macrophage accumulation in adipose tissue is characterized by an increased inflammatory response, risk of insulin resistance, and CVDs in obese people (Xu et al., 2003; Yu et al., 2016). FABP-5 is also expressed in endothelial cells (Iso et al., 2013; Masouye et al., 1997). FABP-5 plays a role in regulating angiogenesis-related endothelial cells functions, including proliferation, migration, and survival. FABP4 and FABP5 are expressed in microvessels of different organs with higher fatty acids metabolic activity (Antohe et al., 1998; Yu et al., 2016). Endothelial cells can regulate the expression of FABP-5 depending on tissue types and microvasculature (Bae et al., 2020).

Transgenic mice, FABP-5 in adipose tissue showed enhanced lipolysis and reduced insulin sensitivity after feeding a high-fat diet (Hertzel et al., 2002; Maeda et al., 2003). Deletion of *FABP-5* gene reduces inflammatory response (Babaev et al., 2011). As circulating FABP-5 level is lower than FABP-4, circulating FABP-5 doesn't always correlate with metabolic disorders associated with FABP-4 (Ishimura et al., 2013; Yeung et al., 2008). However, since inflammation is an essential pathogenic mediator of metabolic syndrome, the role of FABP-4 and FABP-5 in metabolic disease are evident. In animal models, combined knockout of FABP-4 and FABP-5 ensures efficient defense against insulin resistance, obesity, and atherosclerosis than mice deficient for either FABP4 or FABP5 (Boord et al., 2004; Maeda et al., 2005). Along with LCFAs, FABP-5 binds with PPARδ and retinoic acid; upon binding these ligands, it mobilizes to the nucleus to induce PPARδ and regulates cell growth and development. Regulation of endothelial cells homeostasis and vascular integrity is vital to organ physiology and tissue repair, regeneration, and tumor growth. Based on the additive functions of FABP-4 and FABP-5 in other cell types and their co-expression in microvascular endothelial cells, FABP-5 played a role in regulating angiogenesis-related endothelial cell functions, including proliferation, migration, and survival. As a downstream inducer of FABP5-related effects in endothelial cells, the critical role of PPARδ was demonstrated. Endothelial cells homeostasis and vascular integrity regulation are vital to normal physiology and tissue repair, regeneration, and tumor growth.

However, in FABP4 deficiency, the conversion of glucose to fatty acids is stimulated by twofold (Shaughnessy, Smith, Kodukula, Storch, & Fried, 2000). This suggests that FABP4 may suppress lipogenesis in adipocytes, thereby contributing to the observed improvements in fasting plasma glucose levels and insulin in obese FABP4-deficient mice (Shaughnessy et al., 2000). Furthermore, the increased plasma levels of fatty acids suppress *Srebp1c* expression in the liver (Maeda et al., 2005), thus contributing to the development of fatty liver in FABP4-knock-out mice. However, both basal and stimulated levels of lipolysis are significantly decreased in adipocytes from *FABP-4 knock-down* mice, despite the compensatory increase in FABP5 levels (Scheja et al., 1999).

The roles of FABP-4 and FABP-5 in fatty acid uptake by the endothelial cells of heart and skeletal muscle are studied (Iso et al., 2013). The double knockout of *FABP-4* and *FABP-5* confers better defense against obesity, insulin resistance, and atherosclerosis than mice deficient for either *FABP-4* or *FABP-5* (Boord et al., 2004; Maeda et al., 2005). Despite its 55% amino acid sequence homology with FABP-4, the function of FABP-5 in endothelial cells is not clearly understood. Based on the functions of FABP-4 and FABP-5 in other cell types and their co-expression in microvascular endothelial cells, it's clear that FABP-5 plays a role in regulating angiogenesis-related endothelial cells functions, including proliferation, migration, and survival. The regulation of endothelial cell homeostasis and vascular integrity is vital to cardiac physiology and tissue repair, regeneration, and tumor growth.. Interestingly, cellular senescence and oxidative stress induce FABP-4 expression in microvascular endothelial cells (Ha et al., 2006; Lee, Wang, & Vanhoutte, 2010). The ectopic expression of FABP-4 in injured arterial endothelial cells demonstrates its role in atherosclerosis plaque formation (Fuseya et al., 2017; Lee et al., 2007).

FABP-4 regulates the functions of various mitogenic pathways, such as stem cell factor/c-kit signaling, which plays a vital role in its pro-angiogenic activity. Vascular endothelial growth factor-A (VEGF-A), basic fibroblast growth factor (bFGF), mammalian endothelial target of rapamycin complex 1 (mTORC1) induce FABP-4 expression in cardiac, skeletal muscle, and renal endothelial cells to stimulate angiogenesis and fatty acid transport (Elmasri et al., 2012, 2009; Harjes, Bridges, McIntyre, Fielding, & Harris, 2014; Iso et al., 2013).

Inhibition of FABP-4 blocks most of the VEGFA effects in endothelial cells. NOTCH signaling system is involved in FABP-4, FABP-5, lipoprotein lipase, FAT/CD36, and angiopoietin-like 4 (ANGPTL4) expression (Jabs et al., 2018). Human recombinant DLL4 mediated stimulation of NOTCH signaling increases the expression of FABP4, without involving VEGFA (Harjes et al., 2014). Both basal and VEGFA and/or the NOTCH pathways mediated stimulated expression of FABP4 gene depends on the transcription factor FOXO1. This may be crucial for angiogenesis in the tumor environment (see Section 4). More information is required on the relationship between angiogenesis and fatty acid transport.

Metabolic disorders play a crucial role in CVDs, abnormal lipid signaling, high lipids storage, trafficking, and signaling capacity of macrophages and adipocytes. FABPs are highly expressed in several metabolically active tissues (Furuhashi et al., 2008). Henceforth, the contribution of FABPs in cardiovascular impacts is significant (Furuhashi et al., 2008). In turn, FABP-4 expression is controlled by VEGF-A and mammalian endothelial target of rapamycin complex 1 (Elmasri et al., 2012, 2009). FABP-4 induces a pro-angiogenic role by promoting cell proliferation, survival, and migration in endothelial cells (Elmasri et al., 2012, 2009; Ghelfi et al., 2013). The pro-angiogenic role of FABP-4 in first trimester placental trophoblast cells has been demonstrated (Basak et al., 2018).

As fatty acids are the fundamental source of energy in diabetic patients, therefore overexpression of *FABP3* is obvious in these patients (Karbek et al., 2011). Whereas FABP-3 is involved in arachidonic acid uptake by last trimester placental trophoblasts supporting fetal growth (Leroy et al., 2017). PPARα agonists also influence *FABP3* expression (Motojima, 2000).

Myocardial overexpression of *FABP-3* is correlated with exacerbated cardiac dysfunction and increased apoptosis (Zhuang et al., 2019). Within half an hour of chest pain due to myocardial injury, blood FABP-3 level rises and clears by renal filtration (Haastrup et al., 2000; Knowlton, Apstein, Saouf, & Brecher, 1989). FABP-3 has been proposed as an effective biomarker for myocardial infarction and heart failure (Thygesen et al., 2012). Elevated levels of FABP-3 with peripheral arterial disease suggest the significance of utilizing FABP-3 as the biomarker (Syed et al., 2020). FABP-3 and FABP-5 are expressed in endothelial cells of the heart and skeletal muscles, while FABP-4 and FABP-5 are expressed in the endothelial cells of other tissues with active fatty active metabolism (Antohe et al., 1998; Yu et al., 2016).

4. FABPs in various cancers

Cancer cells need lipids to synthesize membrane biomass for their uncontrollable growth and proliferation. Cancer cells can reprogram their lipid metabolism that favors producing more energy and macromolecules, which are essential for their rapid growth, proliferation, division, and invasion in metastasis. To this end, membrane lipid bilayers provide a suitable platform for molecular crosstalk and signaling in oncogenesis (Soteriou, Kalli, Connell, Tyler, & Thorne, 2021). The mechanism by which carcinoma cells reprogram fatty acid metabolism may involve *de novo* synthesis and/or exogenous uptake of fatty acids into the plasma and intracellular pool. The cancer lesions are often inflamed due to the higher presence of infiltrated immune cells.

The FABPs regulate inflammatory response by changing their expression and thereby control fatty acid supply to the cancer cells. The roles of FABP-directed fatty acids metabolism, their supply as an energy source, and their nuclear receptor mediated modulation in inflammation, angiogenesis, membrane biogenesis, and other growth favoring activities in cancer cells are proposed. Therefore, these FABPs are being targeted to reduce cancer metastasis in addition to routine cancer therapy. Incremental reports suggest that altered expression of lipid transport and metabolic proteins exacerbate inflammation-associated ectopic lipid accumulation and induce a favorable microenvironment for the invasion and metastasis of cancer cells (Fig. 2). The dysregulated lipid signaling via fatty acid activation of PPARs leads to activating genes associated with the tumorigenesis microenvironment. FABPs are trafficking these fatty acids to PPARs, promoting their

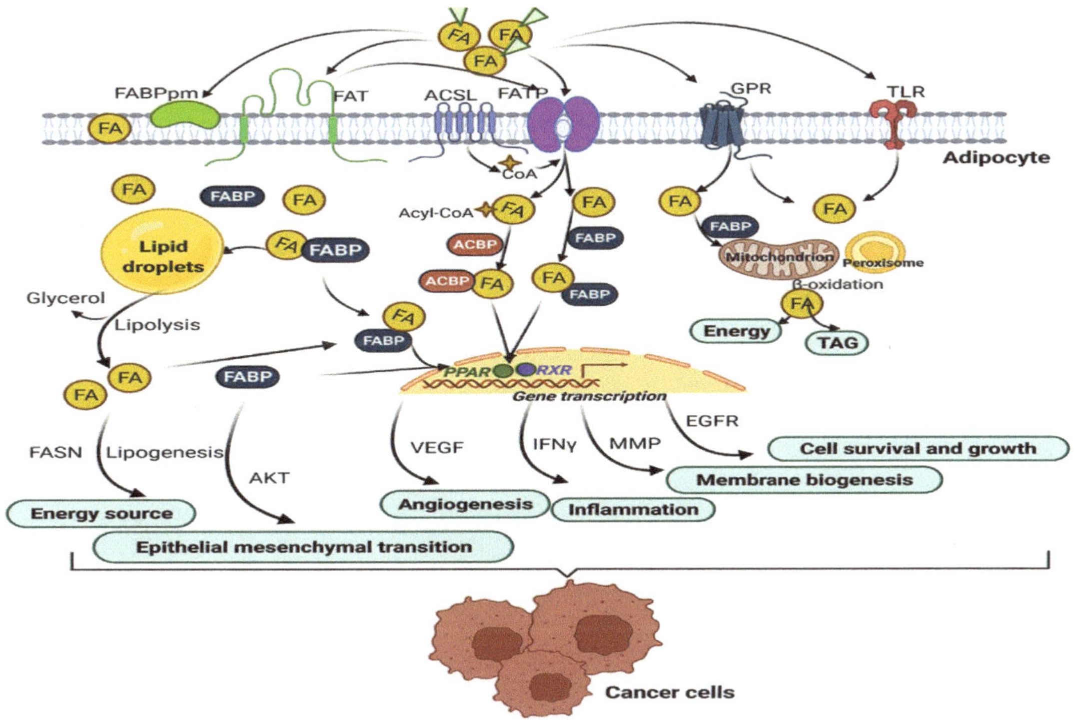

Fig. 2 Endocrine effects of FABPs promote the microenvironment of cancer metastasis. Extracellular albumin-bound free fatty acids can be transported via FATP, FAT, FABPpm, GPR, TLR and released into the cytosol. FABP or ACBP bound intracellular fatty acids are directed to either storage into lipid droplets or transferred to b-oxidation. Increased expression of FABPs triggers inflammation associated with ectopic lipid accumulation in adipocytes by reprogramming lipid metabolism due to altered triglyceride storage of lipid droplets and de novo and/or exogenous supply of fatty acids. The increased FABPs expression can enhance PPAR/RXR mediated transcription of FASN, IFNγ, VEGF, and MMPs, EGFR promoting lipid accumulation, inflammation, angiogenesis, membrane biogenesis, and other growth favoring activities for cancer metastasis. FA, fatty acid; FAT, fatty acid translocase (CD36); FABP pm, plasma membrane fatty acid-binding protein; FATP, fatty acid transporter protein; GPR, G-coupled protein receptor; TLR; toll-like receptor; PPAR, peroxisome proliferator-activated receptor; RXR, retinoid X receptors.

activation (O'Sullivan & Kaczocha, 2020). Activated fatty acids convert to triacylglycerols via a series of catalytic steps, ultimately sequestered into lipid droplets. Lipid droplets (LDs) are dynamic, often controlled by the balance between lipolysis and lipophagy. The altered phenotype of LDs and their metabolism correlates with the stages of the cancers (Petan, Jarc, & Jusović, 2018). Cancer cells promote fatty acid uptake by activating the expression of FABP-7 and FABP-3 in response to hypoxia and stimulating mitochondrial fatty acid oxidation for their growth and proliferation (Bensaad et al., 2014).

Prostate cancer (PCa) is the second most cancer in Men in the US, next to dermal cancer. PCa is the second most common cancer globally in men, accounting for 15% of new cancer cases (Rawla, 2019). FABPs bind different fatty acids and transport them to the intracellular components of different cellular sites. Skewed expression of FABPs occurs widely in various tissue-specific cancers (Gharpure et al., 2018; Kwong, Jamil, Rhodes, Taib, & Chung, 2019; O'Sullivan & Kaczocha, 2020; Zeng, Sauter, & Li, 2020). FABPs coordinate gene expression of several pathways, including inflammation and metabolism (Coe & Bernlohr, 1998; Hertzel & Bernlohr, 2000). The obesity-linked ectopic lipid overloads are associated with high-grade prostate cancer and prostate cancer mortality (Parikesit, Mochtar, Umbas, & Hamid, 2016). However, underlying mechanisms that connect fatty acid overload and prostate cancer progression are poorly understood. A group of FABP genes is co-expressed collectively in metastatic prostate tissues indicate augmented supply of lipids to these sites. A novel FABP-12 subtype is expressed and localized in the chromosomal region of 8q21.13 in prostate cells (Liu & Godbout, 2020). The epidermal fatty acid-binding proteins (E-FABP or FABP-5) are intracellular proteins, highly expressed in adipocytes and macrophages of tumors. FABP-5 is expressed abnormally in various cancers justified to target FABP-5 in treating cancers. Knock-down of FABP5 reduced the cellular invasiveness due to altered lipid metabolism, lipolysis, and de novo FA synthesis in highly aggressive prostate and breast cancer cells mediated by downregulated expression of acyl-CoA synthetase long-chain family member 1 (ACSL1), hormone-sensitive lipase (HSL), monoacylglycerol lipase (MAGL), elongation of long-chain fatty acid member 6 (Elovl6) (Senga, Kobayashi, Kawaguchi, Ando, & Fujii, 2018). The enhanced fatty acid-binding protein 5 (FABP5) expression and increased mitochondrial oxidative metabolism in tumor-infiltrating T cells are used as an immune-metabolic prognostic biomarker of hepatocellular carcinoma (Liu et al., 2020). FABP5 is reported as a

HIF-1α-binding partner. Both FABP5 and HIF-1α expression are upregulated and associated with a poor prognosis of hepatocellular carcinoma (HCC). FABP-5 stimulates HIF-1α activation and lipid metabolism in these cells (Seo et al., 2020). However, more data with longitudinal studies are required to determine FABP5 and its potential as a predictive biomarker for HCC. Thus, targeting FABP-5 led to altering lipid signaling intermediates' profiles and lowered the lipid-mediated invasion in PCa cells, indicating that PCa metastasis may depend on FABP-5 (Carbonetti et al., 2019). By inactivating FABP-5, progression of prostate metastasis was suppressed in PCa cells. The FABP-5-PPARγ-signaling was inhibited by a recombinant FABP-5 inhibitor named dmrFABP-5 indicates FABP-5's role in cancer metastasis (Al-Jameel et al., 2019). Blocking of FABP-5 suppressed the PGE_2 mediated inflammation and thus could be a novel target in inflammation-associated tumors (Bogdan et al., 2018). FABP-5 also promotes cancer invasion by inducing the epithelial-to-mesenchymal (EMT) transition of hepatocellular carcinoma (Ohata et al., 2017). FABP-5 is reported as a biomarker for cervical cancer where FABP-5 gene silencing lowers the MMP2 and MMP9 expression in cervical cancer (Wang et al., 2016).

Epidermal fatty acid-binding protein (E-FABP), also known as FABP-5, was first cloned in psoriatic skin due to its excess presence in keratinocytes (Madsen, Rasmussen, Leffers, Honore, & Celis, 1992). FABP-5 expression in keratinocytes maintains the water permeability barrier in the skin and promotes differentiation of fatty acids mediated keratin-1 expression in these cells (Owada, Suzuki, Noda, & Kondo, 2002). FABP-5 expression prevents induced tumorigenesis in rat skin lesions (Zhang et al., 2018). FABP-5 mRNA levels are upregulated in skin malignancy, including psoriasis vulgaris, atopic dermatitis, Mycosis fungoides, and Sézary syndrome (Takahashi-Shishido et al., 2021). Expression of FABP-5 decreases with the progress of accumulated tumor lesions indicating its dependency on malignancy.

Adipokines are implicated to link obesity with prostate cancer (Cabia, Andrade, Carreira, Casanueva, & Crujeiras, 2016). The FABP4 or ap2 or A-FABP, is a lipid chaperone, highly expressed in adipose, also functions as an adipokine by secreting hormones in regulating lipid and glucose metabolism. Due to these pleiotropic effects, FABP-4 was targeted in controlling obesity-associated cancers in the breast and prostate (Prentice, Saksi, & Hotamisligil, 2019; Zeng et al., 2020). The circulatory FABP-4 levels were significantly increased in postmenopausal obese women independent of breast cancer status. Moreover, FABP-4 did not associate with estrogen levels in these women. Due to these observations, FABP-4 status is predicted as an independent biomarker in assessing the risks of developing breast cancer (Li, Hao, Yan, Kong, & Sauter, 2019). FABP-4 substantially increased the metastatic potential of ovarian cancer by regulating the pathway associated with the metabolism of ovarian cancer cells (Gharpure et al., 2018). FABP-4 promotes lipid accumulation, proliferation, migration, survival of prostate cancer cells. In addition, FABP-4 also express angiogenic modulators in regulating growth of tumor progression in the prostate (Chaudry & Dutta-Roy, 1993). The higher circulatory FABP-4 facilities cancer cells metabolism associated with lipid metabolism, mitogenesis, proliferation, inflammation and insulin resistance (Scifres, Chen, Nelson, & Sadovsky, 2011). FABP-4 plays as a common control point in regulating angiogenesis and metabolic signaling of tumor-associated endothelial cells (Harjes et al., 2014). The FABP-4 was reported to induce by NOTCH directly. The differentiation and stemness of cancer stem cells are controlled by NOTCH, a cell fate-determining transcription factor. It was proposed that the insulin-responsive FOXO1 transcription factor favored DLL4-NOTCH signaling induced by FABP-4. The higher circulatory FABP-4 levels (as it is also found in plasma in obese people) promote the growth, progression, and survival of PCa cells (Uehara, Takahashi, Oha, Ogawa, & Izumi, 2014). The roles of exogenous FABP-4 on prostate cancer progression are not clearly understood.

A higher FABP-4 expression promotes its endocrine function by inducing the production of pro-inflammatory cytokines in prostate stromal cells. The augmented secretion and expression of FABP-4 stimulate the invasiveness of prostate cancer cells by upregulating MMPs expression and activation of MAPK and PI3K meditators (Huang et al., 2017). Exogenous FABP-4 stimulates proliferation, migration and angiogenesis in a concentration and time dependent manner (Basak et al., 2018). FABP-4 affects angiogenesis by targeting VEGF and its receptor in endothelial cells (Elmasri et al., 2012; Ghelfi et al., 2013). However, the mechanism of FABP-4 on angiogenesis and metabolism of prostate cancer survival and progression is still not clear. The increased production of pro-inflammatory cytokine-induced by excess FABP-4 in circulation could affect angiogenesis in cancer cells.

The anti-angiogenesis therapies in cancer starve tumor cells with reduced blood supply. Several such therapies undergo clinical trials in colorectal, ovarian, and other cancers with limited success (Mullard, 2014). The blocking of VEGF and its receptor failed to provide due success as anti-angiogenic therapy against tumor progression due to alternate forms of VEGFs, drug resistance, and others. Lessons from these led to exploring alternate pro-angiogenic molecules targeting angiogenic signaling and controlling lipid metabolism. Cancer cells prefer glucose as an energy substrate over oxidative phosphorylation, which targeted tumor cells with an anti-glycolytic inhibitor (PFKFB3) but achieved little success. Reducing hyper glycolysis instead blocking vis-a-vis by controlling fatty acid metabolism in cancer cells promised a better strategy against tumor progression. Since glycolysis is connected with fatty acid metabolism and fatty acids acts as pro-angiogenic growth factors, targeting mediators that control both fatty acid metabolism, and angiogenesis in cancer is being highlighted recently. In this regard, the roles of FABPs as cancer therapeutic promised potential since FABPs are reported to act as angiogenic and metabolic mediators in endothelial cells and carcinoma cells. The tumor microenvironment is promoted by endothelial cells as these forms new blood vessels from pre-existing one in angiogenesis. Again, fatty acids regulated the secretion and expression of VEGF in choriocarcinoma cells via different mechanisms, including the expression of FABP-4. Therefore, determining the interplay between angiogenic factors, dietary fatty acids, and FABP-4 could be necessary to explore novel targets in mitigating cancer cell progression.

The healthy epithelial cells and cancer cells differentially express FABPs due to the different utilization of fatty acids in these cells. FABP-4 plays a critical role in metabolism and inflammatory process both locally and systematically. FABP-4 could be a potential therapeutic target in the outcome of immune-metabolic diseases such as cancer. A higher expression of FABPs justified its overdoing in trafficking fatty acids to metastasis sites in cancer cells. The roles of FABPs from exogenous and endogenous sources and their targeted inhibition in prostate cancer progression are not clearly understood. The expression or plasma levels of a metabolic marker like FABP-4 could be a potential biomarker in addition to prostate-specific antigen (PSA). The levels of FABP-4 may help in anticipating the disease's prognosis, stage, and aggressiveness, providing advanced treatment and management options in solid tumors.

Diet or hormone-induced obesity is associated with low-grade inflammation that secretes the production of multiple pro-inflammatory cytokines in adipose, including FABP-4. The circulatory form of FABP-4 may have endocrine functions to regulate downstream metabolism and inflammation in adipose. The exogenous circulatory FABP-4 directly interacts with cell adhesion mediators and controls endothelial cells' intracellular lipid flux (Saavedra et al., 2015). FABP-4 contributes to the cellular differentiation and signaling process. FABP-4 seems to couple lipid metabolism with inflammation (Xu et al., 2015). The circulatory FABP-4 is elevated in breast cancer patients, (Hancke, Grubeck, Hauser, Kreienberg, & Weiss, 2010) which is also considered a biomarker of cell injury or inflammation. The roles of circulatory FABP-4 in prostate cancer and its clinical contribution to disease progression is not understood.

Higher expression and excess secretion of FABP-4 promotes invasiveness and angiogenesis of the cancer cells in the prostate through the upregulation of immune-metabolic genes, mitogenic mediators, and collagenases activities. The differentiation of fats cells is impaired in PCa (Kwon, Zhang, Zhang, & Xin, 2016). FABP-4 was used as a survival biomarker in renal cell carcinoma patients (Tölle, Suhail, Jung, Jung, & Stephan, 2011). Angiogenesis drives the blood supply to the tumor tissues and promotes the growth and survival of the malignant cells (Fernandez-Martinez & Lucio-Cazana, 2015; Ku, Liu, Lin, Lu, & Lin, 2016). FABP-4 is also involved in mediating cellular differentiation and signaling in PCa. E-cadherin involves in cell-to-cell adhesion associated with invasion and metastasis of tumor cells. Silencing FABP-1 in liver carcinoma resulted in inhibition of proliferation and VEGF signaling of the HCC cells (Ku et al., 2016).

The fatty acid metabolism of prostate is controlled by prostaglandin metabolism. The FABP-4 can mediate its effects on prostaglandin metabolism in prostate cells. An extensive review of the association between PUFA and cancer incidence suggests no correlation between n-6 PUFA intake and cancer risks. However, blood levels of n-6 PUFAs are negatively associated with cancer risk (Kim & Kim, 2020). The n-6 and n-3 fatty acid derived eicosanoids, and docosanoids act in a different way in angiogenesis and vascularization of endothelial cells of cancer tissue. The n-6 fatty acid promotes tumor angiogenesis by activating the expression and secretion of angiogenic growth factors. Again, n-3 fatty acid metabolites inhibit tumor angiogenesis by lowering VEGF expression and secretion. The proliferation and invasion of PC3 cells were increased by blocking PGE2-EP2 signaling indicating the role of PGE2 in cancer metastasis (Fernandez-Martinez & Lucio-Cazana, 2015). The aldo-keto reductase (AKR)1C is elevated in prostate cancer (Dozmorov et al., 2010). Overexpression of AKR1C3 promotes aggressiveness of PC3 cells by upregulating angiogenesis. However, the roles of FABP-4 on AKR1C3 activity and prostaglandin metabolism in prostate cancer cells are not known. The expression of fatty acid desaturase 2 (FADS2) is profoundly altered in several cancers. It allows selective enrichment of phospholipid membrane with ARA and its metabolites that promote pro-inflammatory activities in the tumor microenvironment (Zhao, Pflug, Lai, & Wang, 2016). It is unknown whether the blocking of FABPs can affect the action of fatty acid desaturase. A systemic review suggests a higher omega-3 intake may decrease prostate cancer mortality, but more data is required to establish the correlation. DHA has been shown to suppress cancer cell viability by suppressing fatty acid synthase in PCa cells (Zhao et al., 2016). However, the mechanism of DHA action is not known. Whether the protective effects of long chain fatty acids on disease progression are mediated by alteration in gene expression is unknown. Fatty acids act as a ligand for PPARγ that directly regulate gene expression in lipid metabolism in prostate cells (Forootan et al., 2016; Wu et al., 2015). The expression of FABP subtypes on solid carcinoma has received tremendous attention in biomarker discovery in cancers (Table 2). The best possible way to manage solid cancers involve an early diagnosis to explore characteristic tissue markers. Altered expression of FABP as a tissue biomarker in a solid tumor is promising for a diagnostic or prognostic target for new therapeutic strategies. The standard tissue composition of FABP subtypes is wholly altered in cancer tissue, leading to a new composition of FABP subtypes in these solid tumors. Since fatty acids play a major role in synthesizing tumor biomass membrane in cancer, a large aggregation of fatty acids at the metastatic site is facilitated by FABPs.

Table 2 Prevalence of fatty acid binding protein expression in various human carcinoma.[a]

Carcinoma type	FABP subtypes						
	FABP-1	**FABP-3**	**FABP-4**	**FABP-5**	**FABP-6**	**FABP-7**	**FABP-9**	**FABP-12**
Prostate	+[1]		+[1]	+[1] +[2]			+[1] +[2]	+[1]
Breast			+[2] +[3] +[4] +[5]	+[3] +[4]		+[1] +[2]		
Colon	+[2] +[3]		+[6] +[7]	+[5]	+[1]			
Bladder			+[8] +[9]	+[6]				
Kidney	+[4]			+[7]		+[3]		
Dermal		+[1]	+[10]			+[4]		
Stomach	+[5]	+[2] +[3]						
Lung		+[4]						
Liver	+[6] +[7] +[8]		+[11]	+[8]				

Continued

Table 2 Prevalence of fatty acid binding protein expression in various human carcinoma.[a]—cont'd

Carcinoma type	FABP subtypes							
	FABP-1	FABP-3	FABP-4	FABP-5	FABP-6	FABP-7	FABP-9	FABP-12
Cervix		+[5]		+[9]				
Ovary			+[12]+[13]					
Oral				+[10]		+[5]		
Glioblastoma						+[6]+[7]		

[a]The studies reported for each FABP subtypes are listed as "+" with superscript numbers.
FABP-1: 1- (Das, Hammamieh, Neill, Melhem, & Jett, 2001); 2-(Lawrie, Dundas, Curran, & Murray, 2004); 3-(Wood et al., 2017); 4-(Tolle et al., 2009); 5-(Hashimoto et al., 2004); 6-(Suzuki, Watanabe, & Ono, 1990); 7-(Inoue, Takahashi, Fujii, Kitagawa, & Fukusato, 2014); 8-(Wang et al., 2014)
FABP-3: 1-(Linge et al., 2012); 2-(Hashimoto et al., 2004); 3-(Chen et al., 2021); 4-(Tang et al., 2016); 5-(Davidson et al., 2017)
FABP-4: 1-(Huang et al., 2017); 2-(Hancke et al., 2010); 3-(Kim, Lee, & Koo, 2015); 4-(Guaita-Esteruelas et al., 2017); 5-(Li et al., 2019); 6-(Zhang et al., 2019); 7-(Tian et al., 2020); 8-(Ohlsson, Moreira, Gromov, Sauter, & Celis, 2005); 9-(Boiteux et al., 2009); 10-(Madsen et al., 1992); 11-(Thompson et al., 2018); 12-(Nieman et al., 2011); 13-(Gharpure et al., 2018);
FABP-5: 1-(Pang et al., 2010); 2-(Carbonetti et al., 2019); 3-(Liu et al., 2011); 4-(Powell et al., 2015); 5-(Kawaguchi et al., 2016); 6-(Ostergaard et al., 1997); 7-(Wu et al., 2019); 8-(Ohata et al., 2017); 9-(Zhang et al., 2020); 10-(Fang, Wong, Chiang, & Chen, 2010);
FABP-6: 1-(Ohmachi et al., 2006); *FABP-7*: 1-(Zhang et al., 2010); 2-(Tang et al., 2010); 3-(Nagao et al., 2018); 4-(Slipicevic et al., 2008); 5-(Phuchareon et al., 2014); 6-(Liang et al., 2005); 7-(Kaloshi et al., 2007); *FABP-9*: 1-(Al Fayi et al., 2016); 2-(Liu & Godbout, 2020); *FABP-12*: 1-(Liu et al., 2020).

5. Conclusions

FABPs regulate metabolic and inflammatory pathways, its inhibition can improve type 2 diabetes mellitus and atherosclerosis. The signaling responsive elements of FABPs and determinants of FABP-mediated functions may be exploited in preventing or treating these diseases. FABPs are involved in cardiac dysfunction, angiogenesis and cancers. FABPs are promising tissue biomarker and can be used as target for novel therapeutic strategies.

Acknowledgments

This work is supported in part by grants from Throne Holst Foundation and the Faculty of Medicine, University Oslo, Norway.

Declarations

Funding: Not applicable.
Conflicts of interest: The author declares no conflicts of interest.
Ethics approval: Not applicable

Consent to participate: Not applicable.
Consent for publication: Not applicable.
Availability of data and material: Not applicable.
Code availability: Not applicable.

References

Al Fayi, M. S., Gou, X., Forootan, S. S., Al-Jameel, W., Bao, Z., Rudland, P. R., et al. (2016). The increased expression of fatty acid-binding protein 9 in prostate cancer and its prognostic significance. *Oncotarget*, 7(50), 82783–82797. https://doi.org/10.18632/oncotarget.12635.

Al-Jameel, W., Gou, X., Jin, X., Zhang, J., Wei, Q., Ai, J., et al. (2019). Inactivated FABP5 suppresses malignant progression of prostate cancer cells by inhibiting the activation of nuclear fatty acid receptor PPARγ. *Genes Cancer*, *10*(3-4), 80–96. https://doi.org/10.18632/genesandcancer.192.

Amri, E. Z., Bertrand, B., Ailhaud, G., & Grimaldi, P. (1991). Regulation of adipose cell differentiation. I. Fatty acids are inducers of the aP2 gene expression. *Journal of Lipid Research*, *32*(9), 1449–1456. Retrieved from https://www.ncbi.nlm.nih.gov/pubmed/1753215.

Antohe, F., Popov, D., Radulescu, L., Simionescu, N., Borchers, T., Spener, F., et al. (1998). Heart microvessels and aortic endothelial cells express the 15 kDa heart-type fatty acid-binding proteins. *European Journal of Cell Biology*, *76*(2), 102–109. https://doi.org/10.1016/S0171-9335(98)80022-8.

Aragones, G., Saavedra, P., Heras, M., Cabre, A., Girona, J., & Masana, L. (2012). Fatty acid-binding protein 4 impairs the insulin-dependent nitric oxide pathway in vascular endothelial cells. *Cardiovascular Diabetology*, *11*, 72. https://doi.org/10.1186/1475-2840-11-72.

Babaev, V. R., Runner, R. P., Fan, D., Ding, L., Zhang, Y., Tao, H., et al. (2011). Macrophage Mal1 deficiency suppresses atherosclerosis in low-density lipoprotein receptor-null mice by activating peroxisome proliferator-activated receptor-gamma--regulated genes. *Arteriosclerosis, Thrombosis, and Vascular Biology*, *31*(6), 1283–1290. https://doi.org/10.1161/ATVBAHA.111.225839.

Bae, H., Hong, K. Y., Lee, C. K., Jang, C., Lee, S. J., Choe, K., et al. (2020). Angiopoietin-2-integrin alpha5beta1 signaling enhances vascular fatty acid transport and prevents ectopic lipid-induced insulin resistance. *Nature Communications*, *11*(1), 2980. https://doi.org/10.1038/s41467-020-16795-4.

Basak, S., Sarkar, A., Mathapati, S., & Duttaroy, A. K. (2018). Cellular growth and tube formation of HTR8/SVneo trophoblast: Effects of exogenously added fatty acid-binding protein-4 and its inhibitor. *Molecular and Cellular Biochemistry*, *437*(1–2), 55–64. https://doi.org/10.1007/s11010-017-3095-9.

Baxa, C. A., Sha, R. S., Buelt, M. K., Smith, A. J., Matarese, V., Chinander, L. L., et al. (1989). Human adipocyte lipid-binding protein: purification of the protein and cloning of its complementary DNA. *Biochemistry*, *28*(22), 8683–8690. https://doi.org/10.1021/bi00448a003.

Bensaad, K., Favaro, E., Lewis, C. A., Peck, B., Lord, S., Collins, J. M., et al. (2014). Fatty acid uptake and lipid storage induced by HIF-1α contribute to cell growth and survival after hypoxia-reoxygenation. *Cell Reports*, *9*(1), 349–365. https://doi.org/10.1016/j.celrep.2014.08.056.

Bogdan, D., Falcone, J., Kanjiya, M. P., Park, S. H., Carbonetti, G., Studholme, K., et al. (2018). Fatty acid-binding protein 5 controls microsomal prostaglandin E synthase 1

(mPGES-1) induction during inflammation. *The Journal of Biological Chemistry, 293*(14), 5295–5306. https://doi.org/10.1074/jbc.RA118.001593.
Boiteux, G., Lascombe, I., Roche, E., Plissonnier, M. L., Clairotte, A., Bittard, H., et al. (2009). A-FABP, a candidate progression marker of human transitional cell carcinoma of the bladder, is differentially regulated by PPAR in urothelial cancer cells. *International Journal of Cancer, 124*(8), 1820–1828. https://doi.org/10.1002/ijc.24112.
Boord, J. B., Maeda, K., Makowski, L., Babaev, V. R., Fazio, S., Linton, M. F., et al. (2004). Combined adipocyte-macrophage fatty acid-binding protein deficiency improves metabolism, atherosclerosis, and survival in apolipoprotein E-deficient mice. *Circulation, 110*(11), 1492–1498. https://doi.org/10.1161/01.CIR.0000141735.13202.B6.
Cabia, B., Andrade, S., Carreira, M. C., Casanueva, F. F., & Crujeiras, A. B. (2016). A role for novel adipose tissue-secreted factors in obesity-related carcinogenesis. *Obesity Reviews, 17*(4), 361–376. https://doi.org/10.1111/obr.12377.
Campbell, F. M., Bush, P. G., Veerkamp, J. H., & Dutta-Roy, A. K. (1998). Detection and cellular localization of plasma membrane-associated and cytoplasmic fatty acid-binding proteins in human placenta. *Placenta, 19*(5-6), 409–415. https://doi.org/10.1016/s0143-4004(98)90081-9.
Cao, H., Sekiya, M., Ertunc, M. E., Burak, M. F., Mayers, J. R., White, A., et al. (2013). Adipocyte lipid chaperone AP2 is a secreted adipokine regulating hepatic glucose production. *Cell Metabolism, 17*(5), 768–778. https://doi.org/10.1016/j.cmet.2013.04.012.
Carbonetti, G., Wilpshaar, T., Kroonen, J., Studholme, K., Converso, C., d'Oelsnitz, S., et al. (2019). FABP5 coordinates lipid signaling that promotes prostate cancer metastasis. *Science Reports, 9*(1), 18944. https://doi.org/10.1038/s41598-019-55418-x.
Cataltepe, O., Arikan, M. C., Ghelfi, E., Karaaslan, C., Ozsurekci, Y., Dresser, K., et al. (2012). Fatty acid binding protein 4 is expressed in distinct endothelial and non-endothelial cell populations in glioblastoma. *Neuropathology and Applied Neurobiology, 38*(5), 400–410. https://doi.org/10.1111/j.1365-2990.2011.01237.x.
Chaudry, A. A., & Dutta-Roy, A. K. (1993). Purification and characterization of a fatty acid binding protein from human prostatic tissue. *Lipids, 28*(5), 383–388. https://doi.org/10.1007/BF02535934.
Chen, X., Hu, S. L., Feng, Y., Li, P., Mao, Q. S., & Xue, W. J. (2021). Expression of fatty acid-binding protein-3 in gastrointestinal stromal tumors and its significance for prognosis. *The Journal of Surgical Research, 260*, 462–466. https://doi.org/10.1016/j.jss.2020.11.003.
Chen, N. N., Li, Y., Wu, M. L., Liu, Z. L., Fu, Y. S., Kong, Q. Y., et al. (2012). CRABP-II- and FABP5-independent all-trans retinoic acid resistance in COLO 16 human cutaneous squamous cancer cells. *Experimental Dermatology, 21*(1), 13–18. https://doi.org/10.1111/j.1600-0625.2011.01392.x.
Cheng, C., Geng, F., Cheng, X., & Guo, D. (2018). Lipid metabolism reprogramming and its potential targets in cancer. *Cancer Communications (London), 38*(1), 27. https://doi.org/10.1186/s40880-018-0301-4.
Chmurzynska, A. (2006). The multigene family of fatty acid-binding proteins (FABPs): Function, structure and polymorphism. *Journal of Applied Genetics, 47*(1), 39–48. https://doi.org/10.1007/BF03194597.
Coe, N. R., & Bernlohr, D. A. (1998). Physiological properties and functions of intracellular fatty acid-binding proteins. *Biochimica et Biophysica Acta, 1391*(3), 287–306.
Coe, N. R., Simpson, M. A., & Bernlohr, D. A. (1999). Targeted disruption of the adipocyte lipid-binding protein (aP2 protein) gene impairs fat cell lipolysis and increases cellular fatty acid levels. *Journal of Lipid Research, 40*(5), 967–972. Retrieved from http://www.ncbi.nlm.nih.gov/pubmed/10224167.
Cook, J. S., Lucas, J. J., Sibley, E., Bolanowski, M. A., Christy, R. J., Kelly, T. J., et al. (1988). Expression of the differentiation-induced gene for fatty acid-binding protein is activated by glucocorticoid and cAMP. *Proceedings of the National Academy of Sciences*

of the United States of America, 85(9), 2949–2953. Retrieved from http://www.ncbi.nlm.nih.gov/pubmed/2452440.

Corsico, B., Cistola, D. P., Frieden, C., & Storch, J. (1998). The helical domain of intestinal fatty acid binding protein is critical for collisional transfer of fatty acids to phospholipid membranes. *Proceedings of the National Academy of Sciences of the United States of America, 95*(21), 12174–12178. https://doi.org/10.1073/pnas.95.21.12174.

Das, R., Hammamieh, R., Neill, R., Melhem, M., & Jett, M. (2001). Expression pattern of fatty acid-binding proteins in human normal and cancer prostate cells and tissues. *Clinical Cancer Research,* 7(6), 1706–1715. Retrieved from https://www.ncbi.nlm.nih.gov/pubmed/11410510.

Davidson, B., Hellesylt, E., Holth, A., Danielsen, H. E., Skeie-Jensen, T., & Katz, B. (2017). Neuron navigator-2 and cyclin D2 are new candidate prognostic markers in uterine sarcoma. *Virchows Archiv, 471*(3), 355–362. https://doi.org/10.1007/s00428-017-2172-5.

Dozmorov, M. G., Azzarello, J. T., Wren, J. D., Fung, K.-M., Yang, Q., Davis, J. S., et al. (2010). Elevated AKR1C3 expression promotes prostate cancer cell survival and prostate cell-mediated endothelial cell tube formation: Implications for prostate cancer progressioan. *BMC Cancer, 10,* 672. https://doi.org/10.1186/1471-2407-10-672.

Duttaroy, A. K. (2009). Transport of fatty acids across the human placenta: A review. *Progress in Lipid Research, 48*(1), 52–61. https://doi.org/10.1016/j.plipres.2008.11.001.

Dutta-Roy, A. K. (2000). Cellular uptake of long-chain fatty acids: Role of membrane-associated fatty-acid-binding/transport proteins. *Cellular and Molecular Life Sciences, 57*(10), 1360–1372. https://doi.org/10.1007/pl00000621.

Dutta-Roy, A. K., Gopalswamy, N., & Trulzsch, D. V. (1987). Prostaglandin E1 binds to Z protein of rat liver. *European Journal of Biochemistry, 162*(3), 615–619. https://doi.org/10.1111/j.1432-1033.1987.tb10683.x.

Egbuche, O., Biggs, M. L., Ix, J. H., Kizer, J. R., Lyles, M. F., Siscovick, D. S., et al. (2020). Fatty acid binding protein-4 and risk of cardiovascular disease: The cardiovascular health study. *Journal of American Heart Association, 9*(7), e014070. https://doi.org/10.1161/JAHA.119.014070.

Ellinghaus, P., Wolfrum, C., Assmann, G., Spener, F., & Seedorf, U. (1999). Phytanic acid activates the peroxisome proliferator-activated receptor alpha (PPARalpha) in sterol carrier protein 2-/sterol carrier protein x-deficient mice. *The Journal of Biological Chemistry, 274*(5), 2766–2772. https://doi.org/10.1074/jbc.274.5.2766.

Elmasri, H., Ghelfi, E., Yu, C. W., Traphagen, S., Cernadas, M., Cao, H., et al. (2012). Endothelial cell-fatty acid binding protein 4 promotes angiogenesis: role of stem cell factor/c-kit pathway. *Angiogenesis.* https://doi.org/10.1007/s10456-012-9274-0.

Elmasri, H., Karaaslan, C., Teper, Y., Ghelfi, E., Weng, M., Ince, T. A., et al. (2009). Fatty acid binding protein 4 is a target of VEGF and a regulator of cell proliferation in endothelial cells. *FASEB Journal, 23*(11), 3865–3873. https://doi.org/10.1096/fj.09-134882.

Esteves, A., & Ehrlich, R. (2006). Invertebrate intracellular fatty acid binding proteins. *Comparative Biochemistry and Physiology. Toxicology & Pharmacology, 142*(3-4), 262–274. https://doi.org/10.1016/j.cbpc.2005.11.006.

Fang, L. Y., Wong, T. Y., Chiang, W. F., & Chen, Y. L. (2010). Fatty-acid-binding protein 5 promotes cell proliferation and invasion in oral squamous cell carcinoma. *Journal of Oral Pathology & Medicine, 39*(4), 342–348. https://doi.org/10.1111/j.1600-0714.2009.00836.x.

Fernandez-Martinez, A. B., & Lucio-Cazana, J. (2015). Intracellular EP2 prostanoid receptor promotes cancer-related phenotypes in PC3 cells. *Cellular & Molecular Life Sciences, 72*(17), 3355–3373. https://doi.org/10.1007/s00018-015-1891-5.

Forootan, F. S., Forootan, S. S., Gou, X., Yang, J., Liu, B., Chen, D., et al. (2016). Fatty acid activated PPARgamma promotes tumorigenicity of prostate cancer cells by up regulating VEGF via PPAR responsive elements of the promoter. *Oncotarget,* 7(8), 9322–9339. https://doi.org/10.18632/oncotarget.6975.

Furuhashi, M., Fucho, R., Gorgun, C. Z., Tuncman, G., Cao, H., & Hotamisligil, G. S. (2008). Adipocyte/macrophage fatty acid-binding proteins contribute to metabolic deterioration through actions in both macrophages and adipocytes in mice. *Journal of Clinical Investigation*, *118*(7), 2640–2650. https://doi.org/10.1172/JCI34750.

Furuhashi, M., Fuseya, T., Murata, M., Hoshina, K., Ishimura, S., Mita, T., et al. (2016). Local production of fatty acid-binding protein 4 in epicardial/perivascular fat and macrophages is linked to coronary atherosclerosis. *Arteriosclerosis, Thrombosis, and Vascular Biology*, *36*(5), 825–834. https://doi.org/10.1161/ATVBAHA.116.307225.

Furuhashi, M., & Hotamisligil, G. S. (2008). Fatty acid-binding proteins: Role in metabolic diseases and potential as drug targets. *Nature Reviews. Drug Discovery*, 7(6), 489–503. https://doi.org/10.1038/nrd2589.

Furuhashi, M., Ishimura, S., Ota, H., & Miura, T. (2011). Lipid chaperones and metabolic inflammation. *International Journal of Inflammation*, *2011*, 642612. https://doi.org/10.4061/2011/642612.

Furuhashi, M., Saitoh, S., Shimamoto, K., & Miura, T. (2014). Fatty acid-binding protein 4 (FABP4): Pathophysiological insights and potent clinical biomarker of metabolic and cardiovascular diseases. *Clinical Medicine Insights. Cardiology*, *8*(Suppl. 3), 23–33. https://doi.org/10.4137/CMC.S17067.

Fuseya, T., Furuhashi, M., Matsumoto, M., Watanabe, Y., Hoshina, K., Mita, T., et al. (2017). Ectopic fatty acid-binding protein 4 expression in the vascular endothelium is involved in neointima formation after vascular injury. *Journal of American Heart Association*, *6*(9). https://doi.org/10.1161/JAHA.117.006377.

Gharpure, K. M., Pradeep, S., Sans, M., Rupaimoole, R., Ivan, C., Wu, S. Y., et al. (2018). FABP4 as a key determinant of metastatic potential of ovarian cancer. *Nature Communications*, *9*(1), 2923. https://doi.org/10.1038/s41467-018-04987-y.

Ghelfi, E., Yu, C.-W., Elmasri, H., Terwelp, M., Lee, C. G., Bhandari, V., et al. (2013). Fatty acid binding protein 4 regulates VEGF-induced airway angiogenesis and inflammation in a transgenic mouse model: Implications for asthma. *The American Journal of Pathology*, *182*(4), 1425–1433. https://doi.org/10.1016/j.ajpath.2012.12.009.

Girona, J., Rosales, R., Plana, N., Saavedra, P., Masana, L., & Vallve, J. C. (2013). FABP4 induces vascular smooth muscle cell proliferation and migration through a MAPK-dependent pathway. *PLoS One*, *8*(11), e81914. https://doi.org/10.1371/journal.pone.0081914.

Gregor, M. F., & Hotamisligil, G. S. (2011). Inflammatory mechanisms in obesity. *Annual Review of Immunology*, *29*, 415–445. https://doi.org/10.1146/annurev-immunol-031210-101322.

Guaita-Esteruelas, S., Saavedra-Garcia, P., Bosquet, A., Borras, J., Girona, J., Amiliano, K., et al. (2017). Adipose-derived fatty acid-binding proteins plasma concentrations are increased in breast cancer patients. *Oncologist*, *22*(11), 1309–1315. https://doi.org/10.1634/theoncologist.2016-0483.

Ha, M. K., Soo Cho, J., Baik, O. R., Lee, K. H., Koo, H. S., & Chung, K. Y. (2006). Caenorhabditis elegans as a screening tool for the endothelial cell-derived putative aging-related proteins detected by proteomic analysis. *Proteomics*, *6*(11), 3339–3351. https://doi.org/10.1002/pmic.200500395.

Haastrup, B., Gill, S., Kristensen, S. R., Jorgensen, P. J., Glatz, J. F., Haghfelt, T., et al. (2000). Biochemical markers of ischaemia for the early identification of acute myocardial infarction without St segment elevation. *Cardiology*, *94*(4), 254–261. https://doi.org/10.1159/000047326.

Hancke, K., Grubeck, D., Hauser, N., Kreienberg, R., & Weiss, J. M. (2010). Adipocyte fatty acid-binding protein as a novel prognostic factor in obese breast cancer patients. *Breast Cancer Research and Treatment*, *119*(2), 367. https://doi.org/10.1007/s10549-009-0577-9.

Hardman, W. E. (2004). (n-3) Fatty acids and cancer therapy. *The Journal of Nutrition, 134*(12 Suppl), 3427S–3430S. Retrieved from http://www.ncbi.nlm.nih.gov/pubmed/15570049.

Harjes, U., Bridges, E., McIntyre, A., Fielding, B. A., & Harris, A. L. (2014). Fatty acid-binding protein 4, a point of convergence for angiogenic and metabolic signaling pathways in endothelial cells. *Journal of Biological Chemistry, 289*(33), 23168–23176. https://doi.org/10.1074/jbc.M114.576512.

Hashimoto, T., Kusakabe, T., Sugino, T., Fukuda, T., Watanabe, K., Sato, Y., et al. (2004). Expression of heart-type fatty acid-binding protein in human gastric carcinoma and its association with tumor aggressiveness, metastasis and poor prognosis. *Pathobiology, 71*(5), 267–273. https://doi.org/10.1159/000080061.

Hashimoto, T., Kusakabe, T., Watanabe, K., Sugino, T., Fukuda, T., Nashimoto, A., et al. (2004). Liver-type fatty acid-binding protein is highly expressed in intestinal metaplasia and in a subset of carcinomas of the stomach without association with the fatty acid synthase status in the carcinoma. *Pathobiology, 71*(3), 115–122. https://doi.org/10.1159/000076465.

Hertzel, A. V., Bennaars-Eiden, A., & Bernlohr, D. A. (2002). Increased lipolysis in transgenic animals overexpressing the epithelial fatty acid binding protein in adipose cells. *Journal of Lipid Research, 43*(12), 2105–2111. https://doi.org/10.1194/jlr.m200227-jlr200.

Hertzel, A. V., & Bernlohr, D. A. (2000). The mammalian fatty acid-binding protein multigene family: Molecular and genetic insights into function. *Trends in Endocrinology and Metabolism, 11*(5), 175–180. https://doi.org/10.1016/s1043-2760(00)00257-5.

Hotamisligil, G. S. (2006). Inflammation and metabolic disorders. *Nature, 444*(7121), 860–867. https://doi.org/10.1038/nature05485.

Hotamisligil, G. S. (2017). Inflammation, metaflammation and immunometabolic disorders. *Nature, 542*(7640), 177–185. https://doi.org/10.1038/nature21363.

Hotamisligil, G. S., & Bernlohr, D. A. (2015). Metabolic functions of FABPs—Mechanisms and therapeutic implications. *Nature Reviews in Endocrinology, 11*(10), 592–605. https://doi.org/10.1038/nrendo.2015.122.

Hotamisligil, G. S., Johnson, R. S., Distel, R. J., Ellis, R., Papaioannou, V. E., & Spiegelman, B. M. (1996). Uncoupling of obesity from insulin resistance through a targeted mutation in aP2, the adipocyte fatty acid binding protein. *Science, 274*(5291), 1377–1379. Retrieved from http://www.ncbi.nlm.nih.gov/pubmed/8910278.

Hsu, B. G., Chen, Y. C., Lee, R. P., Lee, C. C., Lee, C. J., & Wang, J. H. (2010). Fasting serum level of fatty-acid-binding protein 4 positively correlates with metabolic syndrome in patients with coronary artery disease. *Circulation Journal, 74*(2), 327–331. https://doi.org/10.1253/circj.cj-09-0568.

Huang, M., Narita, S., Inoue, T., Koizumi, A., Saito, M., Tsuruta, H., et al. (2017). Fatty acid binding protein 4 enhances prostate cancer progression by upregulating matrix metalloproteinases and stromal cell cytokine production. *Oncotarget, 8*(67), 111780–111794. https://doi.org/10.18632/oncotarget.22908.

Inoue, M., Takahashi, Y., Fujii, T., Kitagawa, M., & Fukusato, T. (2014). Significance of downregulation of liver fatty acid-binding protein in hepatocellular carcinoma. *World Journal of Gastroenterology, 20*(46), 17541–17551. https://doi.org/10.3748/wjg.v20.i46.17541.

Ishimura, S., Furuhashi, M., Watanabe, Y., Hoshina, K., Fuseya, T., Mita, T., et al. (2013). Circulating levels of fatty acid-binding protein family and metabolic phenotype in the general population. *PLoS One, 8*(11), e81318. https://doi.org/10.1371/journal.pone.0081318.

Iso, T., Maeda, K., Hanaoka, H., Suga, T., Goto, K., Syamsunarno, M. R., et al. (2013). Capillary endothelial fatty acid binding proteins 4 and 5 play a critical role in fatty acid uptake in heart and skeletal muscle. *Arteriosclerosis, Thrombosis, and Vascular Biology, 33*(11), 2549–2557. https://doi.org/10.1161/ATVBAHA.113.301588.

Jabs, M., Rose, A. J., Lehmann, L. H., Taylor, J., Moll, I., Sijmonsma, T. P., et al. (2018). Inhibition of endothelial notch signaling impairs fatty acid transport and leads to metabolic and vascular remodeling of the adult heart. *Circulation, 137*(24), 2592–2608. https://doi.org/10.1161/CIRCULATIONAHA.117.029733.

Jump, D. B., & Clarke, S. D. (1999). Regulation of gene expression by dietary fat. *Annual Review of Nutrition, 19*, 63–90. https://doi.org/10.1146/annurev.nutr.19.1.63.

Kaloshi, G., Mokhtari, K., Carpentier, C., Taillibert, S., Lejeune, J., Marie, Y., et al. (2007). FABP7 expression in glioblastomas: Relation to prognosis, invasion and EGFR status. *Journal of Neurooncology, 84*(3), 245–248. https://doi.org/10.1007/s11060-007-9377-4.

Kamijo-Ikemori, A., Sugaya, T., & Kimura, K. (2014). L-type fatty acid binding protein (L-FABP) and kidney disease. *Rinsho Byori, 62*(2), 163–170. Retrieved from https://www.ncbi.nlm.nih.gov/pubmed/24800492.

Karbek, B., Ozbek, M., Bozkurt, N. C., Ginis, Z., Gungunes, A., Unsal, I. O., et al. (2011). Heart-type fatty acid binding protein (H-FABP): Relationship with arterial intima-media thickness and role as diagnostic marker for atherosclerosis in patients with impaired glucose metabolism. *Cardiovascular Diabetology, 10*, 37. https://doi.org/10.1186/1475-2840-10-37.

Kato, I., Land, S., Majumdar, A. P., Barnholtz-Sloan, J., & Severson, R. K. (2010). Functional polymorphisms to modulate luminal lipid exposure and risk of colorectal cancer. *Cancer Epidemiology, 34*(3), 291–297. https://doi.org/10.1016/j.canep.2010.02.010.

Kawaguchi, K., Senga, S., Kubota, C., Kawamura, Y., Ke, Y., & Fujii, H. (2016). High expression of fatty acid-binding protein 5 promotes cell growth and metastatic potential of colorectal cancer cells. *FEBS Open Biology, 6*(3), 190–199. https://doi.org/10.1002/2211-5463.12031.

Khan, S. H., & Sorof, S. (1990). Preferential binding of growth inhibitory prostaglandins by the target protein of a carcinogen. *Proceedings of the National Academy of Sciences of the United States of America, 87*(23), 9401–9405. https://doi.org/10.1073/pnas.87.23.9401.

Kim, Y., & Kim, J. (2020). N-6 polyunsaturated fatty acids and risk of cancer: Accumulating evidence from prospective studies. *Nutrients, 12*(9), 2523. Retrieved from https://www.mdpi.com/2072-6643/12/9/2523.

Kim, S., Lee, Y., & Koo, J. S. (2015). Differential expression of lipid metabolism-related proteins in different breast cancer subtypes. *PLoS One, 10*(3), e0119473. https://doi.org/10.1371/journal.pone.0119473.

Knowlton, A. A., Apstein, C. S., Saouf, R., & Brecher, P. (1989). Leakage of heart fatty acid binding protein with ischemia and reperfusion in the rat. *Journal of Molecular and Cellular Cardiology, 21*(6), 577–583. Retrieved from http://www.ncbi.nlm.nih.gov/pubmed/2778809.

Ku, C. Y., Liu, Y. H., Lin, H. Y., Lu, S. C., & Lin, J. Y. (2016). Liver fatty acid-binding protein (L-FABP) promotes cellular angiogenesis and migration in hepatocellular carcinoma. *Oncotarget*. https://doi.org/10.18632/oncotarget.7571.

Kwon, O. J., Zhang, B., Zhang, L., & Xin, L. (2016). High fat diet promotes prostatic basal-to-luminal differentiation and accelerates initiation of prostate epithelial hyperplasia originated from basal cells. *Stem Cell Research, 16*(3), 682–691. https://doi.org/10.1016/j.scr.2016.04.009.

Kwong, S. C., Jamil, A. H. A., Rhodes, A., Taib, N. A., & Chung, I. (2019). Metabolic role of fatty acid binding protein 7 in mediating triple-negative breast cancer cell death via PPAR-α signaling. *Journal of Lipid Research, 60*(11), 1807–1817. https://doi.org/10.1194/jlr.M092379.

Lamounier-Zepter, V., Look, C., Alvarez, J., Christ, T., Ravens, U., Schunck, W. H., et al. (2009). Adipocyte fatty acid-binding protein suppresses cardiomyocyte contraction:

A new link between obesity and heart disease. *Circulation Research, 105*(4), 326–334. https://doi.org/10.1161/CIRCRESAHA.109.200501.

Lawrie, L. C., Dundas, S. R., Curran, S., & Murray, G. I. (2004). Liver fatty acid binding protein expression in colorectal neoplasia. *British Journal of Cancer, 90*(10), 1955–1960. https://doi.org/10.1038/sj.bjc.6601828.

Leckie, R. L., Manuck, S. B., Bhattacharjee, N., Muldoon, M. F., Flory, J. M., & Erickson, K. I. (2014). Omega-3 fatty acids moderate effects of physical activity on cognitive function. *Neuropsychologia, 59*, 103–111. https://doi.org/10.1016/j.neuropsychologia.2014.04.018.

Lee, M. Y., Tse, H. F., Siu, C. W., Zhu, S. G., Man, R. Y., & Vanhoutte, P. M. (2007). Genomic changes in regenerated porcine coronary arterial endothelial cells. *Arteriosclerosis, Thrombosis, and Vascular Biology, 27*(11), 2443–2449. https://doi.org/10.1161/ATVBAHA.107.141705.

Lee, M. Y., Wang, Y., & Vanhoutte, P. M. (2010). Senescence of cultured porcine coronary arterial endothelial cells is associated with accelerated oxidative stress and activation of NFkB. *Journal of Vascular Research, 47*(4), 287–298. https://doi.org/10.1159/000265563.

Leroy, C., Tobin, K. A. R., Basak, S., Cathrine Staff, A., & Duttaroy, A. K. (2017). Fatty acid-binding protein3 expression in BeWo cells, a human placental choriocarcinoma cell line. *Prostaglandins, Leukotrienes, and Essential Fatty Acids, 120*, 1–7. https://doi.org/10.1016/j.plefa.2017.04.002.

Levi, A. J., Gatmaitan, Z., & Arias, I. M. (1969). Two hepatic cytoplasmic protein fractions, Y and Z, and their possible role in the hepatic uptake of bilirubin, sulfobromophthalein, and other anions. *Journal of Clinical Investigations, 48*(11), 2156–2167. https://doi.org/10.1172/JCI106182.

Li, B., Hao, J., Yan, X., Kong, M., & Sauter, E. R. (2019). A-FABP and oestrogens are independently involved in the development of breast cancer. *Adipocyte, 8*(1), 379–385. https://doi.org/10.1080/21623945.2019.1690827.

Li, Y. T., Tian, X. T., Wu, M. L., Zheng, X., Kong, Q. Y., Cheng, X. X., et al. (2018). Resveratrol suppresses the growth and enhances retinoic acid sensitivity of anaplastic thyroid cancer cells. *International Journal of Molecular Science, 19*(4). https://doi.org/10.3390/ijms19041030.

Liang, Y., Diehn, M., Watson, N., Bollen, A. W., Aldape, K. D., Nicholas, M. K., et al. (2005). Gene expression profiling reveals molecularly and clinically distinct subtypes of glioblastoma multiforme. *Proceedings of the National Academy of Sciences of the United Sttaes of America, 102*(16), 5814–5819. https://doi.org/10.1073/pnas.0402870102.

Linge, A., Kennedy, S., O'Flynn, D., Beatty, S., Moriarty, P., Henry, M., et al. (2012). Differential expression of fourteen proteins between uveal melanoma from patients who subsequently developed distant metastases versus those who did not. *Investigative Ophthalmology & Visual Science, 53*(8), 4634–4643. https://doi.org/10.1167/iovs.11-9019.

Liou, H. L., Kahn, P. C., & Storch, J. (2002). Role of the helical domain in fatty acid transfer from adipocyte and heart fatty acid-binding proteins to membranes: Analysis of chimeric proteins. *The Journal of Biological Chemistry, 277*(3), 1806–1815. https://doi.org/10.1074/jbc.M107987200.

Liu, R.-Z., Choi, W.-S., Jain, S., Dinakaran, D., Xu, X., Han, W. H., et al. (2020). The FABP12/PPARγ pathway promotes metastatic transformation by inducing epithelial-to-mesenchymal transition and lipid-derived energy production in prostate cancer cells. *Molecular Oncology, 14*(12), 3100–3120. https://doi.org/10.1002/1878-0261.12818.

Liu, R. Z., & Godbout, R. (2020). An amplified fatty acid-binding protein gene cluster in prostate cancer: Emerging roles in lipid metabolism and metastasis. *Cancers (Basel), 12*(12). https://doi.org/10.3390/cancers12123823.

Liu, R. Z., Graham, K., Glubrecht, D. D., Germain, D. R., Mackey, J. R., & Godbout, R. (2011). Association of FABP5 expression with poor survival in triple-negative breast cancer: Implication for retinoic acid therapy. *American Journal of Pathology*, *178*(3), 997–1008. https://doi.org/10.1016/j.ajpath.2010.11.075.

Liu, R. Z., Li, X., & Godbout, R. (2008). A novel fatty acid-binding protein (FABP) gene resulting from tandem gene duplication in mammals: Transcription in rat retina and testis. *Genomics*, *92*(6), 436–445. https://doi.org/10.1016/j.ygeno.2008.08.003.

Liu, F., Liu, W., Zhou, S., Yang, C., Tian, M., Jia, G., et al. (2020). Identification of FABP5 as an immunometabolic marker in human hepatocellular carcinoma. *Journal of Immunotherapy Cancer*, *8*(2). https://doi.org/10.1136/jitc-2019-000501.

Liu, R. Z., Monckton, E. A., & Godbout, R. (2012). Regulation of the FABP7 gene by PAX6 in malignant glioma cells. *Biochemical and Biophysical Research Communications*, *422*(3), 482–487. https://doi.org/10.1016/j.bbrc.2012.05.019.

Llaverias, G., Noe, V., Penuelas, S., Vazquez-Carrera, M., Sanchez, R. M., Laguna, J. C., et al. (2004). Atorvastatin reduces CD68, FABP4, and HBP expression in oxLDL-treated human macrophages. *Biochemical and Biophysical Research Communications*, *318*(1), 265–274. https://doi.org/10.1016/j.bbrc.2004.04.021.

Lopes-Coelho, F., Andre, S., Felix, A., & Serpa, J. (2018). Breast cancer metabolic cross-talk: Fibroblasts are hubs and breast cancer cells are gatherers of lipids. *Molecular and Cellular Endocrinology*, *462*(Pt. B), 93–106. https://doi.org/10.1016/j.mce.2017.01.031.

Madsen, P., Rasmussen, H. H., Leffers, H., Honore, B., & Celis, J. E. (1992). Molecular cloning and expression of a novel keratinocyte protein (psoriasis-associated fatty acid-binding protein [PA-FABP]) that is highly up-regulated in psoriatic skin and that shares similarity to fatty acid-binding proteins. *The Journal of Investigative Dermatology*, *99*(3), 299–305. Retrieved from http://www.ncbi.nlm.nih.gov/pubmed/1512466.

Maeda, K., Cao, H., Kono, K., Gorgun, C. Z., Furuhashi, M., Uysal, K. T., et al. (2005). Adipocyte/macrophage fatty acid binding proteins control integrated metabolic responses in obesity and diabetes. *Cell Metabolism*, *1*(2), 107–119. https://doi.org/10.1016/j.cmet.2004.12.008.

Maeda, K., Uysal, K. T., Makowski, L., Gorgun, C. Z., Atsumi, G., Parker, R. A., et al. (2003). Role of the fatty acid binding protein mal1 in obesity and insulin resistance. *Diabetes*, *52*(2), 300–307. https://doi.org/10.2337/diabetes.52.2.300.

Makowski, L., Boord, J. B., Maeda, K., Babaev, V. R., Uysal, K. T., Morgan, M. A., et al. (2001). Lack of macrophage fatty-acid-binding protein aP2 protects mice deficient in apolipoprotein E against atherosclerosis. *Nature Medicine*, 7(6), 699–705. https://doi.org/10.1038/89076.

Makowski, L., Brittingham, K. C., Reynolds, J. M., Suttles, J., & Hotamisligil, G. S. (2005). The fatty acid-binding protein, aP2, coordinates macrophage cholesterol trafficking and inflammatory activity. Macrophage expression of aP2 impacts peroxisome proliferator-activated receptor gamma and IkappaB kinase activities. *Journal of Biological Chemistry*, *280*(13), 12888–12895. https://doi.org/10.1074/jbc.M413788200.

Makowski, L., & Hotamisligil, G. S. (2005). The role of fatty acid binding proteins in metabolic syndrome and atherosclerosis. *Current Opinion in Lipidology*, *16*(5), 543–548. https://doi.org/10.1097/01.mol.0000180166.08196.07.

Mallick, R., Basak, S., & Duttaroy, A. K. (2021). Fatty acids and evolving roles of their proteins in neurological, cardiovascular disorders and cancers. *Progress in Lipid Research*, *101116*. https://doi.org/10.1016/j.plipres.2021.101116.

Mallick, R., & Duttaroy, A. K. (2022). Modulation of endothelium function by fatty acids. *Molecular and Cellular Biochemistry*, *477*(1), 15–38. https://doi.org/10.1007/s11010-021-04260-9.

Masouye, I., Hagens, G., Van Kuppevelt, T. H., Madsen, P., Saurat, J. H., Veerkamp, J. H., et al. (1997). Endothelial cells of the human microvasculature express epidermal fatty acid-binding protein. *Circulation Research*, *81*(3), 297–303. https://doi.org/10.1161/01.res.81.3.297.

Melki, S. A., & Abumrad, N. A. (1993). Expression of the adipocyte fatty acid-binding protein in streptozotocin-diabetes: Effects of insulin deficiency and supplementation. *Journal of Lipid Research*, *34*(9), 1527–1534. Retrieved from http://www.ncbi.nlm.nih.gov/pubmed/7693843.

Miyoshi, T., Onoue, G., Hirohata, A., Hirohata, S., Usui, S., Hina, K., et al. (2010). Serum adipocyte fatty acid-binding protein is independently associated with coronary atherosclerotic burden measured by intravascular ultrasound. *Atherosclerosis*, *211*(1), 164–169. https://doi.org/10.1016/j.atherosclerosis.2010.01.032.

Mohlig, M., Weickert, M. O., Ghadamgadai, E., Machlitt, A., Pfuller, B., Arafat, A. M., et al. (2007). Adipocyte fatty acid-binding protein is associated with markers of obesity, but is an unlikely link between obesity, insulin resistance, and hyperandrogenism in polycystic ovary syndrome women. *European Journal of Endocrinology*, *157*(2), 195–200. https://doi.org/10.1530/EJE-07-0102.

Moradi, A., Forootan, F. S., Hosseini, M., & Pouresmaeili, F. (2019). Leydig cells express the FABP9 in human testis. *Human Antibodies*, *27*(4), 275–278. https://doi.org/10.3233/HAB-190382.

Motojima, K. (2000). Differential effects of PPARalpha activators on induction of ectopic expression of tissue-specific fatty acid binding protein genes in the mouse liver. *The International Journal of Biochemistry & Cell Biology*, *32*(10), 1085–1092. https://doi.org/10.1016/s1357-2725(00)00046-7.

Mullard, A. (2014). New drugs cost US[dollar]2.6 billion to develop. *Nature Reviews Drug Discovery*, *13*(12), 877. https://doi.org/10.1038/nrd4507.

Nagao, K., Shinohara, N., Smit, F., de Weijert, M., Jannink, S., Owada, Y., et al. (2018). Fatty acid binding protein 7 may be a marker and therapeutic targets in clear cell renal cell carcinoma. *BMC Cancer*, *18*(1), 1114. https://doi.org/10.1186/s12885-018-5060-8.

Nieman, K. M., Kenny, H. A., Penicka, C. V., Ladanyi, A., Buell-Gutbrod, R., Zillhardt, M. R., et al. (2011). Adipocytes promote ovarian cancer metastasis and provide energy for rapid tumor growth. *Nature Medicine*, *17*(11), 1498–1503. https://doi.org/10.1038/nm.2492.

Ockner, R. K., Manning, J. A., Poppenhausen, R. B., & Ho, W. K. (1972). A binding protein for fatty acids in cytosol of intestinal mucosa, liver, myocardium, and other tissues. *Science*, *177*(4043), 56–58. https://doi.org/10.1126/science.177.4043.56.

Ohata, T., Yokoo, H., Kamiyama, T., Fukai, M., Aiyama, T., Hatanaka, Y., et al. (2017). Fatty acid-binding protein 5 function in hepatocellular carcinoma through induction of epithelial-mesenchymal transition. *Cancer Medicine*, *6*(5), 1049–1061. https://doi.org/10.1002/cam4.1020.

Ohlsson, G., Moreira, J. M., Gromov, P., Sauter, G., & Celis, J. E. (2005). Loss of expression of the adipocyte-type fatty acid-binding protein (A-FABP) is associated with progression of human urothelial carcinomas. *Molecular Cell Proteomics*, *4*. https://doi.org/10.1074/mcp.M500017-MCP200.

Ohmachi, T., Inoue, H., Mimori, K., Tanaka, F., Sasaki, A., Kanda, T., et al. (2006). Fatty acid binding protein 6 is overexpressed in colorectal cancer. *Clinical Cancer Research*, *12*(17), 5090–5095. https://doi.org/10.1158/1078-0432.ccr-05-2045.

Ostergaard, M., Rasmussen, H. H., Nielsen, H. V., Vorum, H., Orntoft, T. F., & Wolf, H. (1997). Proteome profiling of bladder squamous cell carcinomas: identification of markers that define their degree of differentiation. *Cancer Research*, *57*.

O'Sullivan, S. E., & Kaczocha, M. (2020). FABP5 as a novel molecular target in prostate cancer. *Drug Discovery Today*. https://doi.org/10.1016/j.drudis.2020.09.018.

Owada, Y., Suzuki, I., Noda, T., & Kondo, H. (2002). Analysis on the phenotype of E-FABP-gene knockout mice. *Molecular Cell Biochemistry*, *239*(1-2), 83–86. Retrieved from https://www.ncbi.nlm.nih.gov/pubmed/12479572.

Pang, J., Liu, W.-P., Liu, X.-P., Li, L.-Y., Fang, Y.-Q., Sun, Q.-P., et al. (2010). Profiling protein markers associated with lymph node metastasis in prostate cancer by DIGE-based proteomics analysis. *Journal of Proteome Research*, *9*(1), 216–226. https://doi.org/10.1021/pr900953s.

Parikesit, D., Mochtar, C. A., Umbas, R., & Hamid, A. R. (2016). The impact of obesity towards prostate diseases. *Prostate International*, *4*(1), 1–6. https://doi.org/10.1016/j.prnil.2015.08.001.

Payne, C. M., Bernstein, C., Dvorak, K., & Bernstein, H. (2008). Hydrophobic bile acids, genomic instability, Darwinian selection, and colon carcinogenesis. *Clinical and Experimental Gastroenterology*, *1*, 19–47. https://doi.org/10.2147/ceg.s4343.

Petan, T., Jarc, E., & Jusović, M. (2018). Lipid droplets in cancer: Guardians of fat in a stressful world. *Molecules*, *23*(8). https://doi.org/10.3390/molecules23081941.

Phuchareon, J., Overdevest, J. B., McCormick, F., Eisele, D. W., van Zante, A., & Tetsu, O. (2014). Fatty acid binding protein 7 is a molecular marker in adenoid cystic carcinoma of the salivary glands: Implications for clinical significance. *Translational Oncology*, 7(6), 780–787. https://doi.org/10.1016/j.tranon.2014.10.003.

Powell, C. A., Nasser, M. W., Zhao, H., Wochna, J. C., Zhang, X., Shapiro, C., et al. (2015). Fatty acid binding protein 5 promotes metastatic potential of triple negative breast cancer cells through enhancing epidermal growth factor receptor stability. *Oncotarget*, *6*(8), 6373–6385. https://doi.org/10.18632/oncotarget.3442.

Prentice, K. J., Saksi, J., & Hotamisligil, G. S. (2019). Adipokine FABP4 integrates energy stores and counterregulatory metabolic responses. *Journal of Lipid Research*, *60*(4), 734–740. https://doi.org/10.1194/jlr.S091793.

Queipo-Ortuno, M. I., Escote, X., Ceperuelo-Mallafre, V., Garrido-Sanchez, L., Miranda, M., Clemente-Postigo, M., et al. (2012). FABP4 dynamics in obesity: Discrepancies in adipose tissue and liver expression regarding circulating plasma levels. *PLoS One*, 7(11), e48605. https://doi.org/10.1371/journal.pone.0048605.

Rawla, P. (2019). Epidemiology of prostate cancer. *World Journal of Oncology*, *10*(2), 63–89. https://doi.org/10.14740/wjon1191.

Rhee, E. J., Lee, W. Y., Park, C. Y., Oh, K. W., Kim, B. J., Sung, K. C., et al. (2009). The association of serum adipocyte fatty acid-binding protein with coronary artery disease in Korean adults. *European Journal of Endocrinology*, *160*(2), 165–172. https://doi.org/10.1530/EJE-08-0665.

Saavedra, P., Girona, J., Bosquet, A., Guaita, S., Canela, N., Aragones, G., et al. (2015). New insights into circulating FABP4: Interaction with cytokeratin 1 on endothelial cell membranes. *Biochimica et Biophysica Acta*, *1853*(11 Pt. A), 2966–2974. https://doi.org/10.1016/j.bbamcr.2015.09.002.

Sacchettini, J. C., Gordon, J. I., & Banaszak, L. J. (1989). Crystal structure of rat intestinal fatty-acid-binding protein. Refinement and analysis of the Escherichia coli-derived protein with bound palmitate. *Journal of Molecular Biology*, *208*(2), 327–339. Retrieved from http://www.ncbi.nlm.nih.gov/pubmed/2671390.

Scheja, L., Makowski, L., Uysal, K. T., Wiesbrock, S. M., Shimshek, D. R., Meyers, D. S., et al. (1999). Altered insulin secretion associated with reduced lipolytic efficiency in aP2-/- mice. *Diabetes*, *48*(10), 1987–1994. https://doi.org/10.2337/diabetes.48.10.1987.

Scifres, C. M., Chen, B., Nelson, D. M., & Sadovsky, Y. (2011). Fatty acid binding protein 4 regulates intracellular lipid accumulation in human trophoblasts. *The Journal of Clinical Endocrinology & Metabolism*, *96*(7), E1083–E1091. https://doi.org/10.1210/jc.2010-2084.

Senga, S., Kobayashi, N., Kawaguchi, K., Ando, A., & Fujii, H. (2018). Fatty acid-binding protein 5 (FABP5) promotes lipolysis of lipid droplets, de novo fatty acid (FA) synthesis and activation of nuclear factor-kappa B (NF-κB) signaling in cancer cells. *Biochimica et*

Biophysica Acta: Molecular and Cell Biology of Lipids, 1863(9), 1057–1067. https://doi.org/10.1016/j.bbalip.2018.06.010.

Seo, J., Jeong, D. W., Park, J. W., Lee, K. W., Fukuda, J., & Chun, Y. S. (2020). Fatty-acid--induced FABP5/HIF-1 reprograms lipid metabolism and enhances the proliferation of liver cancer cells. *Communications Biology, 3*(1), 638. https://doi.org/10.1038/s42003-020-01367-5.

Shaughnessy, S., Smith, E. R., Kodukula, S., Storch, J., & Fried, S. K. (2000). Adipocyte metabolism in adipocyte fatty acid binding protein knockout mice (aP2-/-) after short-term high-fat feeding: Functional compensation by the keratinocyte [correction of keritinocyte] fatty acid binding protein. *Diabetes, 49*(6), 904–911. https://doi.org/10.2337/diabetes.49.6.904.

Shum, B. O., Mackay, C. R., Gorgun, C. Z., Frost, M. J., Kumar, R. K., Hotamisligil, G. S., et al. (2006). The adipocyte fatty acid-binding protein aP2 is required in allergic airway inflammation. *Journal of Clinical Investigations, 116*(8), 2183–2192. https://doi.org/10.1172/JCI24767.

Simpson, M. A., LiCata, V. J., Ribarik Coe, N., & Bernlohr, D. A. (1999). Biochemical and biophysical analysis of the intracellular lipid binding proteins of adipocytes. *Molecular and Cellular Biochemistry, 192*(1-2), 33–40. Retrieved from https://www.ncbi.nlm.nih.gov/pubmed/10331656.

Sitar-Taut, A. V., Cozma, A., Fodor, A., Coste, S. C., Orasan, O. H., Negrean, V., et al. (2021). New insights on the relationship between leptin, ghrelin, and leptin/ghrelin ratio enforced by body mass index in obesity and diabetes. *Biomedicines, 9*(11). https://doi.org/10.3390/biomedicines9111657.

Slipicevic, A., Jorgensen, K., Skrede, M., Rosnes, A. K., Troen, G., Davidson, B., et al. (2008). The fatty acid binding protein 7 (FABP7) is involved in proliferation and invasion of melanoma cells. *BMC Cancer, 8*, 276. https://doi.org/10.1186/1471-2407-8-276.

Soteriou, C., Kalli, A. C., Connell, S. D., Tyler, A. I. I., & Thorne, J. L. (2021). Advances in understanding and in multi-disciplinary methodology used to assess lipid regulation of signalling cascades from the cancer cell plasma membrane. *Progress in Lipid Research, 81*, 101080. https://doi.org/10.1016/j.plipres.2020.101080.

Stejskal, D., & Karpisek, M. (2006). Adipocyte fatty acid binding protein in a Caucasian population: a new marker of metabolic syndrome? *European Journal of Clinical Investigations, 36*(9), 621–625. https://doi.org/10.1111/j.1365-2362.2006.01696.x.

Storch, J., & McDermott, L. (2009). Structural and functional analysis of fatty acid-binding proteins. *Journal of Lipid Research, 50*(Suppl), S126–S131. https://doi.org/10.1194/jlr.R800084-JLR200.

Suzuki, T., Watanabe, K., & Ono, T. (1990). Immunohistochemical demonstration of liver fatty acid-binding protein in human hepatocellular malignancies. *Journal of Pathology, 161*(1), 79–83. https://doi.org/10.1002/path.1711610113.

Syed, M. H., Zamzam, A., Khan, H., Singh, K., Forbes, T. L., Rotstein, O., et al. (2020). Fatty acid binding protein 3 is associated with peripheral arterial disease. *JVS-Vascular Science, 1*, 168–175. https://doi.org/10.1016/j.jvssci.2020.08.003.

Takahashi-Shishido, N., Sugaya, M., Morimura, S., Suga, H., Oka, T., Kamijo, H., et al. (2021). Mycosis fungoides and Sezary syndrome tumor cells express epidermal fatty acid-binding protein, whose expression decreases with loss of epidermotropism. *Journal of Dermatology, 48*(5), 685–689. https://doi.org/10.1111/1346-8138.15775.

Tang, Z., Shen, Q., Xie, H., Zhou, X., Li, J., Feng, J., et al. (2016). Elevated expression of FABP3 and FABP4 cooperatively correlates with poor prognosis in non-small cell lung cancer (NSCLC). *Oncotarget*, 7(29), 46253–46262. https://doi.org/10.18632/oncotarget.10086.

Tang, X. Y., Umemura, S., Tsukamoto, H., Kumaki, N., Tokuda, Y., & Osamura, R. Y. (2010). Overexpression of fatty acid binding protein-7 correlates with basal-like subtype of breast cancer. *Pathology, Research and Practice, 206*(2), 98–101. https://doi.org/10.1016/j.prp.2009.06.010.

Thompson, K. J., Austin, R. G., Nazari, S. S., Gersin, K. S., Iannitti, D. A., & McKillop, I. H. (2018). Altered fatty acid-binding protein 4 (FABP4) expression and function in human and animal models of hepatocellular carcinoma. *Liver International, 38*(6), 1074–1083. https://doi.org/10.1111/liv.13639.

Thumser, A. E., & Wilton, D. C. (1996). The binding of cholesterol and bile salts to recombinant rat liver fatty acid-binding protein. *Biochemistry Journal, 320*(Pt. 3), 729–733. Retrieved from http://www.ncbi.nlm.nih.gov/pubmed/9003356.

Thygesen, K., Alpert, J. S., Jaffe, A. S., Simoons, M. L., Chaitman, B. R., White, H. D., et al. (2012). Third universal definition of myocardial infarction. *Journal of the American College of Cardiology, 60*(16), 1581–1598. https://doi.org/10.1016/j.jacc.2012.08.001.

Tian, W., Zhang, W., Zhang, Y., Zhu, T., Hua, Y., Li, H., et al. (2020). FABP4 promotes invasion and metastasis of colon cancer by regulating fatty acid transport. *Cancer Cell International, 20*, 512. https://doi.org/10.1186/s12935-020-01582-4.

Tolle, A., Jung, M., Lein, M., Johannsen, M., Miller, K., Moch, H., et al. (2009). Brain-type and liver-type fatty acid-binding proteins: New tumor markers for renal cancer? *BMC Cancer, 9*, 248. https://doi.org/10.1186/1471-2407-9-248.

Tölle, A., Suhail, S., Jung, M., Jung, K., & Stephan, C. (2011). Fatty acid binding proteins (FABPs) in prostate, bladder and kidney cancer cell lines and the use of IL-FABP as survival predictor in patients with renal cell carcinoma. *BMC Cancer, 11*(1), 1–10. https://doi.org/10.1186/1471-2407-11-302.

Tsai, I. T., Wu, C. C., Hung, W. C., Lee, T. L., Hsuan, C. F., Wei, C. T., et al. (2020). FABP1 and FABP2 as markers of diabetic nephropathy. *International Journal of Medical Sciences, 17*(15), 2338–2345. https://doi.org/10.7150/ijms.49078.

Uehara, H., Takahashi, T., Oha, M., Ogawa, H., & Izumi, K. (2014). Exogenous fatty acid binding protein 4 promotes human prostate cancer cell progression. *International Journal of Cancer, 135*(11), 2558–2568. https://doi.org/10.1002/ijc.28903.

Uysal, K. T., Scheja, L., Wiesbrock, S. M., Bonner-Weir, S., & Hotamisligil, G. S. (2000). Improved glucose and lipid metabolism in genetically obese mice lacking aP2. *Endocrinology, 141*(9), 3388–3396. https://doi.org/10.1210/endo.141.9.7637.

Wang, W., Chu, H. J., Liang, Y. C., Huang, J. M., Shang, C. L., Tan, H., et al. (2016). FABP5 correlates with poor prognosis and promotes tumor cell growth and metastasis in cervical cancer. *Tumour Biology, 37*(11), 14873–14883. https://doi.org/10.1007/s13277-016-5350-1.

Wang, B., Tao, X., Huang, C. Z., Liu, J. F., Ye, Y. B., & Huang, A. M. (2014). Decreased expression of liver-type fatty acid-binding protein is associated with poor prognosis in hepatocellular carcinoma. *Hepatogastroenterology, 61*(133), 1321–1326. Retrieved from https://www.ncbi.nlm.nih.gov/pubmed/25436304.

Wood, S. M., Gill, A. J., Brodsky, A. S., Lu, S., Friedman, K., Karashchuk, G., et al. (2017). Fatty acid-binding protein 1 is preferentially lost in microsatellite instable colorectal carcinomas and is immune modulated via the interferon γ pathway. *Modern Pathology, 30*(1), 123–133. https://doi.org/10.1038/modpathol.2016.170.

Wu, X., Deng, F., Li, Y., Daniels, G., Du, X., Ren, Q., et al. (2015). ACSL4 promotes prostate cancer growth, invasion and hormonal resistance. *Oncotarget, 6*(42), 44849–44863. https://doi.org/10.18632/oncotarget.6438.

Wu, G., Xu, Y., Wang, Q., Li, J., Li, L., Han, C., et al. (2019). FABP5 is correlated with poor prognosis and promotes tumour cell growth and metastasis in clear cell renal cell carcinoma. *European Journal of Pharmacology, 862*, 172637. https://doi.org/10.1016/j.ejphar.2019.172637.

Xiao, Y., Yao, L., Li, X., Zhong, H., Chen, X. Y., Tang, W. L., et al. (2010). Relationship of adipocyte fatty acid-binding protein to adiponectin ratio with femoral intima-media thickness and endothelium-dependent vasodilation in patients with newly-diagnosed type 2 diabetes mellitus. *Zhonghua Yi Xue Za Zhi*, *90*(4), 231–235. Retrieved from https://www.ncbi.nlm.nih.gov/pubmed/20356535.

Xu, H., Barnes, G. T., Yang, Q., Tan, G., Yang, D., Chou, C. J., et al. (2003). Chronic inflammation in fat plays a crucial role in the development of obesity-related insulin resistance. *Journal of Clinical Investigations*, *112*(12), 1821–1830. https://doi.org/10.1172/JCI19451.

Xu, H., Hertzel, A. V., Steen, K. A., Wang, Q., Suttles, J., & Bernlohr, D. A. (2015). Uncoupling lipid metabolism from inflammation through fatty acid binding protein-dependent expression of UCP2. *Molecular and Cellular Biology*, *35*(6), 1055–1065. https://doi.org/10.1128/MCB.01122-14.

Xu, A., Tso, A. W., Cheung, B. M., Wang, Y., Wat, N. M., Fong, C. H., et al. (2007). Circulating adipocyte-fatty acid binding protein levels predict the development of the metabolic syndrome: A 5-year prospective study. *Circulation*, *115*(12), 1537–1543. https://doi.org/10.1161/circulationaha.106.647503.

Xu, A., Wang, Y., Xu, J. Y., Stejskal, D., Tam, S., Zhang, J., et al. (2006). Adipocyte fatty acid-binding protein is a plasma biomarker closely associated with obesity and metabolic syndrome. *Clinical Chemistry*, *52*(3), 405–413. https://doi.org/10.1373/clinchem.2005.062463.

Yeung, D. C., Wang, Y., Xu, A., Cheung, S. C., Wat, N. M., Fong, D. Y., et al. (2008). Epidermal fatty-acid-binding protein: A new circulating biomarker associated with cardio-metabolic risk factors and carotid atherosclerosis. *European Heart Journal*, *29*(17), 2156–2163. https://doi.org/10.1093/eurheartj/ehn295.

Yeung, D. C., Xu, A., Tso, A. W., Chow, W. S., Wat, N. M., Fong, C. H., et al. (2009). Circulating levels of adipocyte and epidermal fatty acid-binding proteins in relation to nephropathy staging and macrovascular complications in type 2 diabetic patients. *Diabetes Care*, *32*(1), 132–134. https://doi.org/10.2337/dc08-1333.

Yoo, H. J., Kim, S., Park, M. S., Choi, H. Y., Yang, S. J., Seo, J. A., et al. (2011). Serum adipocyte fatty acid-binding protein is associated independently with vascular inflammation: analysis with (18)F-fluorodeoxyglucose positron emission tomography. *The Journal of Clinical Endocrinology and Metabolism*, *96*(3), E488–E492. https://doi.org/10.1210/jc.2010-1473.

Yu, C. W., Liang, X., Lipsky, S., Karaaslan, C., Kozakewich, H., Hotamisligil, G. S., et al. (2016). Dual role of fatty acid-binding protein 5 on endothelial cell fate: A potential link between lipid metabolism and angiogenic responses. *Angiogenesis*, *19*(1), 95–106. https://doi.org/10.1007/s10456-015-9491-4.

Zeng, J., Sauter, E. R., & Li, B. (2020). FABP4: A new player in obesity-associated breast cancer. *Trends in Molecular Medicine*, *26*(5), 437–440. https://doi.org/10.1016/j.molmed.2020.03.004.

Zhang, Y., Hao, J., Zeng, J., Li, Q., Rao, E., Sun, Y., et al. (2018). Epidermal FABP prevents chemical-induced skin tumorigenesis by regulation of TPA-induced IFN/p53/SOX2 pathway in keratinocytes. *The Journal of Investigative Dermatology*, *138*(9), 1925–1934. https://doi.org/10.1016/j.jid.2018.02.041.

Zhang, C., Liao, Y., Liu, P., Du, Q., Liang, Y., Ooi, S., et al. (2020). FABP5 promotes lymph node metastasis in cervical cancer by reprogramming fatty acid metabolism. *Theranostics*, *10*(15), 6561–6580. https://doi.org/10.7150/thno.44868.

Zhang, H., Rakha, E. A., Ball, G. R., Spiteri, I., Aleskandarany, M., Paish, E. C., et al. (2010). The proteins FABP7 and OATP2 are associated with the basal phenotype and patient outcome in human breast cancer. *Breast Cancer Research Treatment*, *121*(1), 41–51. https://doi.org/10.1007/s10549-009-0450-x.

Zhang, Y., Zhao, X., Deng, L., Li, X., Wang, G., Li, Y., et al. (2019). High expression of FABP4 and FABP6 in patients with colorectal cancer. *World Journal of Surgical Oncology*, *17*(1), 171. https://doi.org/10.1186/s12957-019-1714-5.

Zhao, H., Pflug, B. R., Lai, X., & Wang, M. (2016). Metabolic and molecular regulation of dietary polyunsaturated fatty acids on prostate cancer. *Proteomics. Clinical Applications*, *10*(3), 267–279. https://doi.org/10.1002/prca.201500066.

Zhuang, L., Li, C., Chen, Q., Jin, Q., Wu, L., Lu, L., et al. (2019). Fatty acid-binding protein 3 contributes to ischemic heart injury by regulating cardiac myocyte apoptosis and MAPK pathways. *American Journal of Physiology. Heart and Circulatory Physiology*, *316*(5), H971–H984. https://doi.org/10.1152/ajpheart.00360.2018.

CHAPTER SIX

Protein expression in exocrine pancreatic diseases. Focus on VMP1 mediated autophagy

Maria I. Vaccaro[a,b,*], Florencia Mitchell[c], Francisco Rivera[c], and Claudio D. Gonzalez[b,c]

[a]University of Buenos Aires, CONICET, Institute for Biochemistry and Molecular Medicine, Department of Pathophysiology, Buenos Aires, Argentina
[b]CEMIC University Hospital, Unit of Translational Medical Research, Buenos Aires, Argentina
[c]CEMIC University Institute, School of Medicine, Department of Pharmacology, Buenos Aires, Argentina
*Corresponding author: e-mail address: mvaccaro@ffyb.uba.ar

Contents

Abstract

The exocrine pancreas produces enzymes involved in the digestive process whereas endocrine pancreas mainly regulates glucose metabolism. Diseases of the exocrine pancreas are characterized by high morbidity and mortality. Acute pancreatitis is a painful disease in which pancreatic secretory proteins are prematurely activated causing the digestion of the gland. Pancreatic adenocarcinoma is one of the most malignant cancers due to its resistance to treatment, its late diagnosis and high capacity for metastasis. Autophagy is a catabolic process that aims at degrading cytoplasmic contents and damaged organelles, to preserve cell viability and homeostasis. VMP1 is a transmembrane protein that plays a key role in triggering autophagy and being part of the autophagosome membrane. A specific type of selective autophagy pathway called zymophagy protects the pancreas against self-digestion in the setting of acute pancreatitis by sequestering intracellularly activated zymogen granules. Mitophagy is also

Advances in Protein Chemistry and Structural Biology, Volume 132
ISSN 1876-1623
https://doi.org/10.1016/bs.apcsb.2022.07.001

responsible for maintaining pancreatitis as a mild disease by preserving mitochondrial function. Dysregulation of these selective autophagic processes by pancreatitis itself constitutes a risk factor for development of severe disease. In pancreatic adenocarcinoma, VMP1 mediated autophagy promotes cancer cell survival and resistance to chemotherapy. Therefore, it is relevant to highlight a role for controlling VMP1 expression and targeting VMP1 molecular pathways to improve exocrine pancreatic diseases prognosis.

1. Introduction

The pancreas is a 6 in. gland with very relevant endocrine and exocrine functions. The exocrine pancreas produces enzymes involved in the digestive process whereas endocrine pancreas mainly regulates glucose metabolism. Autophagy is a catabolic cell process that aims at degrading dysfunctional proteins, damaged organelles and microbes, among others, to preserve cell viability and homeostasis. Normal autophagy is essential to maintain both exocrine and endocrine pancreatic integrity and physiology. Autophagy is a highly regulated process involving a set of biologically preserved transduction pathways. Autophagy has many triggers, mainly stressors such as starvation, hypoxia, inflammation, and microorganism invasion (Klionsky et al., 2021). Abnormalities in autophagy are associated with pancreatic diseases, including pancreatitis and pancreatic adenocarcinoma. Acute pancreatitis is a painful disease in which pancreatic secretory proteins are prematurely activated causing digestion of the gland (Lee & Papachristou, 2019). Pancreatic adenocarcinoma is one of the most malignant cancers due to its resistant to treatment, its late diagnosis and high capacity for metastasis (Tempero, 2019). Looking for new proteins that pancreatic cells produce during the acute phase of pancreatic inflammation we identified a novel pancreatitis associated transmembrane protein (Dusetti et al., 2002). We named this protein Vacuole Membrane Protein (VMP1) and first explored the association of VMP1 with acute pancreatitis (Vaccaro, Grasso, Ropolo, Iovanna, & Cerquetti, 2003). VMP1 is upregulated in acute pancreatitis as well as in chronic pancreatitis (Jiang et al., 2004). Several other investigations also showed a strong association between VMP1 and pancreatic adenocarcinoma. VMP1 regulation is also altered in diabetes mellitus, an endocrine disease associated with pancreatic beta cell dysfunction (Grasso et al., 2009). This body of evidence suggests a strong association between VMP1 and autophagy alterations. Consequently, VMP1 expression, regulation, and role in the pathophysiology of pancreas diseases through dysfunctional autophagy is a clear goal of investigational interest. In this chapter we focus on the role of autophagy mediated by VMP1 expression in exocrine pancreatic diseases such as acute and chronic pancreatitis and pancreatic adenocarcinoma and its potential relevance in the development of more effective and specific therapeutic approaches.

2. The exocrine pancreas

The pancreas acts as an exocrine and endocrine gland. The exocrine pancreas has as main functions secreting digestive enzymes, ions, and water into the gastrointestinal tract, particularly into the duodenum. Digestive enzymes are required for the breakdown of food into smaller, molecular components that can be absorbed via the epithelium that lines the gastrointestinal (GI) tract. Although food digestion begins with participation of enzymes from the salivary glands, stomach, and gastrointestinal tract's surface epithelium, the exocrine pancreas plays a predominant and vital role in the digestive process. This is such that when exocrine pancreatic function is lost, nutritional absorption is severely impaired, resulting in malnutrition.

Another important function of the exocrine pancreas consists of ion and water secretion. This is needed for optimum transport of digestive enzymes from pancreatic acinar cells to the lumen of the intestine where they act. Furthermore, the pH of pancreatic secretions is alkaline due to a large quantity of sodium bicarbonate ($NaHCO_3$) contained (pancreatic secretions can contain up to 140 mM of HCO_3^-). This alkaline composition plays a role in buffering the acidic pH of gastric chyme transported from the stomach to the intestine. A neutral pH in the gut lumen is required for correct epithelial function and proper digestion by enzymes (Ishiguro et al., 2012).

Growing evidence that food consumption has a direct regulatory effect on the release and appropriate supply of digestive enzymes and does not only work through neural circuits and hormone release, is emerging. Neurotransmitters, hormones, and nutrients themselves activate receptors and intracellular regulatory signaling pathways to activate the release of pancreatic enzymes for digestion. Great interest from different scientific fields of research revolves around the exocrine pancreas and its diverse range of processes and functions and has been doing so for a long time. The first evidence of hormone action was found in the pancreas in the early 20th century (Bayliss & Starling, 1902). It has been the most utilized organ for explaining mechanisms such as synthesis and transport of exportable proteins (Palade, 1975), as well as the signaling pathways involved in protein secretion

(Williams, 2001). Furthermore, the mechanisms underlying the exocrine pancreas' ability to secrete extremely high amounts of HCO_3^- have attracted the curiosity of physiologists and are currently being disputed (Pandol, 2010; Williams, 2010).

3. The pancreatic acinar cell

The exocrine pancreas' main functions revolve around creation, storage, and production of digestive enzymes. Pancreatitis is an inflammatory illness of the pancreas of variable severity. It can range from mild and self-limited to cause considerable morbidity and even death. It is understood that pancreatitis is caused by problems or alterations of the exocrine pancreas´ main functions described above. Acute pancreatitis is a frequent gastrointestinal disease, being in fact one of the leading causes of hospitalization in the US among GI disorders. It is widely assumed that the pancreatitis associated pathobiological events are caused by different cellular insults such as alcoholism, smoking, hyperlipidemia, and gallstones, among others. Despite the fact that this condition has been well defined and relatively well understood for a long time, therapy up to date has remained limited to supportive, and no real improvements have been made regarding targeting pathophysiological processes of the disease to improve its outcome (Lugea et al., 2017).

4. The pancreatic proteins

The transcriptional activity in pancreas is mostly related to exocrine proteins, mainly digestive enzymes. (Danielsson et al., 2014). Although most of these proteins are digestive enzymes, pancreatic juice also contains non enzymatic proteins such as PSP (Pancreatic Stone Protein or Litostatin), which helps to maintain the important ionic enzymatic cofactor Ca^{2+} solubilized in pancreatic juice (Caro, De Caro, Multigner, Dagorn, & Sarles, 1988). Furthermore, acute phase proteins such as PAP (Pancreatitis Associated Protein) and VMP1 were described to enhance their expression during pancreas injury by pathological processes. PAP is a secretory protein (Keim, Rohr, Stöckert, & Haberich, 1984) that has anti-apoptotic and anti-inflammatory properties. (Gironella et al., 2007). VMP1 is a transmembrane protein whose expression triggers autophagy in mammalian cells (Ropolo et al., 2007; Grasso & Vaccaro, 2010).

Most digestive enzymes produced and secreted by the pancreatic acinar cells and contained in the pancreatic juice consist of a diverse set of powerful proteolytic enzymes required for nutritional protein breakdown in the intestine. To maintain an adequate supply of these enzymes whenever necessary, both synthesis and secretion undergo strict long- and short-term regulation in the exocrine pancreas. These digestive enzymes, however, have the potential to damage the tissues they interact with given their proteolytic nature. To prevent this from happening, these proteases are synthesized only as precursors or inactive enzymes called zymogens within pancreatic acinar cells and only become activated when they reach the lumen of the duodenum, where they are expected to act.

This activation process of zymogens to active digestive enzymes is mediated by a protease present at the lumen of the duodenum that cleaves trypsinogen selectively and converts it into active trypsin. This newly formed trypsin, in turn, triggers a cascade of digestive enzyme activation within the duodenum, achieving as a result a high proteolytic capacity required for nutrient breakdown. Digestive zymogens are activated prematurely within the pancreatic acinar cell under diverse noxious stimuli giving way to pancreatitis. Even though the pancreas has protective mechanisms, constant and elevated levels of premature zymogen activation overwhelm these defenses, resulting in cell damage and acute pancreatitis and these initial pathobiological events begin in pancreatic acinar cells (Vaccaro, 2012).

5. Autophagy

Autophagy is a fundamental cellular mechanism by which intracytoplasmic components are transferred to lysosomes to maintain cellular function and homeostasis. This mechanism is implicated in diverse physiological processes such as cell development and survival, differentiation and senescence, and it is also triggered as a response to diverse cellular stressors such as starvation, oxidative stress, hypoxia, hormonal imbalance, or DNA damage among other, promoting degradation and recycling of endogenous material to maintain adequate energy levels, therefore playing a cytoprotective role in these contexts, although it has also been shown to induce a cell death program (Trocoli & Djavaheri-Mergny, 2011).

Autophagy is a catabolic process in which the cell organelles and proteins are sequestered and delivered to the lysosomes, where the sequestered materials are degraded. The aim of this process is to degrade dysfunctional proteins, protein aggregates, damaged organelles and microbes and then recycle them back to the cytosol to preserve cell viability and bioenergetic homeostasis.

Furthermore, this process is somehow involved in disease prevention by removing damaged or old cell components that play a role in disruption of homeostasis and disease occurrence, thus contributing to stopping damage at an early stage. In contrast, dysfunction of autophagy has been related to development of diverse diseases such as inflammatory conditions, neurodegenerative disorders, and cancer, among other (Trocoli & Djavaheri-Mergny, 2011).

Autophagy has many triggers which activate the process with the aim of protecting cells and tissues. Some of these stimulating factors include starvation, diseases, or drugs. Autophagy is also related to processes such as intracellular signaling, granule secretion and phagocytosis. Thus, different autophagy processes (canonical and non-canonical), have been described.

Canonical autophagy processes can be classified into three main types:

1. Microautophagy is the direct sequestration and degradation of small-sized macromolecules by the lysosome or mature endosome, either by invagination or protrusion.
2. Chaperon-mediated-autophagy (CMA) targets specific proteins which are then transferred by a chaperone to the lysosome through its LAMP2A receptor for degradation. Thus, there is no need for any vesicles in this process.
3. Macroautophagy is the most prevalent form, in which the substrates are localized in double membrane vesicles called auto-phagosomes. Autophagosomes then fuse with lysosomes and the cargo is degraded by lysosomal enzymes. The main structures involved in this type of autophagy in mammals are: autophagic body (not in mammals), autophagosome, phagophore assembly site, and a phagophore (Feng, He, Yao, & Klionsky, 2014). There are different forms of macroautophagy: In non-selective autophagy, a portion of the cytosol is engulfed and surrounded by autophagosomes (Trocoli & Djavaheri-Mergny, 2011). In selective autophagy, one type of cellular component is recognized and sequestered in autophagosomes (Shaid, Brandts, Serve, & Dikic, 2013). For instance, mitophagy, lipophagy, xenophagy and zymophagy, the selective autophagy identified in pancreatitis.
   a. Mitophagy is defined as a type of autophagy in which the autophagosome envelops the damaged mitochondria and then fuses with the lysosome. This type of autophagy is frequently associated with pathologies that have mitochondrial injuries such as decreased ATP formation, accumulation of Mitochondrial DNA mutation or production of reactive oxygen species in neurodegenerative or

cardiovascular pathologies, among others. Therefore, selective sequestration and degradation of this dysfunctional mitochondria could benefit cell homeostasis (Chen, Kroemer, & Kepp, 2020)

b. Lipophagy is a selective and complex type of autophagy that regulate the lipid droplets metabolism and is relevant in fatty liver, metabolic syndrome, and aging (Singh & Cuervo, 2012)

c. Xenophagy is a selective type of autophagy that sequesters intracellular microorganisms such as fungi, bacteria, and parasites for degradation, and therefore acts as a protective mechanism from cellular infection (Bauckman, Owusu-Boaitey, & Mysorekar, 2015).

d. Zymophagy is an inducible form of selective autophagy that recognizes and sequesters those zymogen granules intracellularly activated by acute pancreatitis, to protect the cell against self-digestion and death (Bauckman et al., 2015; Grasso et al., 2011; Ropolo, Grasso, & Vaccaro, 2019).

A highly regulated and articulated intracellular machinery of remarkable complexity is required to activate autophagy. Several autophagy related (ATG) proteins are recruited and activated during the process. The pancreatitis associated transmembrane protein VMP1 is one important example of them.

Non-canonical autophagy refers to other processes regulated by the autophagy molecular machinery, such as phagocytosis, inflammatory signaling and secretion. Secretory autophagy (SA) is a recently described non-canonical autophagy process. Its main feature is the release of proteins lacking a N-terminal signal peptide, which normally guides the exocytosis pathway from the endoplasmic reticulum (ER), into the Golgi apparatus, and out to the extracellular space. Without this peptide leading the pathway, proteins can be secreted following secretory autophagy. Abnormalities in this process seem to be related to degenerative and immunoinflammatory diseases.

5.1 VMP1 and its role in autophagy

As mentioned above, in the search of novel genes differentially expressed in acinar cells during acute pancreatitis, we identified a transmembrane protein related to autophagy and named VMP1 (Ropolo et al., 2007). VMP1 is a transmembrane protein located in different tissues such as pancreas, appendix, gallbladder, prostate, ovaries, placenta and small intestine, among other (VMP1 vacuole membrane protein 1 [Homo sapiens (human)]—Gene—NCBI, 2002). VMP1 expression is highly activated in pancreatic acinar cells during acute pancreatitis. Also, VMP1 is rapidly expressed under experimental hypoxia in vivo. This protein was reported to be involved into many different physiological processes such as lipid homeostasis, organelle communication and cellular transport, but plays a main role in triggering autophagy, both in cultured cells and tissues, and it is associated with certain diseases, especially pancreatic diseases such as pancreatitis, diabetes and pancreatic cancer. VMP1 could also be involved in pathological processes such as other types of cancer, cancer metastasis (Sauermann et al., 2008), degenerative and inflammatory diseases.

VMP1 gene is located at 17q23.1, it comprises 133 kb of genomic DNA in centromere-to-telomere orientation, and contains 12 exons, with translation initiation codon located to exon 2, and stop codon to exon 12. This gene has 17 transcripts (splice variants), 264 orthologues and 3 paralogues (VMP1 Ensemble ENSG00000062716). It encodes a 406-amino-acid transmembrane protein containing six hydrophobic regions that make up for 6 putative transmembrane domains. This protein has no homologs in low eukaryotes including yeast. The over-expression of VMP1 in mammalian cells, as well as the transgenic expression in mouse pancreatic acinar cells triggers autophagy. The region associated with autophagy involves 20 amino acids in the C terminus hydrophilic domain (Ropolo et al., 2007). This domain is also known as the VMP1-ATG domain and is required for autophagosome biogenesis (Molejon, Ropolo, Re, Boggio, & Vaccaro, 2013).

VMP1, together with TMEM41B, an integral ER membrane protein that works alongside VMP1 in autophagy, was reported to belong to an evolutionarily conserved family of proteins called *Escherichia coli* DNA gene A (DedA) domain superfamily. They share a characteristic transmembrane domain that adopts a structure containing two reentrant loops named the VTT domain (Morita, Hama, & Mizushima, 2019). VMP1 DedA domain is connected to the three transmembrane helices in N terminal and one in C terminal. Due to its low hydrophobicity, the reentrant loop is thought to be rather unstable in the lipid bilayer. Therefore, self-binding of the DedA domain may stabilize the reentrant loops suggesting that VMP1 and TMEM41B might interact with themselves. These proteins have been proposed to have scramblase activity because they may scramble various phospholipids such as phosphatidylserine and phosphatidylcholine and being involved in vesicle and cellular lipid distribution and lipid homeostasis by actively participating in lipid droplet budding and loading and may be involved in different processes which include balancing and transporting of phospholipids between two membrane leaflets. VTT domain containing proteins were also reported to be involved in cholesterol metabolism, which in turn affects different viruses' capacity to enter and infect cells (Li et al., 2021).

There is an intricate and highly complex molecular apparatus behind autophagy regulation. Several proteins with different biological properties and functions interact upon each other for autophagy to take place appropriately. Four main groups of autophagy related (ATG) proteins are involved in autophagosome biogenesis. First, the ULK1/2 complex that initiate the process; second the BECN1 complexes with PI3Kinase activity that phosphorylates the autophagic membrane; thirst, two ubiquitin like systems that promote the membrane elongation and fourth, two transmembrane proteins: ATG9 and VMP1.

The role of VMP1 in regulation and influence over autophagy is of relevance specifically on autophagosome formation and elongation.

Just to summarize this context (Klionsky et al., 2021), the autophagy pathway involves the following steps (Fig. 1):

- Initiation: it is believed that the Endoplasmic Reticulum and associated membranes (Hamasaki et al., 2013; Nascimbeni et al., 2017) start the formation of the pre-autophagosome structure, also called Isolation Membrane (IM), a crescent-shaped structure which later seals to form the autophagosome. Within the structure of the ER, there are PI(3)P-enriched subdomains, in which autophagosome formation will begin. These are known as omegasomes. ER will intimately interact with IM for its formation in the initiation process, and further elongation. This ER-IM contact is crucial for lipid transport in the process of IM expansion.

For this process to start and develop, several autophagy genes (ATG) need to closely interact and activate different molecular pathways and structures. In this direction, the first step implicated in initiation of the autophagosome formation is phosphorylation of ULK and formation of punctate structures between the ER and its tight association with the ULK1/FIP200 Atg1 complex.

When ULK1 becomes active, it phosphorylates the P13KC3 1 complex formed by BECN1, ATG14,15, and vps34, which is essential for autophagosome formation. Through its ATG domain, VMP1 interacts with the BH3 domain of BECN1, resulting in the recruitment of the P13KC3 1 complex to the autophagosome membrane. Thus, VMP1 is critical for the proper structure of the ATG conjugation systems involved in the early stages of autophagosome formation.

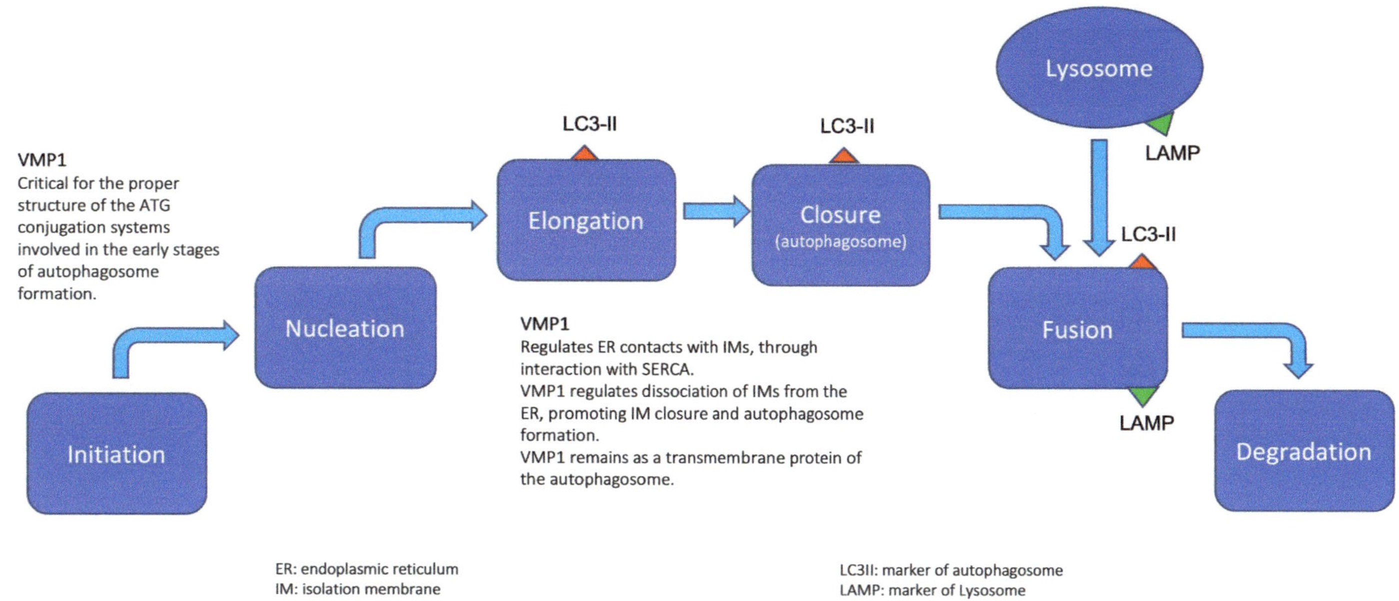

Fig. 1 VMP1 and the canonical autophagy pathway.

Following the previously mentioned steps, BECN1-PI13K complex phosphorylates ATG12 and LC3, resulting in the precise identification of P13P. On the omegasome structure, PI3P is then identified by DFCP1 (double FYVE domain-containing protein). As a result, DFCP1, a PI3P effector, is thought to be a marker of omegasome development. The ATG12-ATG5 complex is then created, which recruits microtubule associated protein 1 light chain 3 (LC3) to the autophagosomal membrane. ATG16L controls this phase by interacting with ATG5 to produce the ATG12-ATG5-ATG16L complex. Then, WIPI protein transfers this complex to the isolation membrane.

Because it is involved in the elongation, maturation, and fusion of the autophagosome-lysosome, Microtubule-associated protein 1A/1B-light chain 3 (LC3) plays an important role in autophagy. Lipidated LC3 (LC3II) is located at both sites of the autophagosome's isolation membrane. Along with other ATG8 family proteins, it is thought to be an autophagosome marker. LC3 is expressed as three isoforms in mammalian cells, LC3A, LC3B and LC3C, but only LC3B-II correlates with increased levels of autophagic vesicles.

Two essential molecules that regulate this stage are: (a) AMPK (an energy sensor), which is involved in regulating metabolism and homeostasis by stimulating ULK phosphorylation, and (b) mTOR (mammalian target of rapamycin) (a nutrient sensor), that acts as an inhibitor of the subsequent process, which regulates cell growth.

- Elongation: the pre autophagosome double membrane expands and transforms into the autophagosome. This stage is stimulated by the ATG12-ATG5-ATG16L1 complex and BECN1 which activates phosphatidylinositol-3 kinase (PI3K) which, when activated, recruits p62, Nbr1 and LC3 proteins resulting in double membrane vacuole that could originate from the ER, the Golgi apparatus, or Mitochondria.
- Maturation and Closure: when the autophagosphore or IM double membrane expands too much, it eventually closes and forms a round vacuole called autophagosome. For this part of the autophagosome formation, VMP1 plays an essential part. ER-IM contact appears to be modulated by changes in calcium levels in mammalian cells. Studies suggest VMP1 regulates ER contacts with IMs, through interaction with SERCA (sarco(endo)plasmic reticulum calcium ATPase), which transports calcium from the cytosol into the ER lumen, and therefore regulates autophagosome formation. Specifically, VMP1 prevents phospholamban and sarcolipin from forming an inhibitory complex with

SERCA2, and this way regulates dissociation of IMs from the ER, promoting IM closure and autophagosome formation. Correct maturation and closure of the autophagosome is essential because it precedes two processes: fusion with the lysosome and the catalytic process.

- Fusion: the movement of the autophagosome across the cytoplasm is carried out via microtubular structures such as LC3. Furthermore, the hydrolases destroy the LC3 reservoir when the autophagosome is delivered to the lysosome; LC3II contained in the outer membrane is cleaved by ATG4b and recycled. Because it is involved in the elongation, maturation, and fusion of the autophagosome-lysosome, LC3 plays an important role in autophagy. Lipidated LC3 (LC3II) is located at both sites of the autophagosome's isolation membrane, functioning as an autophagosome marker, together with other ATG8 proteins. The HOPS complex via STX17 and RAB7 is required for autophagosome-lysosome fusion as well.
- Degradation: the inner membrane and then the content of the autophagosome are degraded by the acidic lumen and lysosomal enzymes within the lysosome. Acidic pH is generated due to the presence of proton pumps and enzymes such as vATPase.

Autophagy is frequently related with exocrine pancreatic diseases. These diseases include acute and chronic pancreatitis, as well as pancreatic cancer. VMP1 involvement in the pathogenesis of these clinical entities has been described in a variety of animals and conditions. The role of VMP1-related pathways in the pathophysiology of exocrine pancreas disorders will be briefly discussed next.

6. VMP1 and pancreatitis

Pancreatitis is a pancreatic inflammatory process defined by pancreas self-digestion in consequence of the activation of zymogens inside the acinar cell resulting in cell structure degradation. It is the most common disease of the pancreas and frequently has a mild or auto limited course, but in a 10% of the cases can be a severe disease involving multiorgan failure and systemic inflammatory response with high mortality. Pancreatitis can be divided into two different chronological types; acute pancreatitis (a sudden short form) and chronic pancreatitis (long standing and progressive form). Both types are caused by the same stressors, but it is believed that many cases of chronic disease result from repetitive episodes of acute pancreatitis.

There are many causes of pancreatitis, but the most prevalent ones are excessive alcohol intake and gallstones. However, there are other important triggering factors for this disease such as hypertriglyceridemia, ERCP procedure, hereditary or genetic, autoimmune, idiopathic, trauma, pancreatic cancer and viruses, i.e., coxsackie virus (Mederos, Reber, & Girgis, 2021; Nesvaderani, Eslick, Vagg, Faraj, & Cox, 2015).

Simultaneous dysregulation of autophagy and persistent inflammation are key factors for developing acute pancreatitis (AP). In addition, the repetitive episodes of AP could cause chronic pancreatitis (CP) because CP is linked to persistent mild inflammation that cannot be resolved. Furthermore, the blockade of autophagy may cause mitochondrial dysfunction, ER stress, p62 accumulation and oxidative stress; this blockade could lead to pancreatitis as a result of trypsinogen activation and dysregulated secretion, fibrosis, cell death and inflammation (Gukovskaya, Gukovsky, Algül, & Habtezion, 2017). In mild pancreatitis, mitophagy and zymophagy become activated, and work as cellular restoration mechanisms preserving mitochondrial bioenergetics (Grasso et al., 2011; Ropolo et al., 2019; Vanasco et al., 2021). In fact, cellular survival, which has as a direct consequence non severity of the disease, depends largely upon adequate mitochondrial function, and effective elimination of damaged mitochondria. Nonetheless, these two autophagy pathways constitute direct targets that might become damaged because of pancreatitis itself. In this direction, we found that during pancreatitis development mitochondrial dysfunction was evidenced by decreased levels of 0_2 consumption and ATP production, with a decreased ATP/0_2 ratio. This dysfunction correlated with changed in mitochondrial dynamics, reflected by optic atrophy 1 (OPA-1) and dynamin-related protein 1 (DRP-1) differential expression, as well as ultrastructural features of mitochondrial fission, elongation, and mitophagy during acute phase of mild pancreatitis in experimental rats.

Mitophagy was evaluated by confocal assay after transfection with the pMITO-RFP-GRP plasmid, a specific label for the process. We demonstrated that VMP1 is required for proper mitophagy during acute pancreatitis, and that VMP1 downregulation avoids damaged mitochondrial clearance. Thus, confirming expression of this protein plays a role in preserving mitophagy and therefore limiting pancreatitis to a mild disease. VMP1 participation in mitophagy was observed by repositioning around specific mitochondrial populations and by detection of mitochondria in autophagosomes isolated with anti-VMP1 antibodies (Vanasco et al., 2021).

A better understanding of the molecular pathways involved in recovering mitochondrial function could be significant and contributory in the development of new therapeutic strategies for acute pancreatitis.

Notably, it has been revealed that zymophagy is triggered in pancreatic acinar cells during pancreatitis in order to sequester and degrade potentially deleterious activated zymogen granules. Thus, autophagy and especially VMP1 could play an essential role protecting the cell from degradation.

When calcium homeostasis in acinar cells is altered by oxidative stress produced by some metabolites (i.e., such as non-oxidative alcohol metabolites, gallstones, or hypertriglyceridemia) there is an increase in intracellular calcium concentration, which is associated with both physiological and pathological responses in AP. Despite the transient spike of calcium concentration in physiological responses, pathological response involves global rise of calcium which maintains in time (Criddle et al., 2007). In addition, prolonged and increased calcium concentration opens the mitochondrial permeability transition pore (MPTP) and, in consequence, ATP formation decreases leading to increased mitochondrial dysfunction and increased calcium concentration again.

The rise in calcium concentration triggers the activation of zymogens inside the cell which can be sequestered and degraded in the autolysosomes by zymophagy. As it was described above, zymophagy is a selective type of autophagy and may play an essential role in protecting the cell from damage produced by the prematurely activated enzymes (Grasso et al., 2011). In pancreatitis, zymophagy has pathophysiological importance, as it controls zymogen activation induced by the disease, and therefore helps to prevent cell death and limit pancreatitis. Pancreatitis-induced zymophagy has been demonstrated in experimental pancreatitis in mice, and cultured pancreatic acinar cells, as well as in humans in the clinical setting. VMP1 has a central role in mediating zymophagy in the setting of acute pancreatitis, together with the ubiquitin-protease USP9x and the ubiquitin-binding protein p62. In fact, there is a close cooperation between the ubiquitin recognition machinery and VMP1 in the formation of the autophagosome (Grasso et al., 2011). During VMP1-selective autophagic pathway, p62 could act as a cargo receptor, shown by colocalization and coexistence of these proteins in isolated autophagosomes by immunofluorescence assays and western blot analyses of proteins from these autophagic organelles. This finding is supported by shown marked colocalization of ubiquitin with activated zymogen granules in cellular models of pancreatitis.

Moreover, colocalization of Ubiquitin and LC3 in VMP1 formed autophagosomes is evidence that the ubiquitinated cargo is sequestered by zymophagy. This suggests that ubiquitination, whether direct or indirect, of activated zymogen granules might take place for recognition by autophagic membranes, acting as a tag for precise engulfment by the VMP1 mediated autophagic pathway. In addition, USP9x has shown to be an essential component for selective zymogen degradation in pancreatitis. The USP9x gene belongs to the peptidase C19 family, encoding for a ubiquitin-specific protease. Our results demonstrate that USP9x interacts with VMP1 early during the process of zymophagy, supporting the role of this protease in the autophagy pathway. To further strengthen this hypothesis, evidence shows that down-regulation of USP9x eliminates activated zymogen granule degradation by autophagy in cellular models of pancreatitis. It is likely that USP9x modulates VMP1 or functions as a recognition signal for altered zymogen granules (Grasso et al., 2011). Interestingly though, these findings allow us to propose that the protective role of selective autophagy could be impaired or decreased in severe pancreatitis. In the setting of an excess of cargo and an overly increased rate of degradation causing the protective role of this process to become disrupted, could lead to a reduced defense mechanism and consequently a more severe form of AP.

Moreover, it has been described that early activation of NF-kB, occurring parallel to trypsinogen activation in time course, is another route to development of AP, and in turn increases damage in acinar cells caused by trypsinogen activation, by a complex cascade of proinflammatory cytokines. Activation of NF-kB responds to different cellular stressors and stimuli and is activated in response to increased calcium concentration in the setting of AP (Sah & Saluja, 2011). Interestingly, NF-kB and autophagy signaling pathways exert a mutual control over each other's activation and inhibition molecular components. Autophagy, on one hand, seems to have a regulatory effect on NF-kB activation, by specifically degrading signaling components of this pathway, and thus having an effect as either termination of the initial NF-kB activation, or inversely, persistently activating this process (Trocoli & Djavaheri-Mergny, 2011). This could have an implication regarding termination of one of the pathways early activated in AP which produces acinar damage. On the other hand, NF-kB can either activate or inhibit signaling components that trigger autophagy, mainly through BECN1 up or down regulation, as well as regulating transcription of certain pro-autophagic regulating genes, in various diseases. Even signaling components of the NF-kB molecular pathway can activate autophagy independently from NF-kB (Trocoli & Djavaheri-Mergny, 2011). Evidence therefore suggests that autophagy alterations could be involved in subsequent alteration of cytokine and pro-inflammatory mediator liberation, such as NF-kB, in the setting of AP, although the specific role of VMP1 in regulating these processes remains unelucidated.

In summary, we thus propose that VMP1 protein can stimulate autophagy and prevent cellular damage by activating zymophagy and regulating mitophagy. Notably, some studies have shown that genetic models lacking ATG proteins had severe complications of pancreatitis (Diakopoulos et al., 2015). Also, when VMP1 is transgenic expressed in mouse pancreas, the mouse model present less trypsinogen activation and mild pancreatitis (Vanasco et al., 2021), thus reinforcing the conclusion that VMP1 could be a key contributing factor to limiting severity in pancreatitis.

7. VMP1 and pancreatic cancer

Pancreatic ductal adenocarcinoma is the most prevalent pancreatic cancer and the third most frequent cause of cancer related death. It is often diagnosed at late stages due to the lack of specific symptoms and there are limited therapeutic options available. It currently has a survival rate of about 8–9% in 5-years (Siegel, Miller, Fuchs, & Jemal, 2022; Survival Rates for Pancreatic Cancer, 2022).

As it is well known, cancer is related to an increased cellular proliferation due to various mechanisms and one of them is apoptosis avoidance (Lowe & Lin, 2000). Autophagy, in turn, can recycle cell organelles and therefore prevent apoptosis, aiding in this mechanism. Fujii et al., have demonstrated that LC3 (a well characterized autophagy biomarker) is increased in pancreatic ductal adenocarcinoma (Fujii et al., 2008). Thus, autophagy could be related to massive cell proliferation by prevention of apoptosis and VMP1 protein could be an important agent in this process. Moreover, it has been demonstrated that autophagy itself is stimulated in pancreatic cancer due to insufficient vascularization of the pancreatic stroma, which promotes cellular hypoxia and nutrient deprivation enhancing cell and ER stress, which activates autophagy.

Some risk factors are associated with the development of pancreatic ductal adenocarcinoma. As it was mentioned above, chronic pancreatitis, in patients with Kras oncogenic mutations, predisposes to pancreatic ductal adenocarcinoma (PDAC). It has been suggested that VMP1 plays a role in the regulation of Kras-induced PDAC initiation. Several studies have demonstrated that Kras mutation requires VMP1 to induce and maintain autophagy levels in cancer cells and those cells expressing VMP1 are associated with poor prognosis (Fujii et al., 2008), making overexpression of VMP1 a possible negative prognostic factor in pancreatic cancer, and thus reinforcing the need to approach its molecular pathway as a possible therapeutic target to improve prognosis. It was demonstrated that VMP1 increases the ability of Kras to induce preneoplastic lesions growth, called pancreatic intraepithelial neoplasia (PanINs). This promoting effect of VMP1 on PanIN formation seems to be related with an increased cell proliferation associated with a decrease in apoptosis. Chloroquine, an inhibitor of autophagy, blocks the effect of VMP1 on PanIN formation. VMP1-mediated autophagy acts cooperatively with mutated Kras to induce PDAC initiation (Loncle et al., 2016). The most frequently detected mutation in pancreatic cancer is Kras that is associated with an energy deficit in acinar cells. Therefore, the cell tries to increase energy production in multiple ways to compensate for this deficit. Autophagy is one of them. The mechanisms by which autophagy is stimulated in this context are poorly understood. It is supposed that the Kras mutation stimulates the PI3k enzyme that then activates some proteins such as AKT1. AKT1 then triggers/stimulates the activity of the transcription factor GLI3 which is an effector of the Hedgehog pathway. GLI3 is highly expressed in cells carrying a mutant KRASG12D allele (Nolan-Stevaux et al., 2009). GLI family is composed of GL1 and GL2 (that are activators) and GLI3 that is predominately a transcriptional repressor. Moreover, GLI3 can interact with other regulator molecules such as histone acetyltransferase p300. The result of the co-regulation of GL3 and p300 histone acetyltransferase is the activation of the VMP1 promoter and therefore, an activation of the autophagy machinery (Lo Ré et al., 2012). Autophagy can be also stimulated by Kras mutation following other pathways such as via ATG 7 and ATG 5 which are present in the autophagosome (Lo Ré et al., 2012).

On the other hand, as briefly mentioned above, ER stress in pancreatic cells can be associated with neoplastic disease. Some studies suggest that ER stress is a defining feature in pancreatic cancer, and it could be an autophagy trigger. As it has been described, VMP1 is a transmembrane protein and its expression and functions in initiation and elongation of the autophagosome is upregulated under cellular stress. Thus, any stressor that could damage the ER could stimulate overexpression of VMP1 and therefore autophagy in pancreatic cancer cells. Autophagy may play an essential role in cell life preservation and therefore cell proliferation, a key process in the development and growth of malignant tumors (Gilabert et al., 2013; Wang, Kou, & Le, 2020). Moreover, some chemotherapeutic drugs, such as gemcitabine, induce ER stress and the overexpression of VMP1 which results in increased drug resistance in some animal models.

8. VMP1 and cancer-related drug resistance

The treatment for pancreatic ductal adenocarcinoma consists mainly of its surgical removal. Also, there are multiple drugs used in agreement with guidelines to treat advanced disease, such as Gemcitabine, Imatinib, Cisplatin and others. However, drug resistance in this type of cancer and its metastasis is very common and chemotherapy still cannot eradicate the tumor or improve survival rate substantially.

It has been shown that downregulation of VMP1 in mice enhances the effects of certain drugs by reducing cell resistance. These studies provided results with rapamycin, cisplatin, gemcitabine, adriamycin, among others. Gilabert et al. also showed that VMP1 expressing tumors grew more, while using gemcitabine, than mice cells which received placebo, suggesting that VMP1 by means of autophagy could be involved in tumor growth and drug resistance.

We have recently found that E2F1, an effector of gemcitabine-induced autophagy, regulates the expression and promoter activity of VMP1. E2F1 is a transcription factor that is involved in the retinoblastoma pathway. The histone acetyltransferase EP300 also modulates the VMP1 promoter activity and seems to work cooperatively with E2F1. We proposed that the E2F1-EP300 activator/coactivator complex controls at least in part the expression and promoter activity of VMP1 induced by gemcitabine in pancreatic cancer stem cell lines such as PANC-1 cells. This mechanism is now being considered as a candidate pathway that mediates gemcitabine-induced autophagy in pancreatic cancer cells carrying oncogenic Kras variants (Ropolo et al., 2020).

Taken together, the experimental and clinical evidence of VMP1 overexpression in pancreatic human cancer elicits an important finding for the role of VMP1 mediated autophagy in cancer. In addition, it has also been suggested by Gilabert et al., that VMP1 could prevent cell death against drugs through pathways other than autophagy. They used a BECN1 binding domain deleted mice and chloroquine to inhibit the autophagy pathway and the result was still an increase of VMP1 expression. This last finding could be

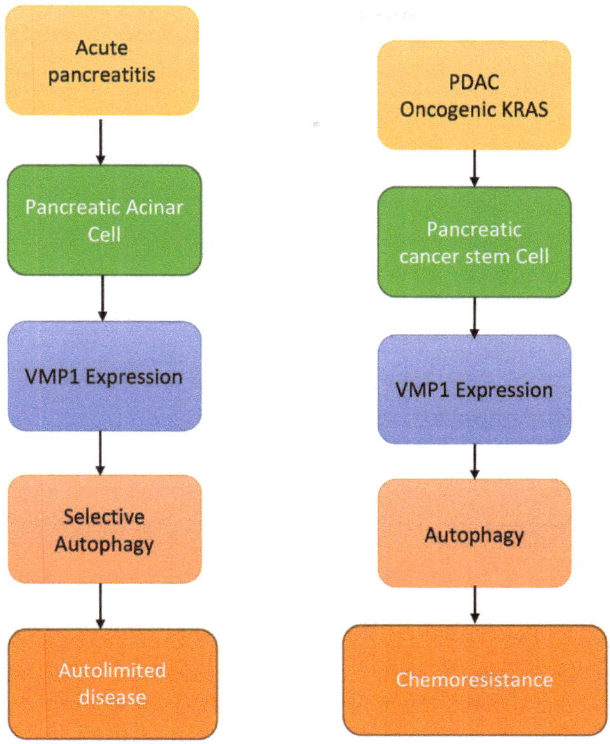

Fig. 2 VMP1 in acute pancreatitis, pancreatic adenocarcinoma and chemoresistance.

also related to the transcriptional or post-transcriptional regulation of VMP1 expression. Nevertheless, VMP1-induced drug resistance is associated with VMP1 overexpression in pancreatic cancer cells. Further research should elucidate autophagy and non-autophagy associated effects on the resistance of pancreatic cancer cells to chemotherapy (Fig. 2).

9. Conclusions

The pancreas is fundamentally an organ that produces, secretes, and exports proteins. Its biology is directly related to the synthesis, storage, and release of essential molecules for intermediary metabolism such as insulin, glucagon, somatostatin and other by the endocrine pancreas; and digestive enzymes by the exocrine pancreas that allow digestion and absorption of essential nutrients. In this chapter we focus on the exocrine pancreas, some specific proteins (VMP1 in particular) and their role in autophagic response to pathological disorders with high morbidity and mortality. The pancreatic acinar cell secretes proteins with enzymatic action capable of hydrolyzing the very tissue that produces them. It is then that these highly differentiated cells must have sophisticated defense mechanisms that protect the tissue from autodigestion, which may include phenotypical changes in the expression of these harming proteins shifting to the activation of molecules able to mediate protection. In the search for new molecules that are differentially expressed during inflammatory disease of the pancreas, we found a transmembrane protein that turned out to be one of the only two transmembrane proteins involved in autophagy, VMP1 (Vaccaro, 2008).

VMP1 is an essential transmembrane protein in the initial steps of autophagosome formation, but also it is an acute phase response protein that mediates selective autophagy as a cell response to disease. In acute pancreatitis, VMP1 expression might account for the most frequent mild and auto-limited disease mediating zymophagy and mitophagy, avoiding the spreading of zymogen activation and allowing energetic recovery. We proposed that the more capacity of VMP1 mediated selective autophagy, the less severity of the disease (Vaccaro, 2012).

On the other hand, in the pancreatic cell trans-differentiation and transformation driven by oncogenic KRAS in the pancreatic ductal adenocarcinoma, the expression of VMP1 mediating autophagy in cancer stems cells is part of the complex's mechanisms for the high cell resistance to chemotherapy seen in those tumors. We summarized here several studies on the transcriptional regulation of VMP1 expression, which is mediated by the critical oncologic gen Kras and P300 as a coactivated transcription factor (Lo Ré et al., 2012; Ropolo et al., 2020). Although these studies may be of great clinical relevance, further studies on both post-transcriptional and post-translational regulation, as well as unconventional secretory mechanisms that would be involved in VMP1 metabolism will be necessary for a better understanding of VMP1 pathways and the elaboration of new and rational therapeutic strategies.

References

Bauckman, K. A., Owusu-Boaitey, N., & Mysorekar, I. U. (2015). Selective autophagy: Xenophagy. *Methods*, *75*, 120–127. https://doi.org/10.1016/j.ymeth.2014.12.005.

Bayliss, W. M., & Starling, E. H. (1902). The mechanism of pancreatic secretion. *The Journal of Physiology*, *28*(5), 325–353.

Caro, A. D., De Caro, A., Multigner, L., Dagorn, J.-C., & Sarles, H. (1988). The human pancreatic stone protein. *Biochimie*, *70*(9), 1209–1214. https://doi.org/10.1016/0300-9084(88)90186-1.

Chen, G., Kroemer, G., & Kepp, O. (2020). Mitophagy: An emerging role in aging and age-associated diseases. *Frontiers in Cell and Development Biology*, *8*, 200.

Criddle, D. N., Gerasimenko, J. V., Baumgartner, H. K., Jaffar, M., Voronina, S., Sutton, R., et al. (2007). Calcium signalling and pancreatic cell death: Apoptosis or necrosis? *Cell Death and Differentiation*, *14*(7), 1285–1294. https://doi.org/10.1038/sj.cdd.4402150.

Danielsson, A., Pontén, F., Fagerberg, L., Hallström, B. M., Schwenk, J. M., Uhlén, M., et al. (2014). The human pancreas proteome defined by transcriptomics and antibody-based profiling. *PLoS One*, *9*(12), e115421.

Diakopoulos, K. N., Lesina, M., Wörmann, S., Song, L., Aichler, M., Schild, L., et al. (2015). Impaired autophagy induces chronic atrophic pancreatitis in mice via sex- and nutrition-dependent processes. *Gastroenterology*, *148*(3), 626–638. e17.

Dusetti, N. J., Jiang, Y., Vaccaro, M. I., Tomasini, R., Azizi Samir, A., Calvo, E. L., et al. (2002). Cloning and expression of the rat vacuole membrane protein 1 (VMP1), a new gene activated in pancreas with acute pancreatitis, which promotes vacuole formation. *Biochemical and Biophysical Research Communications*, *290*(2), 641–649.

Feng, Y., He, D., Yao, Z., & Klionsky, D. J. (2014). The machinery of macroautophagy. *Cell Research*, *24*(1), 24–41.

Fujii, S., Mitsunaga, S., Yamazaki, M., Hasebe, T., Ishii, G., Kojima, M., et al. (2008). Autophagy is activated in pancreatic cancer cells and correlates with poor patient outcome. *Cancer Science*, *99*(9), 1813–1819. https://doi.org/10.1111/j.1349-7006.2008.00893.x.

Gilabert, M., Vaccaro, M. I., Fernandez-Zapico, M. E., Calvo, E. L., Turrini, O., Secq, V., et al. (2013). Novel role of VMP1 as modifier of the pancreatic tumor cell response to chemotherapeutic drugs. *Journal of Cellular Physiology*, *228*(9), 1834–1843.

Gironella, M., Folch-Puy, E., LeGoffic, A., Garcia, S., Christa, L., Smith, A., et al. (2007). Experimental acute pancreatitis in PAP/HIP knock-out mice. *Gut*, *56*(8), 1091–1097.

Grasso, D., Ropolo, A., Lo Ré, A., Boggio, V., Molejón, M. I., Iovanna, J. L., et al. (2011). Zymophagy, a novel selective autophagy pathway mediated by VMP1-USP9x-p62, prevents pancreatic cell death. *The Journal of Biological Chemistry*, *286*(10), 8308–8324.

Grasso, D., Sacchetti, M. L., Bruno, L., Lo Ré, A., Iovanna, J. L., Gonzalez, C. D., et al. (2009). Autophagy and VMP1 expression are early cellular events in experimental diabetes. *Pancreatology*, *9*(1-2), 81–88. https://doi.org/10.1159/000178878.

Grasso, D., & Vaccaro, M. I. (2010). VMP1. *Pancreapedia: Exocrine Pancreas Knowledge Base* (pp. 1–4). American Pancreatic Association, Molecular Page. https://doi.org/10.3998/panc.2010.11.

Gukovskaya, A. S., Gukovsky, I., Algül, H., & Habtezion, A. (2017). Autophagy, inflammation, and immune dysfunction in the pathogenesis of pancreatitis. *Gastroenterology*, *153*(5), 1212–1226.

Hamasaki, M., Furuta, N., Matsuda, A., Nezu, A., Yamamoto, A., Fujita, N., et al. (2013). Autophagosomes form at ER–mitochondria contact sites. *Nature*, *495*(7441), 389–393. https://doi.org/10.1038/nature11910.

Ishiguro, H., Yamamoto, A., Nakakuki, M., Yi, L., Ishiguro, M., Yamaguchi, M., et al. (2012). Physiology and pathophysiology of bicarbonate secretion by pancreatic duct epithelium. *Nagoya Journal of Medical Science*, *74*, 1–18.

Jiang, P. H., Motoo, Y., Vaccaro, M. I., Iovanna, J. L., Okada, G., & Sawabu, N. (2004). Expression of vacuole membrane protein 1 (VMP1) in spontaneous chronic pancreatitis in the WBN/Kob rat. *Pancreas*, *29*(3), 225–230. https://doi.org/10.1097/00006676-200410000-00008.

Keim, V., Rohr, G., Stöckert, H. G., & Haberich, F. J. (1984). An additional secretory protein in the rat pancreas. *Digestion*, *29*(4), 242–249. https://doi.org/10.1159/000199041.

Klionsky, D. J., et al. (2021). Guidelines for the use and interpretation of assays for monitoring autophagy (4th edition). *Autophagy*, *17*(1), 1–382. https://doi.org/10.1080/15548627.2020.1797280.

Lee, P. J., & Papachristou, G. I. (2019). New insights into acute pancreatitis. *Nature Reviews. Gastroenterology & Hepatology, 16*(8), 479–496. https://doi.org/10.1038/s41575-019-0158-2.

Li, Y. E., Wang, Y., Du, X., Zhang, T., Mak, H. Y., Hancock, S. E., et al. (2021). TMEM41B and VMP1 are scramblases and regulate the distribution of cholesterol and phosphatidylserine. *The Journal of Cell Biology, 220*(6). https://doi.org/10.1083/jcb.202103105.

Lo Ré, A. E., Fernández-Barrena, M. G., Almada, L. L., Mills, L. D., Elsawa, S. F., Lund, G., et al. (2012). Novel AKT1-GLI3-VMP1 pathway mediates KRAS oncogene-induced autophagy in cancer cells. *The Journal of Biological Chemistry, 287*(30), 25325–25334.

Loncle, C., Molejon, M. I., Lac, S., Tellechea, J. I., Lomberk, G., Gramatica, L., et al. (2016). The pancreatitis-associated protein VMP1, a key regulator of inducible autophagy, promotes Kras(G12D)-mediated pancreatic cancer initiation. *Cell Death & Disease, 7*, e2295.

Lowe, S. W., & Lin, A. W. (2000). Apoptosis in cancer. *Carcinogenesis, 21*(3), 485–495.

Lugea, A., Waldron, R. T., Mareninova, O. A., Shalbueva, N., Deng, N., Su, H.-Y., et al. (2017). Human pancreatic acinar cells: Proteomic characterization, physiologic responses, and organellar disorders in ex vivo pancreatitis. *The American Journal of Pathology, 187*(12), 2726–2743.

Mederos, M. A., Reber, H. A., & Girgis, M. D. (2021). Acute pancreatitis: A review. *JAMA: The Journal of the American Medical Association, 325*(4), 382–390.

Molejon, M. I., Ropolo, A., Re, A. L., Boggio, V., & Vaccaro, M. I. (2013). The VMP1-Beclin 1 interaction regulates autophagy induction. *Scientific Reports, 3*, 1055.

Morita, K., Hama, Y., & Mizushima, N. (2019). TMEM41B functions with VMP1 in autophagosome formation [Review of TMEM41B functions with VMP1 in autophagosome formation]. *Autophagy, 15*(5), 922–923.

Nascimbeni, A. C., Giordano, F., Dupont, N., Grasso, D., Vaccaro, M. I., Codogno, P., et al. (2017). ER-plasma membrane contact sites contribute to autophagosome biogenesis by regulation of local PI3P synthesis. *The EMBO Journal, 36*(14), 2018–2033.

Nesvaderani, M., Eslick, G. D., Vagg, D., Faraj, S., & Cox, M. R. (2015). Epidemiology, aetiology and outcomes of acute pancreatitis: A retrospective cohort study. *International Journal of Surgery, 23*(Pt A), 68–74.

Nolan-Stevaux, O., Lau, J., Truitt, M. L., Chu, G. C., Hebrok, M., Fernández-Zapico, M. E., et al. (2009). GLI1 is regulated through Smoothened-independent mechanisms in neoplastic pancreatic ducts and mediates PDAC cell survival and transformation. *Genes & Development, 23*(1), 24–36.

Palade, G. (1975). Intracellular aspects of the process of protein synthesis. *Science, 189*(4206), 867.

Pandol, S. J. (2010). Introduction. In *The exocrine pancreas* Morgan & Claypool Life Sciences.

Ropolo, A., Catrinacio, C., Renna, F. J., Boggio, V., Orquera, T., Gonzalez, C. D., et al. (2020). A novel E2F1-EP300-VMP1 pathway mediates gemcitabine-induced autophagy in pancreatic cancer cells carrying oncogenic KRAS. *Frontiers in Endocrinology, 11*, 411.

Ropolo, A., Grasso, D., Pardo, R., Sacchetti, M. L., Archange, C., Lo Re, A., et al. (2007). The pancreatitis-induced vacuole membrane protein 1 triggers autophagy in mammalian cells. *The Journal of Biological Chemistry, 282*(51), 37124–37133.

Ropolo, A., Grasso, D, & Vaccaro, M. I. (2019). Measuring autophagy in pancreatitis. *Methods in Molecular Biology, 1880*, 541–554. https://doi.org/10.1007/978-1-4939-8873-0_35.

Sah, R. P., & Saluja, A. (2011). Molecular mechanisms of pancreatic injury. *Current Opinion in Gastroenterology, 27*(5), 444–451. https://doi.org/10.1097/mog.0b013e328349e346.

Sauermann, M., Sahin, O., Sültmann, H., Hahne, F., Blaszkiewicz, S., Majety, M., et al. (2008). Reduced expression of vacuole membrane protein 1 affects the invasion capacity of tumor cells. *Oncogene, 27*(9), 1320–1326.

Shaid, S., Brandts, C. H., Serve, H., & Dikic, I. (2013). Ubiquitination and selective autophagy. *Cell Death and Differentiation*, *20*(1), 21–30.
Siegel, R. L., Miller, K. D., Fuchs, H. E., & Jemal, A. (2022). Cancer statistics, 2022. *CA: A Cancer Journal for Clinicians*, *72*(1), 7–33.
Singh, R., & Cuervo, A. M. (2012). Lipophagy: Connecting autophagy and lipid metabolism. *International Journal of Cell Biology*, *2012*, 282041.
Survival Rates for Pancreatic Cancer. (2022). Retrieved April 27, 2022, from https://www.cancer.org/cancer/pancreatic-cancer/detection-diagnosis-staging/survival-rates.html.
Tempero, M. A. (2019). NCCN guidelines updates: Pancreatic cancer. *Journal of the National Comprehensive Cancer Network*, *17*(5.5), 603–605. https://doi.org/10.6004/jnccn.2019.5007.
Trocoli, A., & Djavaheri-Mergny, M. (2011). The complex interplay between autophagy and NF-κB signaling pathways in cancer cells. *American Journal of Cancer Research*, *1*(5), 629–649.
Vaccaro, M. I. (2008). Autophagy and pancreas disease. *Pancreatology*, *8*(4-5), 425–429. https://doi.org/10.1159/000151480.
Vaccaro, M. I. (2012). Zymophagy: Selective autophagy of secretory granules. *International Journal of Cell Biology*, *2012*, 396705.
Vaccaro, M. I., Grasso, D., Ropolo, A., Iovanna, J. L., & Cerquetti, M. C. (2003). VMP1 expression correlates with acinar cell cytoplasmic vacuolization in arginine-induced acute pancreatitis. *Pancreatology*, *3*(1), 69–74. https://doi.org/10.1159/000069150.
Vanasco, V., Ropolo, A., Grasso, D., Ojeda, D. S., García, M. N., Vico, T. A., et al. (2021). Mitochondrial dynamics and VMP1-related selective mitophagy in experimental acute pancreatitis. *Frontiers in Cell and Development Biology*, *9*, 640094.
VMP1 vacuole membrane protein 1 [Homo sapiens (human)]—Gene—NCBI. (2002). Retrieved February 22, 2022, from https://www.ncbi.nlm.nih.gov/gene/81671#gene-expression.
Wang, P., Kou, D., & Le, W. (2020). Roles of VMP1 in autophagy and ER–membrane contact: Potential implications in neurodegenerative disorders. *Frontiers in Molecular Neuroscience*, *13*. https://doi.org/10.3389/fnmol.2020.00042.
Williams, J. A. (2001). Intracellular signaling mechanisms activated by cholecystokinin-regulating synthesis and secretion of digestive enzymes in pancreatic acinar cells. *Annual Review of Physiology*, *63*, 77–97.
Williams, J. A. (2010). Regulation of acinar cell function in the pancreas. *Current Opinion in Gastroenterology*, *26*(5), 478–483. https://doi.org/10.1097/mog.0b013e32833d11c6.

Further reading

ENSG00000062716—Summary—Homo_sapiens—Ensemble genome browser 106. (n.d.). Retrieved May 8, 2022, from https://www.ensembl.org/Homo_sapiens/Gene/Summary?g=ENSG00000062716;r=17:59707192-59842255.
Cooley, M. M., Jones, E. K., Gorelick, F. S., & Groblewski, G. E. (2020). Pancreatic Acinar Cell Protein Synthesis, Intracellular Transport, and Export. *Pancreapedia: Exocrine Pancreas Knowledge Base*, https://doi.org/10.3998/panc.2020.15.
Sans, M. D. (2021). Pancreatic Digestive Enzyme Synthesis and its Translational Control. *Pancreapedia: Exocrine Pancreas Knowledge Base*, https://doi.org/10.3998/panc.2021.09.
Gaisano, H. Y., Dolai, S., & Takahashi, T., (2020). Physiologic Exocytosis in Pancreatic Acinar Cells and Pathologic Fusion Underlying Pancreatitis. *Pancreapedia: Exocrine Pancreas Knowledge Base*. https://doi.org/10.3998/panc.2020.04.
Hart, P. A., & Conwell, D. L. (2021). Secretion of the Human Exocrine Pancreas in Health and Disease. *Pancreapedia: Exocrine Pancreas Knowledge Base*, https://doi.org/10.3998/panc.2021.02.

CHAPTER SEVEN

Deciphering the effect of mutations in MMAA protein causing methylmalonic acidemia—A computational approach

N. Madhana Priya[a,†], S. Udhaya Kumar[b,†], D. Thirumal Kumar[c], R. Magesh[a], R. Siva[b], R. Gnanasambandan[b], and C. George Priya Doss[b,*]

[a]Department of Biotechnology, Sri Ramachandra Institute of Higher Education and Research (DU), Chennai, Tamil Nadu, India
[b]Laboratory of Integrative Genomics, Department of Integrative Biology, School of Bio Sciences and Technology, Vellore Institute of Technology, Vellore, Tamil Nadu, India
[c]Meenakshi Academy of Higher Education and Research (Deemed to be University), Chennai, India
*Corresponding author: e-mail address: georgepriyadoss@vit.ac.in

Contents

† Authors contributed equally to this work.

Advances in Protein Chemistry and Structural Biology, Volume 132
ISSN 1876-1623
https://doi.org/10.1016/bs.apcsb.2022.07.003

Abstract

Methylmalonic acidemia (MMA) is a rare genetic disorder affecting multiple body systems. We aimed to investigate the pathogenic mutations in MMAA that are associated with isolated methylmalonic acidemia to identify the structural behavior of MMAA upon mutation. The algorithms such as PredictSNP, iStable, ConSurf, and Align GVGD were employed to analyze the consequence of the mutations. Molecular docking was carried out for the native MMAA, L89P, G274D, and R359G to interpret its interactions with the GDP substrate. The docked complexes were simulated for 200ns aiding GROMACS in apprehending the behavior of MMAA upon mutation and GDP binding. After simulation, cα disruptions were observed using the RMSF plot, which indicated that several regions of mutant MMAAs have highly fluctuated. The gyration and H-bond plots were used to understand the compactness and intermolecular interaction with the GDP molecule. The MDS analysis showed that the mutations L89P, G274D, and R359G are highly unstable even after GDP binding, with the least compactness, fewer H-bonds, and larger conformational cα motions. Our study provided structural and dynamic insights into MMAA protein, which further helps to characterize these mutants and provide potential treatment strategies for MMA patients.

Abbreviations

ED essential dynamics
HGMD human gene mutation database
MDS molecular dynamics simulation
MMA methylmalonic acidemia
PCA principal component analysis

1. Introduction

Methylmalonic acidemia is a condition of aberrant metabolism that is fatal, severe and causes multi-system injury. The earliest reports occurred in 1967 (Zhou et al., 2018). Two primary manifestations of the condition have been distinguished based on phenotype: solo methylmalonic acidurias, mixed methylmalonic aciduria, and homocystinuria (Lindblad et al., 1968). In contrast to paired methylmalonic aciduria and homocystinuria, which is characterized by increased plasma homocysteine and reduced levels of the coenzymes adenosylcobalamin and methylcobalamin, isolated methylmalonic acidurias is caused by defects of methylmalonyl-CoA mutase or even the synthesis of the MUT coenzyme adenosyl (MeCbl). Certain benign MMA subtypes have also been reported (Li et al., 2014).

The enzyme classified as methylmalonyl CoA mutase requires the molecule adoCbl, which is generated from vitamin B12 (commonly known as cobalamin), to operate normally. Several proteins, lipids, and cholesterol can be broken down with the aid of this enzyme (L. Han et al., 2015). One of the final steps in producing AdoCbl is the transportation of vitamin B12 into mitochondria, which appears to be mediated by the MMAA protein. B12 vitamin is subsequently converted into AdoCbl by further chemical processes. The making of the protein adenosylcobalamin (AdoCbl) is made possible by the *MMAA* gene. According to recent research, the MMAA protein can assist and maintain the activity of methylmalonyl-CoA mutase and prevent it from being switched off or inactivated (B. Han et al., 2016). Cobalamin (Cbl) metabolism within cells depends on the *MMAA* gene product. Disorders of the cblA class characterized by isolated methylmalonic aciduria are caused by mutations in this gene (Dempsey-Nunez et al., 2012).

A wide range of clinical manifestations of MMA exists, from a benign illness to fetal neonatal syndrome. The MMA symptoms can appear at any age, from infancy to adulthood. Anorexia, failure to grow, hypotonia, growth retardation, progressive kidney failure, functional immune dysfunction, optic nerve atrophy, as well as hematologic abnormalities are typical symptoms in affected children (Chandler et al., 2015). It was recently revealed that most of the symptoms in children having combined MMA and homocysteinemia were late-onset diffuse lung illnesses (DLD) (Liu et al., 2017). It is still unknown if those with atypical and "benign"/ adult MMA may experience symptoms, even when there is a slight increase in methylmalonate excretion in the urine (Manoli et al., 2013). The prognosis of long-term survival is poor, and patients with MMA suffer from significant morbidity and death (Matsui et al., 1983). The mortality rate of MUT MMA ranges from 60% to 88% in the early studies from the 1980s and 1990s, but by the first decade of the 2000s, it had somewhat decreased to 40% (Harrington et al., 2016). The results of prospective cohort research on 45 Chinese pediatric patients with homocystinuria and MMA between 2006 and 2012 demonstrated that increased urea and urine methylmalonic acid are indicators of worse consequences for the patients (Qiliang et al., 2015).

The primary symptoms of MMA include ketosis, acidosis, hyperammonemia, hyperglycemia, hypoglycemia, and neutropenia (Han et al., 2015). Developmental delay, tubulointerstitial nephritis along with progressive renal failure, metabolic stroke (involvement of the acute and chronic basal ganglia), disabling movement disorder with choreoathetosis, dystonia, pancreatitis, functional immune impairment, growth failure, and optic nerve atrophy are some of the most significant secondary complications of MMA (Liu et al., 2010). Several studies utilized computational algorithms to determine enzymes' structural and behavioral mechanisms upon mutation in cancer and genetic disorders (Kumar et al., 2021; Kumar, Thirumal Kumar, Siva, Doss, & Zayed, 2020; Udhaya Kumar & George Priya Doss, 2021; Udhaya Kumar et al., 2020, 2021). These studies help clinical researchers to comprehend the distinct mutational mechanisms and provide precision therapy for the patients. Similarly, our research utilized computational algorithms to comprehend the mutations and their structural behavior in MMAA protein.

2. Materials and methods

2.1 Data retrieval and collection of mutations

The biological information and the protein sequence of MMAA were retrieved from UniProt. Missense mutations were collected from the UniProt (UniProt ID-Q8IVH4), HGMD, and Research Articles (PubMed and Google Scholar). The search terms used for collecting the mutations were "MMAA Mutations" and "*Homo sapiens*" (Consortium, T. U., & The UniProt Consortium, 2019; Cooper, 1998), The 3-D structure of MMAA protein was retrieved from the RCSB database with the PDB ID: 2WWW (Froese et al., 2010).

2.2 Prediction of pathogenicity

PredictSNP is a compilation of various algorithms, including MAPP, PhD-SNP, PolyPhen1, PolyPhen2, SIFT, SNAP, PANTHER, and nsSNP analyzer (https://loschmidt.chemi.muni.cz/predictsnp/). These tools offer information regarding the type of mutation as to whether they are deleterious or neutral (Bendl et al., 2014). MAPP stands for Multivariate Analysis of Protein Polymorphism and estimates the protein's function on amino acid substitutions. PhD-SNP is used to compute the effect of the SNPs in the gene. The effect of the amino acid substitution on the protein's function was evaluated by PolyPhen (Polymorphism Phenotyping) and Sorting Intolerant from Tolerant (SIFT). The functional significance of the SNPs is predicted by SNAP and other computational tools such as nsSNPAnalyser for the substitutions' phenotypic effect (Bendl et al., 2016).

2.3 Stability analysis for SNPs

Predicting the stability of the mutations is essential as the mutations alter the protein's function by altering the amino acid's chemical entities. The stability was calculated using the iStable web server. iStable computes the stability of the structure upon SNPs as they alter the protein structure and its function. iStable tool predictors such as I-Mutant 2.0, MUpro, and iStable (Capriotti, Fariselli, & Casadio, 2005; Chen, Lin, Liao, Chang, & Chu, 2020; Cheng, Randall, & Baldi, 2006). The combination of these three tools increases the efficiency of the predictions. The fasta sequence of MMAA protein was given as input, and steadiness variations up on each mutation were checked. The results ranged from increase to decrease, which signifies that the protein stability was compromised when the decrease condition occurred.

2.4 Conservational analysis

Mutations that fall in the conserved regions play a crucial role in disease mechanism as it has an increasingly deleterious effect. The fasta sequence was given as input to test the degree of evolutionary conservation of the amino acid with the aid of the ConSurf web server (Glaser et al., 2003). The output of the tool provides the conservational scores from highly conserved (score 9) to the least conserved (score 0) amino acid at its position (Ashkenazy, Erez, Martz, Pupko, & Ben-Tal, 2010, Ashkenazy et al., 2016; Glaser et al., 2003).

2.5 Biophysical property analysis

The Align GVGD is a quantitative measure to identify the mutations that interfere with the protein's function by using the amino acids' protein MSAs and biophysical features. GV measures the perceived range of variation, and GD computes the distance between missense substitutions (Tavtigian et al., 2006).

2.6 Mutant structure models

The most prevalent mutations were selected and imposed on the native structure using UCSF Chimera and Discovery Studio. Energy minimization was carried out for the mutants using a SwissPDB viewer (Guex & Peitsch, 1997).

2.7 Docking

Docking was performed to determine how the amino acid alteration influences the binding of the substrate GDP with the receptor structures. MMAA exhibits GTPase activity modulated by MUT, and the two proteins interact *in vitro* and *in vivo*. The formation of a stable MMAA-MUT complex is nucleotide-selective for MMAA (GMPPNP over GDP) and apoenzyme-dependent for MUT (Froese et al., 2010). The receptor was prepared by adding polar regions followed by Kolmann & Compute Gasteiger charges and AD4 Atoms. The ligand was imported, and the associated torsions were detected, defined, and saved in the pdbqt format. The grid parameter file was created by fixing the active site using a grid of 60points in x, y, and z dimensions (Morris et al., 2009). The Lamarckian GA parameters used in the analysis were 150 for the population size, 25,000,000 for the maximum number of energy assessments, 27,000 for the number of generations, 0.02 for the mutation rate, and 0.8 for the boundary level (Morris et al., 1998, 2009). Docking was performed in triplicate, and the average scores were computed from the triple run, and the resulting complex structures, both native and mutant, were visualized using the UCSF Chimera and Discovery Studio academic visualizer v20 (BIOVIA, Dassault Systèmes, 2020). The complex structures were further energy minimized as a preliminary step for molecular simulation.

2.8 Molecular dynamics

MDS was performed for the native and mutant complexes using GROMACS Package. The topology of the structures was built, implying a GROMOS 54A7 force field (Schmid et al., 2011). The structure was placed in the middle of a dodecahedron box at a distance of 1nm from the edges filled with SPC as the solvent. In order to neutralize the system few of the solvent molecules were replaced with counter ions (Na or Cl) (Sharma, Kumar, & Chandra, 2019). The system was energy minimized by up to 5000 steps to repair the asymmetrical geometries in the structure. Further, the system was equilibrated by NVT and NPT for 50,000 ps to stabilize the temperature and pressure, closely resembling experimental conditions (Miyamoto & Kollman, 1992). Upon the completion of equilibration, MD was run for 200 ns, and the trajectories were analyzed (Junghans, Phillips, & Wall, 2014; Lindahl, Hess, & van der Spoel, 2020). Principal Component Analysis also known as ED was performed to analyze the conformational space taken by the complexes. The gmx anaeig command was used to build the 2D projection of eigenvalues and eigenvectors maintaining the matrix abd covariance (Amadei, Linssen, & Berendsen, 1993; David & Jacobs, 2014).

2.9 Trajectory analysis

MDS was analyzed for RMSF, Radius of Gyration, PCA, and H-bond with gmx rmsf, gmx gyrate, gmx covar, and gmx hbond plugins used, respectively (Lindahl, Hess, & van der Spoel, 2001). Post-analysis, the graphs were plotted using Xmgrace software (Moorhouse & Barry, 2005).

3. Results

3.1 Retrieval of data

A total of 22 mutations from UniProt, 21 from HGMD, and 27 from the literature were taken. After removing the overlapping ones from the selected mutations, 33 mutants were finalized. The 33 mutations were taken for further analysis, implying computational algorithms. The retrieved PDB structure 2WWW was identified by X-ray diffraction method with resolution 2.64 Å, a Homo-2-mer with the substrate GDP co-crystallized along with it.

3.2 Pathogenicity and stability analysis

Out of 33 mutations, 11 of them were identified to be deleterious in all the pathogenicity prediction analysis tools (Supplementary Table 1 in the online version at https://doi.org/10.1016/bs.apcsb.2022.07.003). Initially, out of the 33 mutations, the PredictSNP computed 92% of the mutants to be deleterious; similarly, other tools such as PhD SNP, PolyPhen, SIFT, SNAP, and Panther predicted 75%, 84%, 69%, 72%, 78% and 45% of the mutants to be deleterious, respectively (Table 1). Finally, 33% of the total mutations were predicted to be highly deleterious after considering

Table 1 Pathogenicity results of the identified MMAA mutations.

Accession ID	Mutation	Predict SNP	Ph-D SNP	PolyPhen 1	PolyPhen 2	SIFT	SNAP	Panther
VAR_020835	L89P	D	D	D	D	D	D	D
VAR_020836	R145Q	D	D	D	D	D	D	D
VAR_080015	G147E	D	D	D	D	D	D	D
VAR_017202	Y207C	D	D	D	D	D	D	D
VAR_080020	V220M	D	D	D	D	D	D	D
VAR_080021	I241F	D	D	D	D	D	D	D
VAR_071921	G274D	D	D	D	D	D	D	D
VAR_080025	A287D	D	D	D	D	D	D	D
VAR_038804	R359G	D	D	D	D	D	D	D
VAR_020838	R359Q	D	D	D	D	D	D	D
VAR_080030	G399V	D	D	D	D	D	D	D

D—denotes deleterious.

Table 2 Stability analysis of mutations identified from MMAA protein.

Accession ID	Mutation	I-Mutant	MUpro	iStable
VAR_020835	L89P	D	D	D
VAR_017202	Y207C	D	D	D
VAR_080021	I241F	D	D	D
VAR_071921	G274D	D	D	D
VAR_038804	R359G	D	D	D
VAR_020838	R359Q	D	D	D

D—denotes destabilizing.

all the tools in common. These 11 mutations were computed as deleterious and subjected to the stability analysis (iStable). Further, the iStable predicted six mutations to decrease the protein stability (Table 2 and Supplementary Table 2 in the online version at https://doi.org/10.1016/bs.apcsb.2022.07.003). The predicted six destabilizing mutations were VAR_020835 (L89P), VAR_017202 (Y207C), VAR_080021 (I241F), VAR_071921 (G274D), VAR_038804 (R359G), VAR_020838 (R359Q).

3.3 Conservation and biophysical analysis

Conservation analysis was done for the six pathogenic destabilizing mutations, and all the six mutations were noted to fall in the conserved region with scores of 9 and 7. The three mutations (G274D, R359G, and R359Q) with a score of 9, and the mutations (L89P, Y207C, and I241F) with a score of 7 were identified from ConSurf analysis (Table 3 and Fig. 1). Align GVGD was performed for the above six mutations that were considered to be in the conserved region, with highly deleterious and decreased stability properties. The results showed four mutations (L89P, Y207C, G274D, and R359G) to fall under Class 65 by interfering with the protein's function (Table 4). From this, the prevalence of L89P mutations was higher than the Y207C mutation; therefore, to determine the structural differences *via* docking and simulation, we selected L89P, G274D, and R359G mutations from MMAA.

Table 3 Conservation scores of shortlisted mutations from MMAA protein.

Accession ID	Mutation	ConSurf
VAR_020835	L89P	7
VAR_017202	Y207C	7
VAR_080021	I241F	7
VAR_071921	G274D	9
VAR_038804	R359G	9
VAR_020838	R359Q	9

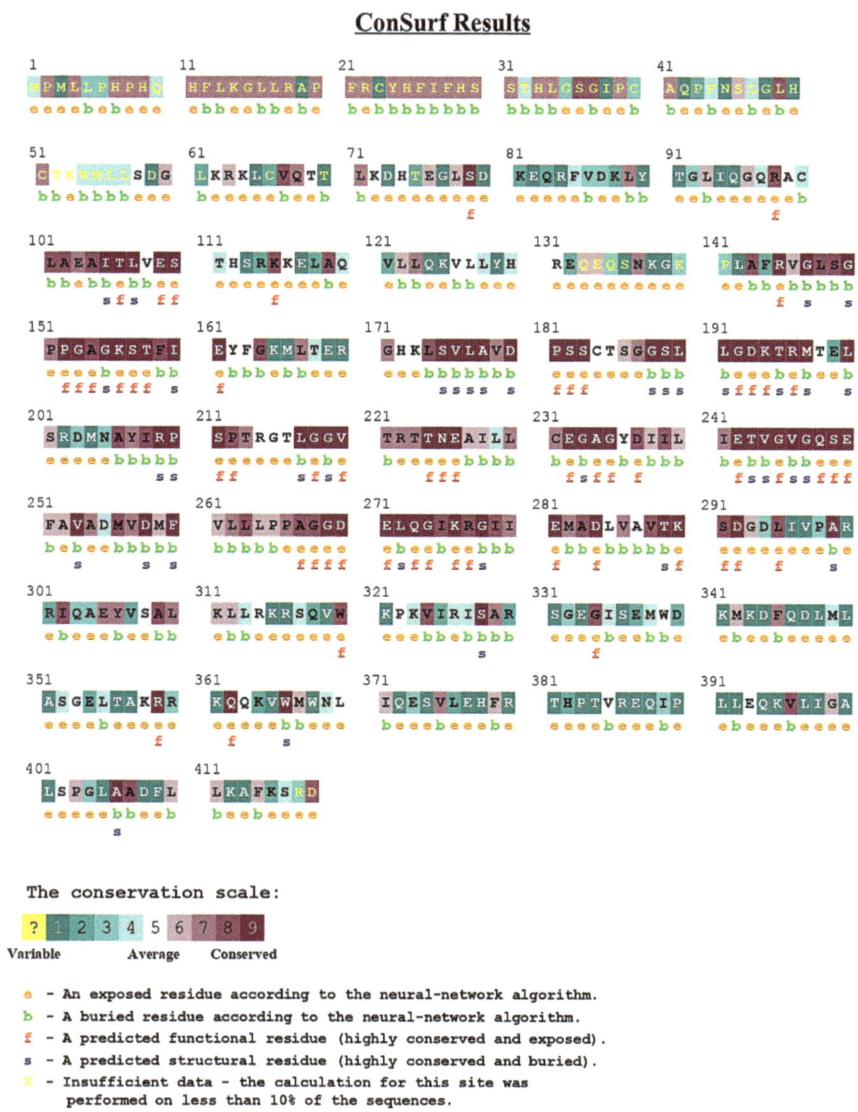

Fig. 1 ConSurf analysis depicting conserved and variable regions. The conservation scores range from 1 to 9, with the most conserved residues with a score of 7 to 9; variable residues from scores 1–4 and neutral residues fall under the scores 5 and 6.

Table 4 Biophysical property analysis of mutations identified from MMAA protein.

Accession ID	Mutation	GVGD class
VAR_020835	L89P	Class 65
VAR_017202	Y207C	Class 65
VAR_071921	G274D	Class 65
VAR_038804	R359G	Class 65

3.4 Receptor and mutant model preparation

The structure 2WWW was taken from PDB, and the query coverage was 7–418 amino acids. Since there were many missing residues, the structure was refined, and missing residues were filled using the SwissModel tool. The selected three mutations, VAR_020835 (L89P), VAR_071921 (G274D), and VAR_038804 (R359G), were imposed on the native structure, and energy was minimized. The refined energy minimized structures were taken for docking with the substrate GDP. The simulation was performed further to determine the less or large hindering effect on MMAA protein upon mutation.

3.5 GDP interaction profiling

The L89P, G274D, R359G, and the native were docked with the GDP to comprehend the binding affinity and interaction patterns. The docking score was −5.98 (kcal/mol) for the native and higher in mutants, with scores (kcal/mol) -10.14, -10.56, and -10.69 for L89P, G274D, and R359G, respectively (Table 5). Minor changes were observed in the receptor structure due to substrate binding. The 2D interaction represented in Fig. 2 explains the several interactions, including van der Waals, salt bridge, H-bonds, and Pi-alkyl, formed between the GDP and MMAA. The residue interactions with GDP are consistent with the docking results for native and mutant MMAA proteins.

3.6 Molecular dynamics

The native and mutant structures were simulated for 200 ns each to observe the interaction and dynamical behavior changes during the course of the

Table 5 Computational binding affinity of GDP molecule with the native and mutant structures.

S. No.	Protein	Docking score (kcal/mol)
1.	Native	−5.98
2.	L89P	−10.14
3.	G274D	−10.56
4.	R359G	−10.69

The binding affinity is calculated by the unit kcal/mol.

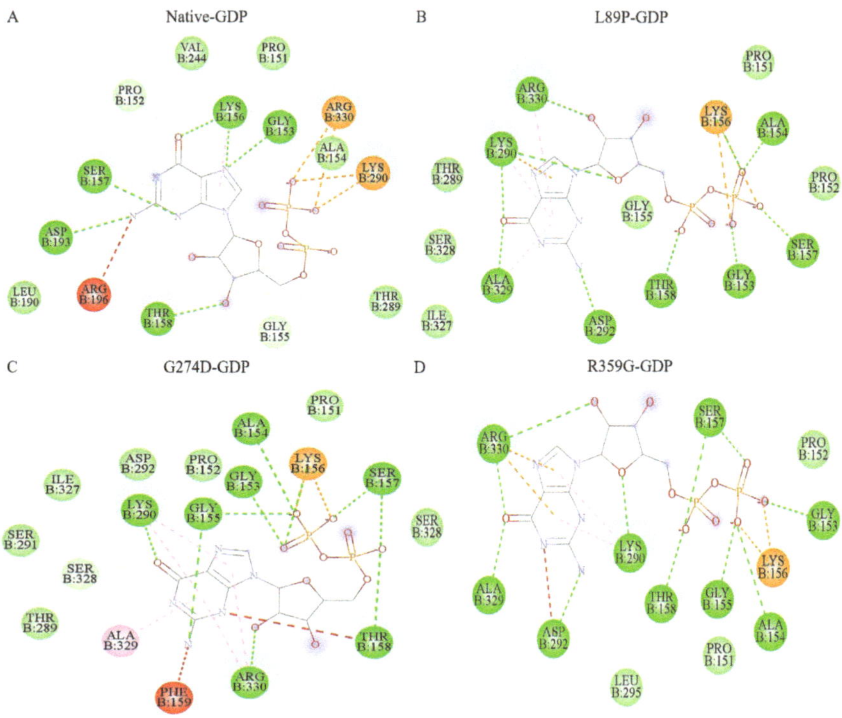

Fig. 2 2D interactions of docked MMAA complexes. The figure shows the interacting residues and the surrounding residues, which can have influence on the binding of the GDP substrate. (A) Shows 8 interacting residues for the native; (B) 9 interacting residues for L89P; (C) 10 interacting residues with G274D; (D) 9 interacting residues for R359G. Among the identified interactions, the dark green dotted lines are the conventional H-bonds, light green dotted lines depict the van der Waals, orange represents salt bridges, and red signifies unfavorable donor/acceptor interactions, respectively.

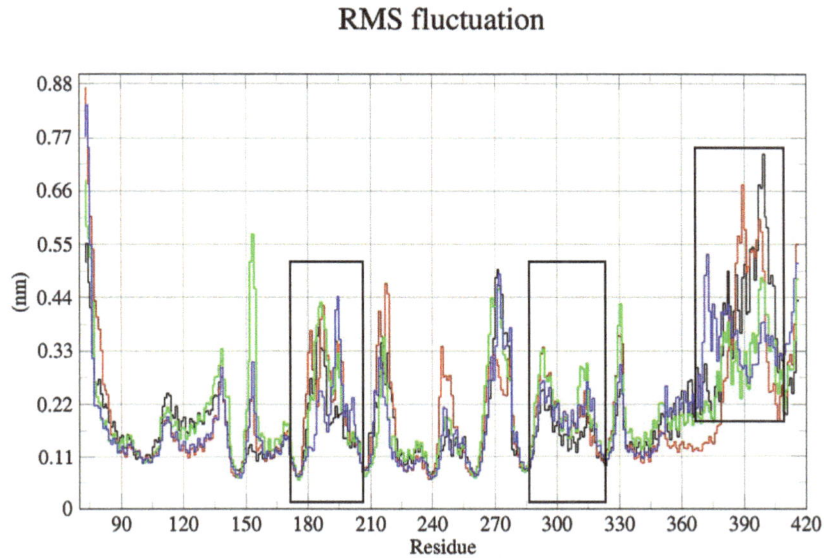

Fig. 3 RMSF plot of the complexes with GDP substrate. The graph colors are as follows: black, red, green, and blue indicate the fluctuation is seen in native, L89P, G274D, and R359G complexes, respectively. Y-axis indicates the fluctuation in nm, and the X-axis shows the time in ns.

simulation. The RMSF analysis is mainly done to observe the structure fluctuation on binding with GDP. The RMSF clearly shows very minimal fluctuations between the native complex in comparison to mutant ones (Fig. 3). The regions marked with the black box depict the changes in the secondary structure (α-helix and β-sheets) of MMAA native and mutants. Fluctuations were observed between 180 and 210, 280–320, and 371–410 residues of all the mutant MMAA proteins. The compactness and H-bond interaction further confirm the RMS fluctuation by utilizing gmx gyrate and hbond command from MDS trajectories. The L89P and G274D mutation showed higher deviation than native and R359G. Similarly, L89P and G274D had lesser H-bonds than native and R359G (Fig. 4). Further, PCA predicts the protein dynamics (Cα motions) occurred due to the influence of the mutations. As predicted in PCA, G274D and R359G showed high motion changes, whereas native and L89P showed lesser motion changes (Fig. 5).

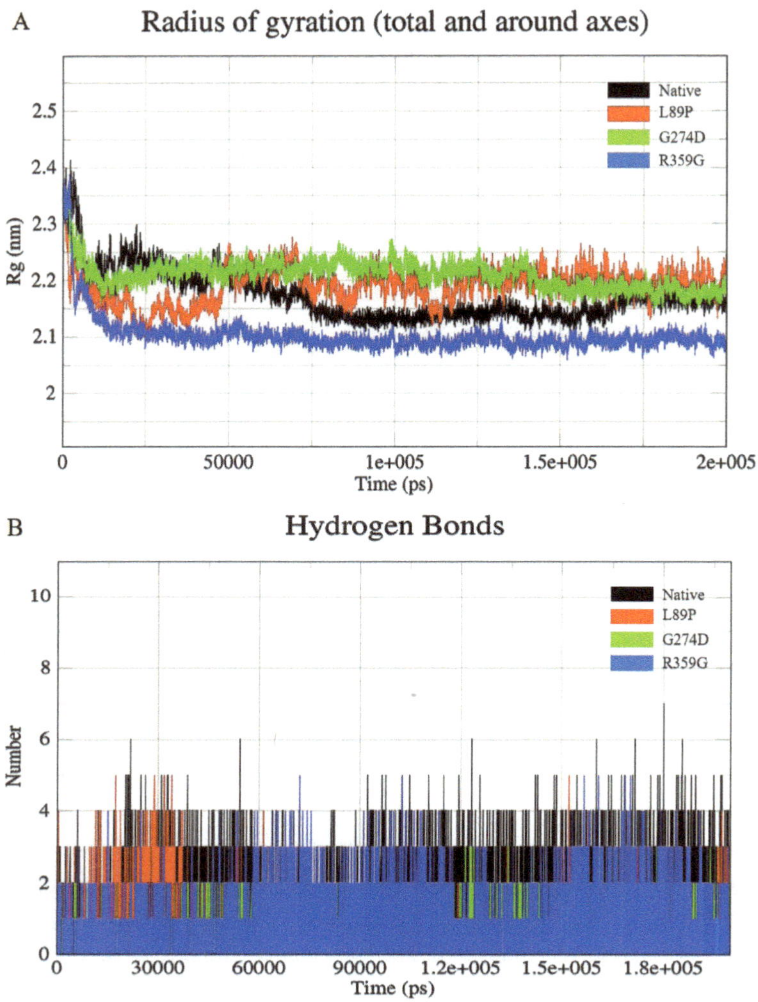

Fig. 4 (A) The Rg plot indicates the compactness of the MMAA complexes. (B) The intermolecular H-bond graph of the MMAA complexes during the simulation period. The graph colors are as follows: black, red, green, and blue indicate the native, L89P, G274D, and R359G complexes, respectively.

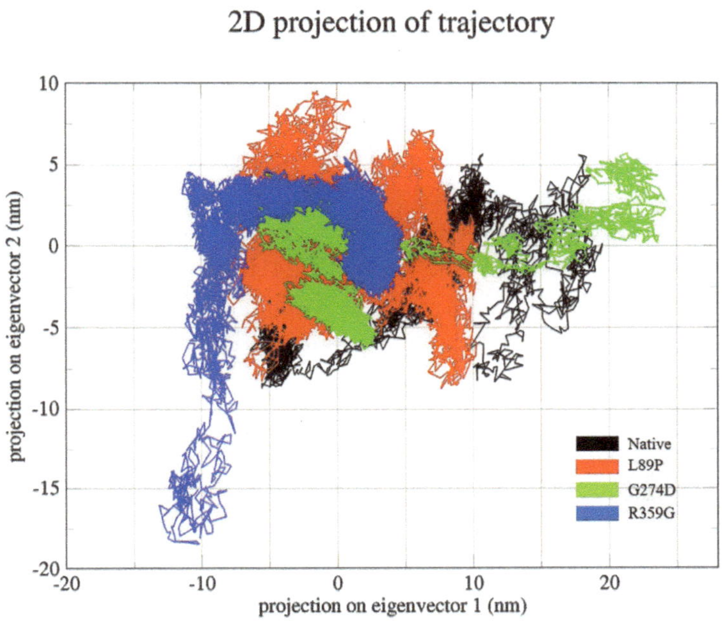

Fig. 5 PCA analysis in the 2D projection of the MMAA complexes. The conformational space in terms of stability is depicted as a graph for native MMAA and mutants. The graph colors are as follows: black, red, green, and blue indicate the native, L89P, G274D, and R359G complexes, respectively.

4. Discussion

Methylmalonic acidemia (MMA) is a category of hereditary disorders wherein the body seems unable to metabolize various proteins and fats (lipids) effectively. People with this illness suffer from converting, or "metabolizing," methymalonyl-coenzyme A (Keyfi, Talebi, & Varasteh, 2016; Matsui, Mahoney, & Rosenberg, 1983). As a result, methylmalonic acid accumulates in the body. The *MMAA* gene codes for a protein involved in synthesizing a chemical known as adenosylcobalamin (AdoCbl) (Tang, Chen, & Liu, 2020; Yang et al., 2004). A total of 33 mutations were retrieved and analyzed using computational algorithms. The analysis of these mutations provides a significant understanding mutant's role in phenotypic and genotypic disease conditions (Sohail, Berg, & Vecchyo, 2022). The most effective mutation can be identified and validated by aiding certain computational tools in combination.

In this study, the pathogenicity of the mutations using the PredictSNP tool initially resulted in deleterious or neutral. PredictSNP is a combination of prediction tools such as PredictSNP, PhD-SNP, PolyPhen1, PolyPhen2, SIFT, SNAP, and PANTHER (Adzhubei et al., 2013; Bendl et al., 2014; Capriotti et al., 2006; Johnson et al., 2008; Pers et al., 2015; Sim et al., 2012). This analysis predicted 11 mutations [VAR_020835 (L89P), VAR_020836 (R145Q), VAR_080015 (G147E), VAR_017202 (Y207C), VAR_080020 (V220M), VAR_080021 (I241F), VAR_071921 (G274D), VAR_080025 (A287D), VAR_038804 (R359G), VAR_020838 (R359Q), and VAR_080030 (G399V)] to be deleterious. These mutations, with their deleterious characteristic, influence the enzyme function, directly triggering the disease condition (Table 1). I-Mutant 2.0, MUpro, and iStable were employed to predict the effects of the pathogenic mutations on the stability of the structure, with the result being increased or decreased (Capriotti et al., 2005; Chen et al., 2020; Cheng et al., 2006). Among the 11 pathogenic mutations, six were predicted [VAR_020835 (L89P), VAR_017202 (Y207C), VAR_080021 (I241F), VAR_071921 (G274D), VAR_038804 (R359G), and VAR_020838 (R359Q)] to show decreased stability in the entire stability analysis tools (Table 2). As a result, these mutations with decreased stability disturb the protein's function, altering its structure by degradation and misfolding.

The evolutionary conservation analysis helps determine the mutations' severity and effects. The mutations located at the conserved region are predicted to be with severe effect than the mutations located at the variable region (Ashkenazy et al., 2010, 2016). ConSurf analysis revealed that all six destabilizing mutations were located in the conserved region (Table 3). On analysis of the ConSurf results, the mutations such as [VAR_020835 (L89P), VAR_017202 (Y207C), VAR_080021 (I241F) had a score of 7 with the conservancy, and VAR_071921 (G274D), VAR_038804 (R359G), VAR_020838 (R359Q) had score 9 with the high conservancy. Analysis of biophysical properties of amino acid substitution shows the phenotype appearance. The results range from Class 15 to Class 65, with Class 65 showing high phenotypic significance (Yadegari & Majidzadeh, 2019). The four mutations such as VAR_020835 (L89P), VAR_017202 (Y207C), VAR_071921 (G274D), and VAR_038804 (R359G), show phenotypic effects with a score of C65. In the case of these mutations; in L89P: Leucine, a non-polar residue is substituted by another non-polar residue proline making the charge neutral; in Y207C: polar tyrosine is substituted by non-polar cysteine being a neutral mutation change; in G274D: glycine is non-polar substituted by aspartic acid a polar residue inducing negative charge, and in R359G: arginine is polar, and the substituted glycine is also polar making the structure positive.

Among the four mutations, three of them [VAR_020835 (L89P), VAR_071921 (G274D), and VAR_038804 (R359G)] were shortlisted based on their prevalence and susceptibility to the disease from the literature (Brennerová et al., 2021; Dempsey-Nunez et al., 2012; Lerner-Ellis et al., 2004; Plessl et al., 2017; Yang et al., 2004). The selected mutations were prepared and docked with the GDP. The GDP molecule is highly essential to stabilize the MMAA-MUT complex. The physiological importance of this interaction is highlighted by a recently identified homoallelic patient mutation of MMAA (Froese et al., 2010). On this basis, we investigated the involvement of GDP with native and mutant MMAA proteins to comprehend their interaction pattern *via* docking and simulation studies. Docking studies reveal the interaction of the receptors with the GDP molecule. The interactions were strong with all four receptors. The native had five H-bonds, van der Waals interaction, one Pi-cation, Pi-alkyl, attractive charge, and unfavorable interaction. Similarly, the mutants also had ~8H-bonds, attractive charge, donor-donor, acceptor-acceptor, Pi-alkyl, Pi-cation, and carbon H-bond (Fig. 2). Compared to the native-GDP complex, all three mutants had distinct interactions (Edwards et al., 2015; Padovani et al., 2006). The dynamics results showed that the native and mutations had slight fluctuations, depicting the interaction with GDP. The fluctuation seen in the RMSF plot might be the potential reason for the differences in the binding of the molecule with the receptor, The Rg did not show very distinct differences but the compactness varied among the receptors, especially in the mutants (L89P and G274D). The H-bonds also play a significant role in understanding the interaction profile that depicts the conformational changes in substrate interaction. Lesser H-bonds were calculated for the L89P and G274D complexes. Furthermore, with the PCA, it was identified that G274D and R359G exhibited more cα motions. The complex with high conformational space is said to be less stable. In this case, the G274D and R359G are predicted to be unstable with large cα motions, and more compactness was seen with L89P even after GDP binding. Therefore, our study computed that L89P, G274D, and R359G have a high efficiency with severe phenotypic disease conditions in methylmalonic acidemia.

5. Conclusion

Methylmalonic acidemia is an inherited disorder that refers to the condition where the body is unable to break down proteins or lipids. This study is the first to predict the potentially damaging mutation using *in silico* tools. We report three highly pathogenic and destabilizing mutations that influence the function of MMAA. The mutants were identified to have a better binding affinity with GDP than the native did. These interactions could possibly affect the GTP hydrolysis process and disrupt the MMAA-MUT complex formation. We also performed 200 ns MD simulations to explore mutations that interact with GDP, and their results revealed several significant structural alterations, particularly for L89P and the G274D mutation. These findings may be considered in designing experimental studies and provide novel insights into MMAA structural mechanisms.

Acknowledgments

The authors would like to thank the Vellore Institute of Technology (VIT), Vellore, Tamil Nadu, India, and Sri Ramachandra Institute of Higher Education and Research, Chennai, India, for providing the necessary research facilities and encouragement to carry out this work.

Author contributions

MPN, UKS, TKD, GR, and GPDC, contributed to designing the study and data acquisition, analysis, and interpretation. MPN, UKS, TKD, GR, and MR are involved in the acquisition, analysis, interpretation of the results, and manuscript drafting. GPDC, SR, and MR supervised the entire study and edited the manuscript. The manuscript was reviewed and approved by all the authors.

Role of funding source

No funding agency is involved in the present study.

References

Adzhubei, I., Jordan, D. M., & Sunyaev, S. R. (2013). Predicting functional effect of human missense mutations using PolyPhen-2. *Current Protocols in Human Genetics*. Editorial Board, Jonathan L. Haines ... et Al. Chapter 7, Unit 7.20.

Amadei, A., Linssen, A. B. M., & Berendsen, H. J. C. (1993). Essential dynamics of proteins. *Proteins: Structure, Function, and Genetics*, *17*(4), 412–425. https://doi.org/10.1002/prot.340170408.

Ashkenazy, H., Abadi, S., Martz, E., Chay, O., Mayrose, I., Pupko, T., et al. (2016). ConSurf 2016: An improved methodology to estimate and visualize evolutionary conservation in macromolecules. *Nucleic Acids Research*, *44*(W1), W344–W350.

Ashkenazy, H., Erez, E., Martz, E., Pupko, T., & Ben-Tal, N. (2010). ConSurf 2010: Calculating evolutionary conservation in sequence and structure of proteins and nucleic acids. *Nucleic Acids Research*, *38*(Issue Web Server), W529–W533. https://doi.org/10.1093/nar/gkq399.

Bendl, J., Musil, M., Štourač, J., Zendulka, J., Damborský, J., & Brezovský, J. (2016). PredictSNP2: A unified platform for accurately evaluating SNP effects by exploiting the different characteristics of variants in distinct genomic regions. *PLoS Computational Biology*, *12*(5), e1004962.

Bendl, J., Stourac, J., Salanda, O., Pavelka, A., Wieben, E. D., Zendulka, J., et al. (2014). PredictSNP: Robust and accurate consensus classifier for prediction of disease-related mutations. *PLoS Computational Biology*, *10*(1), e1003440.

BIOVIA, Dassault Systèmes. (2020). *Discovery studio client. V.20.1.0.19295*. English, San Diego: Dassault Systèmes. https://discover.3ds.com/discovery-studio-visualizer-download.

Brennerová, K., Škopková, M., Ostrožlíková, M., Šaligová, J., Staník, J., Bzdúch, V., et al. (2021). Genetic testing is necessary for correct diagnosis and treatment in patients with isolated methylmalonic aciduria: A case report. *BMC Pediatrics*, *21*(1), 578.

Capriotti, E., Calabrese, R., & Casadio, R. (2006). Predicting the insurgence of human genetic diseases associated to single point protein mutations with support vector machines and evolutionary information. *Bioinformatics*, *22*(22), 2729–2734.

Capriotti, E., Fariselli, P., & Casadio, R. (2005). I-Mutant2.0: Predicting stability changes upon mutation from the protein sequence or structure. *Nucleic Acids Research*, *33*(Issue Web Server), W306–W310. https://doi.org/10.1093/nar/gki375.

Chandler, R. J., LaFave, M. C., Varshney, G. K., Trivedi, N. S., Carrillo-Carrasco, N., Senac, J. S., et al. (2015). Vector design influences hepatic genotoxicity after adeno-associated virus gene therapy. *The Journal of Clinical Investigation*, *125*(2), 870–880.

Chen, C.-W., Lin, M.-H., Liao, C.-C., Chang, H.-P., & Chu, Y.-W. (2020). iStable 2.0: Predicting protein thermal stability changes by integrating various characteristic modules. *Computational and Structural Biotechnology Journal*, *18*, 622–630. https://doi.org/10.1016/j.csbj.2020.02.021.

Cheng, J., Randall, A., & Baldi, P. (2006). Prediction of protein stability changes for single-site mutations using support vector machines. *Proteins*, *62*(4), 1125–1132.

Consortium, T. U., & The UniProt Consortium. (2019). UniProt: A worldwide hub of protein knowledge. *Nucleic Acids Research*, *47*(D1), D506–D515. https://doi.org/10.1093/nar/gky1049.

Cooper, D. (1998). The human gene mutation database. *Nucleic Acids Research*, *26*(1), 285–287. https://doi.org/10.1093/nar/26.1.285.

David, C. C., & Jacobs, D. J. (2014). Principal component analysis: A method for determining the essential dynamics of proteins. *Protein Dynamics (Pembroke, Ont.)*, 193–226. https://doi.org/10.1007/978-1-62703-658-0_11.

Dempsey-Nunez, L., Illson, M. L., Kent, J., Huang, Q., Brebner, A., Watkins, D., et al. (2012). High resolution melting analysis of the MMAA gene in patients with cblA and in those with undiagnosed methylmalonic aciduria. *Molecular Genetics and Metabolism*, *107*(3), 363–367.

Edwards, T. E., Baugh, L., Bullen, J., Baydo, R. O., Witte, P., Thompkins, K., et al. (2015). Crystal structures of mycobacterial MeaB and MMAA-like GTPases. *Journal of Structural and Functional Genomics*, *16*(2), 91–99.

Froese, D. S., Kochan, G., Muniz, J. R. C., Wu, X., Gileadi, C., Ugochukwu, E., et al. (2010). Structures of the human GTPase MMAA and vitamin B12-dependent methylmalonyl-CoA mutase and insight into their complex formation. *The Journal of Biological Chemistry*, *285*(49), 38204–38213.

Glaser, F., Pupko, T., Paz, I., Bell, R. E., Bechor-Shental, D., Martz, E., et al. (2003). ConSurf: Identification of functional regions in proteins by surface-mapping of phylogenetic information. *Bioinformatics*, *19*(1), 163–164.

Guex, N., & Peitsch, M. C. (1997). SWISS-MODEL and the Swiss-Pdb viewer: An environment for comparative protein modeling. *Electrophoresis*, *18*(15), 2714–2723. https://doi.org/10.1002/elps.1150181505.

Han, B., Cao, Z., Tian, L., Zou, H., Yang, L., Zhu, W., et al. (2016). Clinical presentation, gene analysis and outcomes in young patients with early-treated combined

methylmalonic acidemia and homocysteinemia (cblC type) in Shandong province, China. *Brain & Development*, *38*(5), 491–497.
Han, L., Han, F., Ye, J., Qiu, W., Zhang, H., Gao, X., et al. (2015). Spectrum analysis of common inherited metabolic diseases in Chinese patients screened and diagnosed by tandem mass spectrometry. *Journal of Clinical Laboratory Analysis*, *29*(2), 162–168.
Han, L.-S., Huang, Z., Han, F., Ye, J., Qiu, W.-J., Zhang, H.-W., et al. (2015). Clinical features and MUT gene mutation spectrum in Chinese patients with isolated methylmalonic acidemia: Identification of ten novel allelic variants. *World Journal of Pediatrics: WJP*, *11*(4), 358–365.
Harrington, E. A., Sloan, J. L., Manoli, I., Chandler, R. J., Schneider, M., McGuire, P. J., et al. (2016). Neutralizing antibodies against adeno-associated viral capsids in patients with mut methylmalonic acidemia. *Human Gene Therapy*, *27*(5), 345–353.
Johnson, A. D., Handsaker, R. E., Pulit, S. L., Nizzari, M. M., O'Donnell, C. J., & de Bakker, P. I. W. (2008). SNAP: A web-based tool for identification and annotation of proxy SNPs using HapMap. *Bioinformatics*, *24*(24), 2938–2939. https://doi.org/10.1093/bioinformatics/btn564.
Junghans, C., Phillips, J., & Wall, M. (2014). *Gromacs IC Tutorial*. https://doi.org/10.2172/1136453.
Keyfi, F., Talebi, S., & Varasteh, A.-R. (2016). Methylmalonic acidemia diagnosis by laboratory methods. *Reports of Biochemistry & Molecular Biology*, *5*(1), 1.
Kumar, S. U., Sankar, S., Kumar, D. T., Younes, S., Younes, N., Siva, R., et al. (2021). Molecular dynamics, residue network analysis, and cross-correlation matrix to characterize the deleterious missense mutations in GALE causing galactosemia III. *Cell Biochemistry and Biophysics*, *79*(2), 201–219.
Kumar, S. U., Thirumal Kumar, D., Siva, R., Doss, C. G. P., & Zayed, H. (2020). An extensive computational approach to analyze and characterize the functional mutations in the galactose-1-phosphate uridyl transferase (GALT) protein responsible for classical galactosemia. *Computers in Biology and Medicine*, *117*, 103583. https://doi.org/10.1016/j.compbiomed.2019.103583.
Lerner-Ellis, J. P., Dobson, C. M., Wai, T., Watkins, D., Tirone, J. C., Leclerc, D., et al. (2004). Mutations in the MMAA gene in patients with the cblA disorder of vitamin B12 metabolism. *Human Mutation*, *24*(6), 509–516.
Li, Y., Peng, T., Wang, X., Duan, R., Gao, H., Guan, W., et al. (2014). A primary study on down-regulated miR-9-1 and its biological significances in methylmalonic acidemia. *Journal of Molecular Neuroscience: MN*, *53*(2), 280–286.
Lindahl, E., Hess, B., & van der Spoel, D. (2001). GROMACS 3.0: A package for molecular simulation and trajectory analysis. *Journal of Molecular Modeling*, *7*(8), 306–317. https://doi.org/10.1007/s008940100045.
Lindahl, A., Hess, B., & van der Spoel, D. (2020). *GROMACS 2020 source code*. https://doi.org/10.5281/zenodo.3562495.
Lindblad, B., Lindblad, B. S., Olin, P., Svanberg, B., & Zetterström, R. (1968). Methylmalonic acidemia. A disorder associated with acidosis, hyperglycinemia, and hyperlactatemia. *Acta Paediatrica Scandinavica*, *57*(5), 417–424.
Liu, J., Peng, Y., Zhou, N., Liu, X., Meng, Q., Xu, H., et al. (2017). Combined methylmalonic acidemia and homocysteinemia presenting predominantly with late-onset diffuse lung disease: A case series of four patients. *Orphanet Journal of Rare Diseases*, *12*(1), 58.
Liu, M.-Y., Yang, Y.-L., Chang, Y.-C., Chiang, S.-H., Lin, S.-P., Han, L.-S., et al. (2010). Mutation spectrum of MMACHC in Chinese patients with combined methylmalonic aciduria and homocystinuria. *Journal of Human Genetics*, *55*(9), 621–626.
Manoli, I., Sysol, J. R., Li, L., Houillier, P., Garone, C., Wang, C., et al. (2013). Targeting proximal tubule mitochondrial dysfunction attenuates the renal disease of methylmalonic

acidemia. *Proceedings of the National Academy of Sciences of the United States of America, 110*(33), 13552–13557.

Matsui, S. M., Mahoney, M. J., & Rosenberg, L. E. (1983). The natural history of the inherited methylmalonic acidemias. *The New England Journal of Medicine, 308*(15), 857–861.

Miyamoto, S., & Kollman, P. A. (1992). Settle: An analytical version of the SHAKE and RATTLE algorithm for rigid water models. *Journal of Computational Chemistry, 13*(8), 952–962. https://doi.org/10.1002/jcc.540130805.

Moorhouse, M., & Barry, P. (2005). *Bioinformatics biocomputing and Perl: An introduction to bioinformatics computing skills and practice.* John Wiley & Sons.

Morris, G. M., Goodsell, D. S., Halliday, R. S., Huey, R., Hart, W. E., Belew, R. K., et al. (1998). Automated docking using a Lamarckian genetic algorithm and an empirical binding free energy function. *Journal of Computational Chemistry, 19*(14), 1639–1662. https://doi.org/10.1002/(sici)1096-987x(19981115)19:14<1639::aid-jcc10>3.0.co;2-b.

Morris, G. M., Huey, R., Lindstrom, W., Sanner, M. F., Belew, R. K., Goodsell, D. S., et al. (2009). AutoDock4 and AutoDockTools4: Automated docking with selective receptor flexibility. *Journal of Computational Chemistry, 30*(16), 2785–2791.

Padovani, D., Labunska, T., & Banerjee, R. (2006). Energetics of interaction between the G-protein chaperone, MeaB, and B12-dependent methylmalonyl-CoA mutase. *The Journal of Biological Chemistry, 281*(26), 17838–17844.

Pers, T. H., Timshel, P., & Hirschhorn, J. N. (2015). SNPsnap: A web-based tool for identification and annotation of matched SNPs. *Bioinformatics, 31*(3), 418–420. https://doi.org/10.1093/bioinformatics/btu655.

Plessl, T., Bürer, C., Lutz, S., Yue, W. W., Baumgartner, M. R., & Froese, D. S. (2017). Protein destabilization and loss of protein-protein interaction are fundamental mechanisms in cblA-type methylmalonic aciduria. *Human Mutation, 38*(8), 988–1001.

Qiliang, L., Wenqi, S., Quan, W., Xinying, Y., Jiuwei, L., Qiang, S., et al. (2015). Predictors of survival in children with methymalonic acidemia with homocystinuria in Beijing, China: A prospective cohort study. *Indian Pediatrics, 52*(2), 119–124.

Schmid, N., Eichenberger, A. P., Choutko, A., Riniker, S., Winger, M., Mark, A. E., et al. (2011). Definition and testing of the GROMOS force-field versions 54A7 and 54B7. *European Biophysics Journal: EBJ, 40*(7), 843–856.

Sharma, S., Kumar, P., & Chandra, R. (2019). Introduction to molecular dynamics. In *Molecular dynamics simulation of Nanocomposites using BIOVIA materials studio* (pp. 1–38). Lammps and Gromacs. https://doi.org/10.1016/b978-0-12-816954-4.00001-2.

Sim, N.-L., Kumar, P., Hu, J., Henikoff, S., Schneider, G., & Ng, P. C. (2012). SIFT web server: Predicting effects of amino acid substitutions on proteins. *Nucleic Acids Research, 40*(Web Server issue), W452–W457.

Sohail, M., Berg, J., & Vecchyo, D. O.-D. (2022). *Genetic architecture and evolution of complex traits and diseases in diverse human populations.* Frontiers Media SA.

Tang, D., Chen, G., & Liu, S. (2020). Identification of a novel mutation in the MMAA gene in a Chinese boy with isolated methylmalonic acidemia. *Acta Endocrinologica, 16*(2), 242–244.

Tavtigian, S. V., Deffenbaugh, A. M., Yin, L., Judkins, T., Scholl, T., Samollow, P. B., et al. (2006). Comprehensive statistical study of 452 BRCA1 missense substitutions with classification of eight recurrent substitutions as neutral. *Journal of Medical Genetics, 43*(4), 295–305.

Udhaya Kumar, S., & George Priya Doss, C. (2021). Computational investigation to identify potent inhibitors of the GTPase-Kirsten RAt sarcoma virus (K-Ras) mutants G12C and G12D. *Computers in Biology and Medicine, 139*(104946). https://doi.org/10.1016/j.compbiomed.2021.104946.

Udhaya Kumar, S., Rajan, B., Kumar, D. T., Cathryn, R. H., Samprita Das, V., Zayed, H., et al. (2021). Comparison of potential inhibitors and targeting fat mass and obesity-associated protein causing diabesity through docking and molecular dynamics strategies. *Journal of Cellular Biochemistry*, *122*(11), 1625–1638. https://doi.org/10.1002/jcb.30109.

Udhaya Kumar, S., Sankar, S., Younes, S., Thirumal Kumar, D., Ahmad, M. N., Okashah, S. S., et al. (2020). Deciphering the role of Filamin B calponin-homology domain in causing the Larsen syndrome, boomerang dysplasia, and atelosteogenesis type I spectrum disorders via a computational approach. *Molecules*, *25*(23), 5543. https://doi.org/10.3390/molecules25235543.

Yadegari, F., & Majidzadeh, K. (2019). In silico analysis for determining the deleterious non-synonymous single nucleotide polymorphisms of genes. *Molecular Biology Research Communications*, *8*(4), 141–150.

Yang, X., Sakamoto, O., Matsubara, Y., Kure, S., Suzuki, Y., Aoki, Y., et al. (2004). Mutation analysis of the MMAA and MMAB genes in Japanese patients with vitamin B12-responsive methylmalonic acidemia: Identification of a prevalent MMAA mutation. *Molecular Genetics and Metabolism*, *82*(4), 329–333. https://doi.org/10.1016/j.ymgme.2004.05.002.

Zhou, X., Cui, Y., & Han, J. (2018). Methylmalonic acidemia: Current status and research priorities. *Intractable & Rare Diseases Research*, 7(2), 73–78.

CHAPTER EIGHT

Intrinsically disordered proteins in viral pathogenesis and infections

Chandrabose Selvaraj[a,*], Muthuraja Arun Pravin[a], Wardah A. Alhoqail[b], Anuraj Nayarisseri[c], and Sanjeev Kumar Singh[a,*]

[a]Computer Aided Drug Design and Molecular Modeling Lab, Department of Bioinformatics, Science Block, Alagappa University, Karaikudi, Tamil Nadu, India

[b]Department of Biology, College of Education, Majmaah University, Al Majma'ah, Saudi Arabia

[c]*In Silico* Research Laboratory, Eminent Biosciences, Indore, Madhya Pradesh, India

*Corresponding authors: e-mail address: drbioselva@gmail.com; skysanjeev@gmail.com

Contents

Abstract

Disordered proteins serve a crucial part in many biological processes that go beyond the capabilities of ordered proteins. A large number of virus-encoded proteins have extremely condensed proteomes and genomes, which results in highly disordered proteins. The presence of these IDPs allows them to rapidly adapt to changes in their biological environment and play a significant role in viral replication and down-regulation of host defense mechanisms. Since viruses undergo rapid evolution and have a high rate of mutation and accumulation in their proteome, IDPs' insights into viruses are critical for understanding how viruses hijack cells and cause disease. There are many conformational changes that IDPs can adopt in order to interact with different protein partners and thus stabilize the particular fold and withstand high

Advances in Protein Chemistry and Structural Biology, Volume 132
ISSN 1876-1623
https://doi.org/10.1016/bs.apcsb.2022.06.002

mutation rates. This chapter explains the molecular mechanism behind viral IDPs, as well as the significance of recent research in the field of IDPs, with the goal of gaining a deeper comprehension of the essential roles and functions played by viral proteins.

1. Introduction

Several studies have shown the importance of protein functional entities in maintaining the protein's three-dimensional stability (Deller, Kong, & Rupp, 2016). Molecular, structural, and biochemical studies have determined that functional proteins fall into two categories: rigid and semi-rigid. NMR and X-ray crystallographic methods are used to determine protein structure. Proteins have been found to have an aperiodic crystal structure with atoms and backbones instead of a regular periodic structure (Wlodawer, Minor, Dauter, & Jaskolski, 2013). Ramachandran angles remain stable around their equilibrium position due to low-amplitude thermal fluctuations. Scientific thought has been dominated for more than a century by structure-function thinking, however many protein functions don't require a distinctive structure to function (Hollingsworth & Karplus, 2010). Proteins are clearly able to carry out their biological functions because their amino acid sequence encodes a three-fold conformational change that is energy-stable. IDPs are found in the proteome in addition to the orderly protein. Because of their relatively flat-free energy surface, intrinsically disordered proteins, or IDPs, have no well-defined equilibrium structure and appear as highly dynamic and heterogeneous conformers (Uversky, 2020). The lack of a fixed structure distinguishes intrinsically disordered proteins (IDPs) from other proteins and gives them the property of inherent flexibility. This dynamic behavior of IDPs has been referred to as "Dancing Proteins."

Structure-independent functions are carried out by proteins with intrinsically disordered regions (IDPRs) or protein hybrids including both IDPRs and ordered domains (Uversky, 2013). There are a number of distinct IDPs and IDRs that are rapidly becoming more prevalent. A wide variety of biological processes rely on the abundance of these IDP and IDRs in nature. A number of studies have examined the various aspects of IDPs and their rapid rise in popularity. IDPs proteins have recently attracted more attention in the scientific literature as a result of this growth (Wright & Dyson, 2015). Though these proteins are discovered quite recently, owing to their biologically active proteins and their more complex nature, and stability were studied multiple times that shown, the less stable structure and biological functionality have been periodically studied for more than half a century. Owing to the unavailability of the common terminology IDPs constitute an important extension of the protein kingdom. In recent days the advancement of bioinformatics has been widely used to conclude the important naturally flexible proteins and IDPs are recognized as a very broad class of proteins (Kuhlman & Bradley, 2019). Also, several factors can influence the great attention of IDPs, this class of proteins plays an important role in the insights into the molecular mechanism of protein action and in reshaping the protein structure and relationship. Most physiologically active proteins have been found to be extremely flexible, casting doubt on the long-held belief that proteins require specific three-dimensional shapes to operate.

As other organized proteins, IDPs do not perform critical biological tasks due to their lack of a stable structure. Functionally diverse and widely distributed, IDPs are abundant in the proteomes of living things, and their long-term survival in the evolutionary process serves as further proof of their significance and role in biological processes (Marasco & Scognamiglio, 2015). IDPs abundantly occur in the eukaryotic proteome. IDPRs frequently contain signal motifs and posttranslational modification sites, allowing alternative splicing to modify protein signaling and function. Theoretical and experimental methods to IDPs are still lacking. Among IDPs, it is not uncommon to see a shift from functional disorder to order. In many cases, IDPs are coupled to diverse structures in complexes; some of the ordered proteins require partial local unfolding and undergo order-to-disorder transitions to become functional (Darling & Uversky, 2018). In order to develop new experimental and theoretical ways to evaluate IDPs, we need to understand the structural features of IDPs/IDPRs and their conformational variations, flexible state, interaction with other proteins, and functional involvement in biological processes. IDPs/IDPRs have received a lot of attention in the biomedical approach since they are strongly linked to a wide range of human disorders. Many small compounds recognize IDPs to be an attractive target for IDP therapy (DeForte & Uversky, 2016). They also have a significant impact on a wide range of human disorders since they are found among the many proteins that play a critical role in the body's major activities. Because various small compounds have been found to operate as potential inhibitors of protein-protein interactions that involve IDPRs of one of the partners, these IDPs act as potent enticing therapeutic targets to generate novel small molecules. Even while these

IDPs have a key function in regulating, signaling, and recognizing cells, they are also frequently implicated in a variety of pathogenic illnesses (Uversky, Oldfield, & Dunker, 2008). The present chapter discusses the impact of intrinsically disordered proteins in viral entry and virulence.

2. Intrinsically disordered proteins (IDPs)

For more than a century, it was commonly acknowledged that the three-dimensional structure of proteins has a significant influence on their biological activity. It is essential for the normal function of most proteins like enzymes to have many functional groups of amino acids arranged in protein structure to support the various chemical reactions that eventually lead to other related functions. Though recent research states that the fraction of genome-encoded proteins of various organisms lacks definite three-dimensional structures, they play crucial roles in cellular functionality such groups of proteins are termed intrinsically disordered proteins (IDPs) (Wang, Cao, Zhao, & Li, 2011). These include protein clouds, naturally denatured proteins, natively unfolded and partially folded proteins, inherently unstructured proteins, and partially unfolded and mobile proteins. IDPs have been given a variety of names based on the features they exhibit in different studies at different points in time. In eukaryotic proteins, one-third of the proteins anchor the intrinsically disordered regions. In solution, if the IDPs are kept alone, they lack their unique three-dimensional structure (Rezaei-Ghaleh, Blackledge, & Zweckstetter, 2012). IDPs play an essential role in signaling, control, and recognition by interacting with many partners with high specificity and low affinity. One-to-many connections can facilitate this disordered protein-based signaling relationship. The tuning of IDPs comprehensively activates the post-translational modifications which alter the splicing and induced folding (Uversky et al., 2009). According to the significant incidence of IDPs in various clinical states, the root reason is not only protein misfolding, but also the result of incorrect signals or identification. Since IDPs are known for their distinct characteristics, they have been a focus on as prospective therapeutic targets for deteriorating protein-protein interactions (Huang & Liu, 2009).

3. Techniques involved in the prediction of IDPs

IDPs have been found to exist in a variety of structural states, from fully stretched to compact, as demonstrated by a variety of conformational and spectroscopic investigations. Keith Dunker's hypothesis states that the three most recognized conformations of a protein molecule in a functional framework, which proposed that the physiologically active protein can exist in three different native forms, such as ordered form, collapsed disorder state, and extended disorder state (Dunker et al., 2008). Biophysical approaches can be employed to guarantee the conformational study and structure determination of IDP molecules. In some biophysical methods, additional quantitative structure data can be obtained using an indirect method. Quantitative structural data can be obtained through the use of Nuclear Magnetic Resonance (NMR) (Kosol, Contreras-Martos, Cedeno, & Tompa, 2013). IDP structural transitions can be studied using electron paramagnetic resonance spectroscopy (EPR). The development of improved new generation EPR (spin-labels EPR) targets residues other than cysteine broadened the scope of this technique's application. Small-angle X-ray scattering, and small-angle neutron scattering are used for the extraction of quantitative information, to investigate the transient intermediates, and provide detailed information on IDP's nature (Le Breton et al., 2015). Single-molecule techniques like FERT (fluorescence resonance energy transfer), HS-FAS (high-speed atomic force microscopy), and AFM-based force spectroscopy are frequently used to study the dynamics and structure of IDPs. In addition, FERT was used to assess the changes in distance between two residues and the conformation equilibria in less than a second based on the intramolecular distance distribution. Secondary structure information and temporal scaling can be studied using HS-AFM, whereas SM-FS is utilized to quantify secondary structure information (Mazal & Haran, 2019).

Hs-AFM has also been utilized to display several dynamic processes to date. Several other complimentary techniques, including SDS-PAGE-sodium dodecyl sulfate-polyacrylamide gel electrophoresis, gel filtration, and size exclusion chromatography-based analysis, were utilized to explore the unique behavior of analysis in acidic and high-temperature conditions. IDP mobility was also found in SDS-PAGE; this phenomenon is solely explained by the less efficient binding of SDS molecules to highly charged IDPs in comparison to other globular proteins (Goetz et al., 2004). Gel filtration and size exclusion chromatography techniques reveal that IDPs have high apparent molecular weights. The unique behavior of various IDPs such as microtubule-associated protein-2, caldesmon, and -synuclein in different environmental conditions such as acidic and high temperature has been studied, and these environmental conditions lead to the denaturation and precipitation of globular proteins in solution. In contrast to globular proteins, IDPs display a distinct lack of conformational rigidity

(Mishra, Verma, Rao, Uversky, & Nandi, 2020). Although various approaches have been created, it is difficult to compare them in order to discover the most accurate method for predicting protein disorder. This is because the sensitivity and bias of each methodology will vary when applied to different protein sequences. When it comes to studying disordered proteins, the DisProt database has recently been found to be a useful resource for both getting experimental data and comparing the performance of disorder predictors. Over 3500 proteins have been annotated in the DisProt database for intrinsic disorder, 126 of which are virus-encoded proteins (de Brevern, 2020). Evidence was shown by Tokuriki, Oldfield, Uversky, Berezovsky, and Tawfik (2009) when compared to non-viruses and viral proteins. Disruption areas of viral proteins are shorter than in non-viral proteins (Tokuriki et al., 2009). Table 1 depicted the currently available tools to predict the IDPs.

Table 1 Depicted the various IDPs prediction tools.

Tools	Web address	Reference
SPOT-Disorder2	http://sparks-lab.org/server/spot-disorder2/	Hanson, Paliwal, Litfin, and Zhou (2019)
NetSurfP-2	http://www.cbs.dtu.dk/services/NetSurfP-2.0/	Klausen et al. (2019)
SPOT-Disorder-Single	http://sparks-lab.org/server/spot-disorder-single/	Hanson et al. (2019)
IUPred	https://iupred2a.elte.hu/	Dosztanyi (2018)
MobiDB-Lite	http://old.protein.bio.unipd.it/mobidblite/	Necci, Piovesan, Dosztanyi, and Tosatto (2017)
Disopred3	http://bioinf.cs.ucl.ac.uk/disopred	Jones and Cozzetto (2015)
DisPredict_v1.0	https://github.com/tamjidul/DisPredict_v1.0	Iqbal and Hoque (2015)
MFDp2	http://biomine.cs.vcu.edu/servers/MFDp2	Mizianty, Peng, and Kurgan (2013)
ESpritz	http://old.protein.bio.unipd.it/espritz/	Walsh, Martin, Di Domenico, and Tosatto (2012)

Table 1 Depicted the various IDPs prediction tools.—cont'd

Tools	Web address	Reference
GeneSilico Metadisorder	http://genesilico.pl/metadisorder/	Kozlowski and Bujnicki (2012)
PONDR-FIT	https://www.disprot.org/	Xue, Dunbrack, Williams, Dunker, and Uversky (2010)
NORSnet	https://predictprotein.org/	Lieutaud et al. (2016)
PrDOS	http://prdos.hgc.jp/cgi-bin/top.cgi	Ishida and Kinoshita (2007)
Spritz	http://distill.ucd.ie/spritz/	Vullo, Bortolami, Pollastri, and Tosatto (2006)
GlobPlot	http://globplot.embl.de	Linding, Russell, Neduva, and Gibson (2003)

4. Disordered proteins regions in viral proteomes

In a recent paper, the protein sections are classified as disordered, intermediate, or well-folded and stable regions with advanced methods. The pH-dependent approaches identify disordered regions by providing a numerical cutoff. The slope level less than 0.4 was identified as ordered protein, whereas the slope level greater than 0.4 was identified as an intermediately ordered or entirely disordered protein area (van der Lee et al., 2014). The average slope value for these wholly disordered regions is roughly 0.8, while the average slope value for the stable section and completely ordered regions is close to zero. This information may aid in the numerical quantification, differentiation, and classification of intrinsically disordered regions, as well as the creation of novel ways to classify the elemental study in the field of IDPs/IDPRs research. The DENV capsid protein was tested using pH-dependent methods to predict disordered proteins. For example, the type of IDP and conformational ensembles that do not have a single structure are essential in determining how common IDPs are in the proteomes of various animals (Ota et al., 2013). However, IDPs and IDPRs contribute to the majority of proteomes. Viruses have a substantial proportion of disordered proteins, and multicellular organisms may have more than 50% of their amino acids disordered. It is possible to increase both the size and complexity of the interactome by exploiting the binding promiscuity of disordered protein regions. The most effective IDP/IDPRs are found in higher-order organisms, which may benefit the majority of cells in their interactions with various ligands. Using these IDPs/IDPRs as potential binding partners, the cells, tissues, and internal systems of the organism can communicate with one other (Uversky, 2014). Viruses can infect multicellular organisms including humans and mammals, though they have small proteomes and has a big level of disorders. They can evolve new functionalities to enhance the number of biological processes and hence increase their survival rate in this disorderly region. As an example, flaviviruses, which have a very tiny genome, include a high number of disordered proteins (Zheng, Tan, Sugrue, & Tang, 2012).

5. IDPs in diseases

Intrinsically disordered proteins are playing a crucial role in several biological functions. During undesirable conditions, IDPs are engaged with adverse interactions and lead to various pathological conditions. Several pathological proteins associated with diseases like cardiovascular disease, neurodegeneration, genetic disorders, and also some of the human cancer-related proteins are having IDPs or intrinsically disordered regions in their proteome (Kulkarni & Uversky, 2019). Exothermic events in CSF fluid and the brain proteome have been linked to IDPs, according to recent studies. It is possible that the aggregation and fibrillization of proteins is reflected in these exothermic transitions, which are strongly influenced by neurodegenerative diseases. IDPs such as -synuclein and tau protein, for example, are strongly linked to neurodegenerative illnesses, particularly Parkinson's disease. synuclein's amyloid fibrils are linked to Parkinson's disease etiology. These two IDPs are recognized as appropriate targets for SARS-CoV-2, which forms a virion incorporating its genomic RNA bundles in a particle formed of structural proteins (Rotunno et al., 2020). Other proteins, such as p53 and BRCA1, are also recognized as appropriate targets. SARS-CoV-2 infection is connected with intrinsic disorder in the viral proteins, since it is known that the viral proteomes feature a noteworthy level IDPs and proteins that exploit intrinsic disorder during host cell invasion. Virus virulence and pathogenicity have been greatly influenced by IDPs and IDRs, which are essential in DNA and RNA binding mechanisms. An important part of the process of protein-RNA recognition involves RNA and protein conformational modifications that result in structural flexibility for disordered residues (Tenchov & Zhou, 2022).

Knowledge of IDPs in the coronavirus proteome can help us better comprehend viral pathogenicity. SARS-IDPs CoV-2's are of particular relevance because they've been linked to viral infectivity in the past, thus it's important to study them. Although the SARS-CoV-2 proteome's N protein has fewer intrinsically disordered protein regions than the other structural and non-structural proteins, the N protein's IDPs could play a significant role in the virus' functioning and virulence, which makes it a potential target for antiviral drug development (Hoffmann et al., 2020). Disparate disordered prediction methods have predicted the presence of highly disordered sections in N proteins, as well as the average percentage of these proteins. It has been calculated that approximately 65% of IDPs are anticipated in N proteins (Tenchov & Zhou, 2022). One of the protein disordered prediction systems, PrDOS, analyzes the disordered propensity and reports the significant degree of the disordered area in these three segments. There are other reports on the ID linker sections, which are crucial for connecting the N protein's NTD and CTD and are abundant in serine and arginine residues. The IDR's middle region is strongly linked to the RNA binding mechanism, and the IDR in CTD is responsible for the connection of dimer-dimer pairs in human coronaviruses. Conformational alterations in both RNA and protein may lead to the structural flexibility of IDRs, and these findings indicate the role of N protein, which is critical for RNA encapsulation, in ribonucleoprotein core synthesis via genomic RNA interaction (Chauhan et al., 2021).

6. Roles of IDPs in protein interaction and PPI networks

There are three ways in which IDP and IDR contribute to binding diversity: first, they serve as the structural basis for hub protein promiscuity. Alternatively, it may bind to well-structured hub proteins, and finally, IDR can act as a flexible linker that connects the functional domains and induces bidding diversity through the linker enabling mechanism. In addition, it has a variety of functional implications and affects a wide range of biological processes. IDPs, for example, include a posttranslational modification site that includes methylation, glycosylation, ADP-ribosylation, and phosphorylation (Bondos, Dunker, & Uversky, 2022). IDRs have unique qualities such as entropic springs and also contain auto-inhibitory domains, DNA recognition sites, three types of RNAs, and other proteins. It also contains a regulating protease digesting site, flexible linkers, and mediates cell control. IDRs have unique qualities such as entropic springs and also contain auto-inhibitory domains, DNA recognition sites, three types of RNAs, and other proteins. It also contains a regulating protease digesting site, flexible linkers, and mediates cell control (Alston, Soranno, & Holehouse, 2021).

7. Involvement of IDPs in the pathogen-host mediated regulation of cell cycle

In host cells, IDPs are involved in several pathways to control the cell cycle. Generally, the helical portion of the CD kinases inhibitors 1B (p27Kip1) protein is associated with both positive and negative regulation of cell cycles, whereas, the performed helical structure in disordered N-terminal transactivation domain (TAD) of p53 finds regulates the interaction of this protein with mouse double minute 2 homolog (Mdm2) and any mutation in the key residues of this molecular domain may affect the interaction of Mdm2 and subsequently leads the cell regulation and apoptosis (Yoon, Mitrea, Ou, & Kriwacki, 2012). The presence of IDPs in EMI-1 (early mitotic inhibitor protein affects the anaphase-promoting complex/cyclosome that is important for cell division by activating the proteins involved in ubiquitin-mediated degradation of cyclins. The connection between EMI-1 and the anaphase-promoting complex/cyclosome is regulated by kinase phosphorylation. Through molecular mimicry, the unstructured viral protein component could infiltrate host IDPs engaged in various cell regulatory processes and hijack host cell machinery. Furthermore, it disrupts the cell cycle control network in host cells via histone mimicry. IDRs in RNA binding proteins also play an important function in viral cellular machinery on the host-pathogen battleground (Reimann, Gardner, Margottin-Goguet, & Jackson, 2001). The imitation of host cells by affection cellular components disturbs their normal activities. For example, cancer-inducing viruses' aberrant viral oncoprotein targets Retinoblastoma protein and the E2F complex, affecting the normal cell regulation process. Many biological components, including nucleic acids, membranes, and proteins from host cells, interact with viral proteins. The presence of IDPs improves their interactions with host cell components. Because of their lack of three-dimensional structure and tremendous flexibility, they can interact with several binding partners at the same time. The connection of functional domains also facilitates this contact with partners, whilst the flexibility of the IDPs is also important in linking two or more domains to fulfill a specific function, and these linkers provide advantages for evading the host immune system (Dyson & Wright, 2018). The viral proteins interact with the host cell in such a way that it is able to locate the host immune system and interact with the host protein. The flexible area of viral proteins has a lower mutation rate than other viral proteins; hence, it is resistant to mutation. In paramyxoviruses, IDPs and their role in the replicative complex were investigated for the first time. Recently, improved bioinformatics tools have been routinely employed to determine the quantity of IDPs in viruses. Alternative splicing and overlapping reading frames may be impacted by a virus intrinsic disease protein. Viruses have evolved a sophisticated genetic organization for the expression and regulation of their proteins in a regulated manner, despite having a restricted genome for regulatory activities (Maginnis, 2018). Viruses have also adapted their genomic architecture and structure, as well as their transcription method, to employ both positive and negative transcription. Human T-cell lymphotropic virus type 1 (HTLV-1) and delta-retroviruses, for instance, are associated with myelopathy and adult T-cell leukemia (ATL). This extensive utilization of HTLV-1's genetic material is strongly linked to the buildup of intrinsically disordered proteins in its proteome (Zhang, Wei, Wang, Huang, & Chen, 2017).

8. Intrinsically disordered proteins in a SARS-CoV-2

Viruses with a single strand of RNA, such as SARS-Cov-2, have infected millions of people and caused the deaths of many. Finding new treatment targets that are both potent and efficacious will be a crucial element of treating COVID-19 (Selvaraj et al., 2022). During viral infection, the particle containing genomic RNA is composed of four structural proteins: spike glycoprotein, membrane protein, ion channel small envelope protein, and nucleocapsid protein. Spike proteins bind to the human angiotensin-converting enzyme 2 and mediate viral entry into the host cell and the 3D structure is shown in Fig. 1 (Smak, Chandrabose, Tvaroska, & Koca, 2021).

Membrane proteins involved in viral assembly in the endoplasmic reticulum, whereas the ion channel envelops protein and nucleocapsid protein are assembled with viral RNA to form a ribonucleoprotein complex. Using NMR spectroscopy, scientists have uncovered new information on the SARS-CoV-2 proteins that is essential to their understanding of the virus life cycle (Selvaraj et al., 2020). In addition, protocols for the large-scale manufacture of more than 80% of all SARS-CoV-2 proteins are supplied, which is extremely useful for future research. The existence

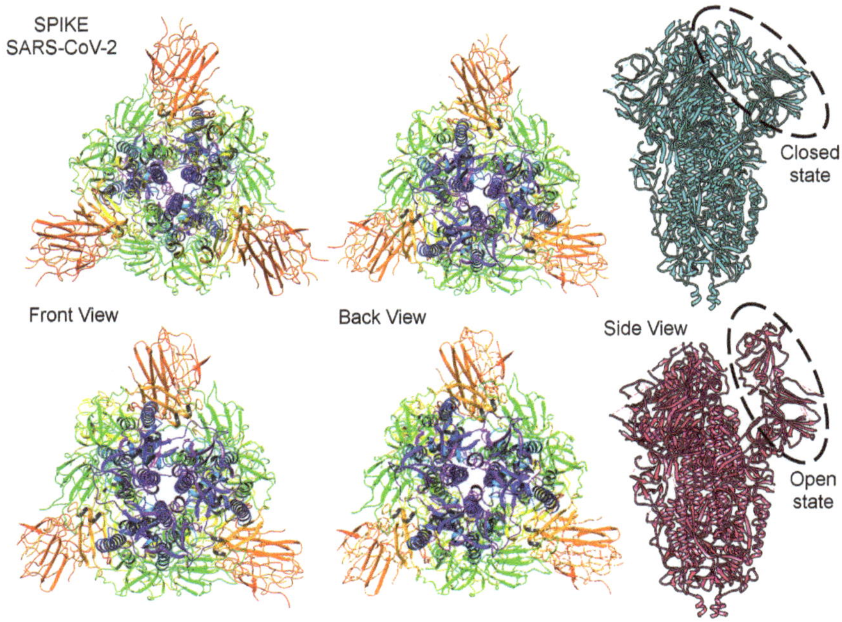

Fig. 1 3D structure of Spike protein from SARS-CoV-2 showing multiple views of protein along with open and closed state.

of disordered regions in the viral genome is strongly linked to viral infectiousness and pathogenicity. According to a recent study, the SARS-CoV-2 proteome has a substantial level of structural order, and few of its protein areas are inherently disordered, which contributes to the virus function and virulence, which classified in to several structural and non-structural proteins as shown in Fig. 2 (Selvaraj, Dinesh, Panwar, Boura, & Singh, 2021).

The RNA binding nucleocapsid strongly stabilizes the genomic RNA within the viral particle, hence regulating transcription, replication, and viral particle packing in the viral genome. N protein consists of three major intrinsically disordered regions and two structural domains, including the N-terminal RNA-binding domain (NTD) and the C-terminal domain. The primary function of RNA-binding N protein is to facilitate the assembly and packaging of viral particles (Selvaraj et al., 2021). The N protein possesses a greater number of IDRs and is essential for stabilizing genomic RNA within the virus particle and controlling viral genome transcription, packaging, and replication. The N protein of SARS-CoV-2 is composed of two structural domains, the N-terminal RNA-binding domain and the

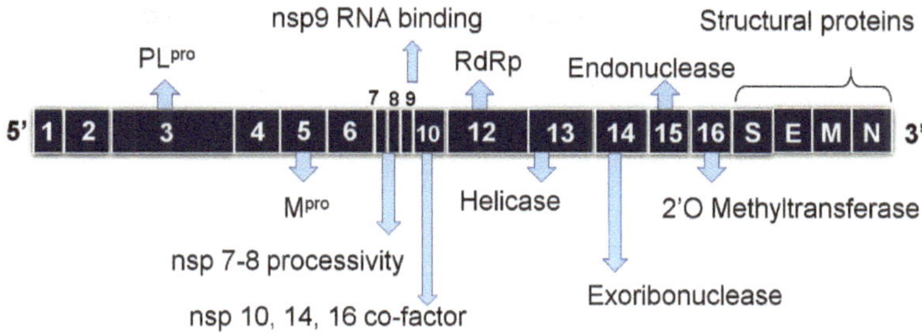

Fig. 2 Protein topology of structural and non-structural proteins in SARS-CoV-2.

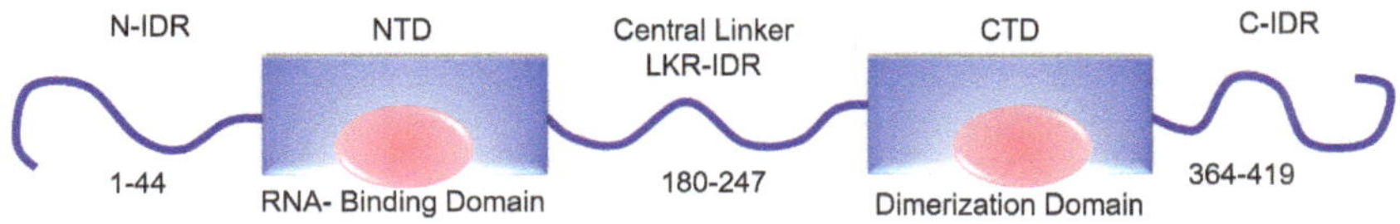

Fig. 3 The IDP region on N-protein of SARS-CoV-2.

C-terminal domain, which are separated by three large IDPRs (Fig. 3). The primary role of this RNA binding N protein is to assemble the genomic RNA into the viral RNA-protein complex and to mediate the packaging of viral particles into 80–90 nm virions.

The highly disordered region of the N protein displays the average proportion of intrinsic disorder predicted by the diverse disorder prediction techniques. Analyzing the disorder propensity of the N protein using disorder prediction methods revealed a considerable degree of disorder in these three domains. The highly flexible linker sections are also present between the N- and C-terminal domains and are rich in disorder-promoting serine and arginine residues. In human coronaviruses, the middle IDRs are responsible for its RNA-binding activity and the IDR located on the boundaries of the C-terminal tail peptides are responsible for dimer-dimer association, whereas the central N-terminal core is primarily responsible for nucleotide binding (Dosztanyi, 2018). In addition, all three IDRs of the N protein have molecular recognition domains. The structural flexibility of disordered areas permits conformational changes in both RNA and protein, which are required for RNA-protein recognition. This information confirms the vital role of the N protein in ribonucleoprotein core production, which is necessary for RNA encapsulation and the life cycle of viruses. Consequently, IDRs serve as excellent therapeutic targets to disrupt the interaction between N-protein and viral genomic RNA (Khan et al., 2020). Table 2

Table 2 Examples of viral proteins with intrinsic disorder according to the DisProt database.

Protein name	Organism	Sequence length	Disorder content
Antitermination protein N	Escherichia phage lambda	107	100.00%
Ribonuclease E	Escherichia coli (strain K12)	1061	53.16%
Protein E7	Human papillomavirus type 16	98	52.04%
Nucleoprotein	Nipah virus	532	25.00%
Nucleoprotein	Hendra virus (isolate Horse/ Autralia/Hendra/1994)	532	25.00%
Phosphoprotein	Nipah virus	709	57.26%
Genome polyprotein	Hepatitis C virus genotype 2a (isolate JFH-1)	3033	6.73%
Large envelope protein	Hepatitis B virus	445	24.27%
Phosphoprotein	Human respiratory syncytial virus A (strain Long)	241	75.10%
Nucleoprotein	Influenza B virus	560	12.50%
Gag polyprotein	Human immunodeficiency virus type 1 group M subtype B (isolate BH10)	512	6.64%

shows important IDPs of various types of virus and their length. Using a technique such as liquid-liquid phase separation to isolate intracellular organelles reveals a number of IDPs with an array of crucial cellular functions. This liquid-liquid phase separation method is commonly used to predict biomolecules like proteins and nucleic acids. There are a number of membrane compartments in eukaryotes that are highly dynamic structures that can rapidly form on demand and lead to the concentration of proteins and biochemical reactions at a specific site where they are required. These membrane-less compartments all include bigger IDPs and lack a well-defined secondary or tertiary structure in addition to a high degree of adaptability to accommodate their dynamic activity. In certain instances, the membranes accumulate stress granules consisting of proteins and RNAs when the cell is under stress and its translation process is limited. In a separate study, the N protein of SARS-CoV-2 was exposed to liquid-liquid phase

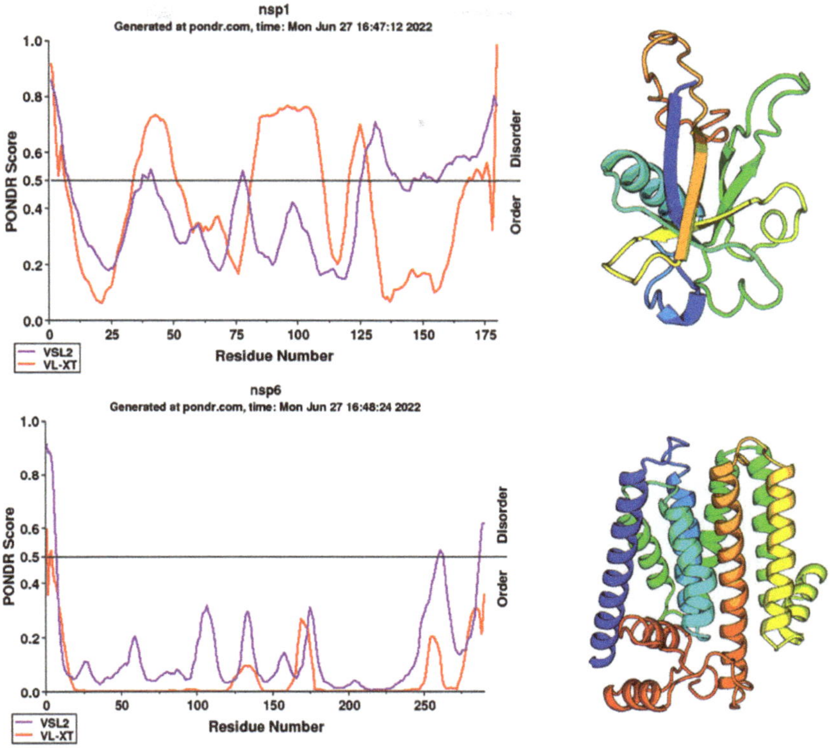

Fig. 4 IDP predicted by PONDR server for the models generated by the I-Tasser models of Nsp1 and Nsp6.

separation to determine if and how stress increases the degree of inherent disorder of IDPs engaged in stress granules (Zhang, Mehta, & Zhang, 2021). The findings demonstrate that the N-protein splits into stress granules via its N-terminal IDR1. This condensing of N protein into stress granules may be a mechanism by which SARS-CoV-2 inhibits the innate immunity of the host cell. Now a days, computational servers strongly predict the IDPS and for example, the two proteins from SARS-CoV-2, namely nsp1 and nsp6 is checked in PONDR VSL2 and VL-XT options. Nsp1 shows more disorder regions than the nsp6 as shown in Fig. 4.

9. Molecular mechanisms of drugs targeting ordered proteins

The majority of pharmacological antagonists and inhibitors are capable of competing with endogenous proteins at identified binding sites

(Selvaraj, Chandra, & Singh, 2022). Several medications are designed with the concept of one drug, one target, and one disease in mind; the primary objective of new compounds targeting proteins is to alter the normal biochemical response in the cell. Small compounds achieve, for instance, the enzyme inhibitory activity or the augmentation of the catalytic activity. The enzyme inhibition experiments could be reversible, competitive, or noncompetitive. Some antimetabolites resemble the endogenous metabolites, and can therefore be utilized as inhibitors. Small compounds that bind to the active site of enzymes can imitate many of the interactions of natural metabolites, resulting in protein function regulation (Colovic, Krstic, Lazarevic-Pasti, Bondzic, & Vasic, 2013). For successful drug development, the active sites of enzymes are typically utilized as targets; however, the molecular mechanism of inhibitory protein-protein interaction also has to advance. Recognizing the significance of the energy of protein-protein interactions, the classic pharmacological targets include protein-protein interactions. However, it will be difficult to develop an effective medication to suppress protein-protein interactions. The difficulty of developing drugs to target protein-protein interactions may stem from the inherent contradiction between organized protein-protein interfaces and the composition of tiny molecules. The available crystal structure of most of the small molecule with greater than 300 Da bind s contiguous epitope of more residues, and also many interfaces between the ordered proteins are much more complex with epitopes consisting of discontinuous or a combination of multiple contiguous epitopes (Mabonga & Kappo, 2019). The scarcity of small-molecule inhibitors of protein-protein interactions may be attributable to the distinct characteristics of the two binding systems. The identification of medications that target enzymes is anticipated to be imprecise (Selvaraj, Omer, Singh, & Singh, 2015). On the other hand, protein interfaces are less conserved, and the lack of inherent specificity is not a problem for protein-protein interactions that represent a promising new route for drug development.

10. Lipid-binding activity of IDPs

One or more immunoreceptor tyrosine-based activation motifs with cytoplasmic intrinsically disordered regions are present in MIRR signaling subunits. In addition, the activation motif contains two tyrosine residues, which become phosphorylated and activate the intracellular signaling cascade when they bind tiny molecules. The presence of IDPs in the immunoreceptor tyrosine-based activation motif (ITAM) facilitates folding and renders the ITAM tyrosines inaccessible for phosphorylation. The evaluation of structural organization using CD and NMR (nuclear magnetic resonance) spectral analysis confirmed the formation of partial helices in ITAM motifs in the presence of LMPG micelles and similar findings were also reported for CD3εcyt. Xu et al. (2008) revealed the presence of DMPG vesicles and bicelles containing 1-polmitoyl-2-oleoyl-phosphatidylglycerol (POPG) and dihexanoyl-phosphatidylcholine (DHPC) as a model for the cell membrane. It was discovered that CD3cyt connects with acidic phospholipid molecules and generates an electrostatic connection between the N-terminal basic residues of CD3cyt and the polar lipid groups, which are essential for binding. The studies with an in vitro kinase assay and CD, NMR spectroscopy analysis revealed that the binding of these molecules may influence the protein folding with partial helix formation in the CD3cyt ITAM region and sequestration of ITAM tyrosine into the hydrophobic lipid area, rendering these residues inaccessible for phosphorylation by tyrosine kinases. The experiments with fluorescence resonance energy transfer and Jurkat cells revealed the close interaction of CD3εcyt with the cell membrane and lead to the αhelical folding transition in ITAM regions which regulates the TCR activation that sequestration of key tyrosine into the lipid bilayers which represent the unrecognized mechanism for receptor activation control (Yao, Flaswinkel, Reth, & Scott, 1995). The conformational changes in the α-helical folding transition while binding of ITAM has a number of IDRs in the cell membrane which regulates the TCR signaling, though no disorder-to-order transition was reported in ζcyt, CD3εcyt, and FcεRIγcyt upon binding to POPG vesicles which suggesting that the mode of binding without folding. These findings suggest that the stable POPG bilayer mimic in the cell membrane than micelles, in addition, the N-terminal cluster basic residues in CD3εcyt and ζcyt are highly crucial for binding of IDPs to phospholipid bilayers, while the ITAM residues are not involved in these kinds of interactions. The residues of ITAM in POPG bind with ζcyt, CD3εcyt, and FcεRIγcyt can remain accessible for phosphorylation. In addition, the fully phosphorylated IDPs of ζcyt have a net charge of −5.5 but still can bind with phospholipid bilayers with no-disorder-to-order transition observed on this binding. This result indicated that the cluster of basic residues and their overall net charge is deriving the interaction of fully phosphorylated ζcyt with POPG lipid bilayers (Cornish, Chamberlain, Owen, & Mott, 2020).

11. Future prospectus

Globally, the emergence of novel and genetically altered viruses is strongly linked to human disease and epidemics, and this is regarded as a major issue. For the development of promising and novel therapeutic medications for the prevention and treatment of viral illnesses, it is crucial to understand the functional mechanism and structural information of viral proteins. Intrinsically disordered proteins in the viral proteome are also closely associated with their infection and pathogenicity. Understanding structural transition and their physiological relationship is crucial for researching the activities of IDPs and their interactions with host proteins engaged in several physiological processes. To locate the inhibitor to limit viral replication, maturation, and pathogenicity, the viral-host protein interactions must be thoroughly investigated. Conformational alterations and flexibility of IDPs allow them to bind structurally distinct molecules with comparable binding affinity, hence their accurate prediction is gaining increasing interest. Current techniques for the effective suppression of IDPs involve either designing small compounds that can bind to and stable the functionally misfolded state of IDPs or directly binding with them and blocking the misfolded interaction. Therefore, it is anticipated that prediction techniques will continue to design IDP-effective inhibitors. Accurate IDP prediction is essential for a comprehensive understanding of protein folding and aids in elucidating the molecular mechanism underlying protein function in order to establish a novel structural and functional hierarchy of proteins. This chapter try to bring the aspects of IDPs and its information's, that relates to the viral mechanism in host invading mechanisms. These information's will be aid to understand the future drug design aspects of viral pathogenesis and human cell invading mechanisms.

Acknowledgment

The authors CS and SKS thankfully acknowledge the Tamil Nadu State Council for Higher Education (TANSCHE) for the research grant (Au/S.o. (P&D): TANSCHE Projects: 117/2021). The author would like to thank Deanship of Scientific Research at Majmaah University for supporting this work.

References

Alston, J. J., Soranno, A., & Holehouse, A. S. (2021). Integrating single-molecule spectroscopy and simulations for the study of intrinsically disordered proteins. *Methods*, *193*, 116–135. https://doi.org/10.1016/j.ymeth.2021.03.018.

Bondos, S. E., Dunker, A. K., & Uversky, V. N. (2022). Intrinsically disordered proteins play diverse roles in cell signaling. *Cell Communication and Signaling: CCS*, *20*(1), 20. https://doi.org/10.1186/s12964-022-00821-7.

Chauhan, A., Avti, P., Shekhar, N., Prajapat, M., Sarma, P., Bhattacharyya, A., et al. (2021). Structural and conformational analysis of SARS CoV 2 N-CTD revealing monomeric and dimeric active sites during the RNA-binding and stabilization: Insights towards potential inhibitors for N-CTD. *Computers in Biology and Medicine*, *134*, 104495. https://doi.org/10.1016/j.compbiomed.2021.104495.

Colovic, M. B., Krstic, D. Z., Lazarevic-Pasti, T. D., Bondzic, A. M., & Vasic, V. M. (2013). Acetylcholinesterase inhibitors: Pharmacology and toxicology. *Current Neuropharmacology*, *11*(3), 315–335. https://doi.org/10.2174/1570159X11311030006.

Cornish, J., Chamberlain, S. G., Owen, D., & Mott, H. R. (2020). Intrinsically disordered proteins and membranes: A marriage of convenience for cell signalling? *Biochemical Society Transactions*, *48*(6), 2669–2689. https://doi.org/10.1042/BST20200467.

Darling, A. L., & Uversky, V. N. (2018). Intrinsic disorder and posttranslational modifications: The darker side of the biological dark matter. *Frontiers in Genetics*, *9*, 158. https://doi.org/10.3389/fgene.2018.00158.

de Brevern, A. G. (2020). Analysis of protein disorder predictions in the light of a protein structural alphabet. *Biomolecules*, *10*(7). https://doi.org/10.3390/biom10071080.

DeForte, S., & Uversky, V. N. (2016). Order, disorder, and everything in between. *Molecules*, *21*(8). https://doi.org/10.3390/molecules21081090.

Deller, M. C., Kong, L., & Rupp, B. (2016). Protein stability: A crystallographer's perspective. *Acta Crystallographica Section F Structural Biology Communications*, *72*(Pt 2), 72–95. https://doi.org/10.1107/S2053230X15024619.

Dosztanyi, Z. (2018). Prediction of protein disorder based on IUPred. *Protein Science*, *27*(1), 331–340. https://doi.org/10.1002/pro.3334.

Dunker, A. K., Oldfield, C. J., Meng, J., Romero, P., Yang, J. Y., Chen, J. W., et al. (2008). The unfoldomics decade: An update on intrinsically disordered proteins. *BMC Genomics*, *9*(Suppl. 2), S1. https://doi.org/10.1186/1471-2164-9-S2-S1.

Dyson, H. J., & Wright, P. E. (2018). How do intrinsically disordered viral proteins hijack the cell? *Biochemistry*, *57*(28), 4045–4046. https://doi.org/10.1021/acs.biochem.8b00622.

Goetz, H., Kuschel, M., Wulff, T., Sauber, C., Miller, C., Fisher, S., et al. (2004). Comparison of selected analytical techniques for protein sizing, quantitation and molecular weight determination. *Journal of Biochemical and Biophysical Methods*, *60*(3), 281–293. https://doi.org/10.1016/j.jbbm.2004.01.007.

Hanson, J., Paliwal, K. K., Litfin, T., & Zhou, Y. (2019). SPOT-Disorder2: Improved protein intrinsic disorder prediction by ensembled deep learning. *Genomics, Proteomics & Bioinformatics*, *17*(6), 645–656. https://doi.org/10.1016/j.gpb.2019.01.004.

Hoffmann, M., Kleine-Weber, H., Schroeder, S., Kruger, N., Herrler, T., Erichsen, S., et al. (2020). SARS-CoV-2 cell entry depends on ACE2 and TMPRSS2 and is blocked by a clinically proven protease inhibitor. *Cell*, *181*(2), 271–280 e278. https://doi.org/10.1016/j.cell.2020.02.052.

Hollingsworth, S. A., & Karplus, P. A. (2010). A fresh look at the Ramachandran plot and the occurrence of standard structures in proteins. *Biomolecular Concepts*, *1*(3-4), 271–283. https://doi.org/10.1515/BMC.2010.022.

Huang, Y., & Liu, Z. (2009). Kinetic advantage of intrinsically disordered proteins in coupled folding-binding process: A critical assessment of the "fly-casting" mechanism. *Journal of Molecular Biology*, *393*(5), 1143–1159. https://doi.org/10.1016/j.jmb.2009.09.010.

Iqbal, S., & Hoque, M. T. (2015). DisPredict: A predictor of disordered protein using optimized RBF Kernel. *PLoS One*, *10*(10), e0141551. https://doi.org/10.1371/journal.pone.0141551.

Ishida, T., & Kinoshita, K. (2007). PrDOS: Prediction of disordered protein regions from amino acid sequence. *Nucleic Acids Research*, *35*(web server issue), W460–W464. https://doi.org/10.1093/nar/gkm363.

Jones, D. T., & Cozzetto, D. (2015). DISOPRED3: Precise disordered region predictions with annotated protein-binding activity. *Bioinformatics*, *31*(6), 857–863. https://doi.org/10.1093/bioinformatics/btu744.

Khan, A., Tahir Khan, M., Saleem, S., Junaid, M., Ali, A., Shujait Ali, S., et al. (2020). Structural insights into the mechanism of RNA recognition by the N-terminal RNA-binding domain of the SARS-CoV-2 nucleocapsid phosphoprotein. *Computational and Structural Biotechnology Journal*, *18*, 2174–2184. https://doi.org/10.1016/j.csbj.2020.08.006.

Klausen, M. S., Jespersen, M. C., Nielsen, H., Jensen, K. K., Jurtz, V. I., Sonderby, C. K., et al. (2019). NetSurfP-2.0: Improved prediction of protein structural features by integrated deep learning. *Proteins*, *87*(6), 520–527. https://doi.org/10.1002/prot.25674.

Kosol, S., Contreras-Martos, S., Cedeno, C., & Tompa, P. (2013). Structural characterization of intrinsically disordered proteins by NMR spectroscopy. *Molecules*, *18*(9), 10802–10828. https://doi.org/10.3390/molecules180910802.

Kozlowski, L. P., & Bujnicki, J. M. (2012). MetaDisorder: A meta-server for the prediction of intrinsic disorder in proteins. *BMC Bioinformatics*, *13*, 111. https://doi.org/10.1186/1471-2105-13-111.

Kuhlman, B., & Bradley, P. (2019). Advances in protein structure prediction and design. *Nature Reviews. Molecular Cell Biology*, *20*(11), 681–697. https://doi.org/10.1038/s41580-019-0163-x.

Kulkarni, P., & Uversky, V. N. (2019). Intrinsically disordered proteins in chronic diseases. *Biomolecules*, *9*(4). https://doi.org/10.3390/biom9040147.

Le Breton, N., Martinho, M., Mileo, E., Etienne, E., Gerbaud, G., Guigliarelli, B., et al. (2015). Exploring intrinsically disordered proteins using site-directed spin labeling electron paramagnetic resonance spectroscopy. *Frontiers in Molecular Biosciences*, *2*, 21. https://doi.org/10.3389/fmolb.2015.00021.

Lieutaud, P., Ferron, F., Uversky, A. V., Kurgan, L., Uversky, V. N., & Longhi, S. (2016). How disordered is my protein and what is its disorder for? A guide through the "dark side" of the protein universe. *Intrinsically Disord Proteins*, *4*(1), e1259708. https://doi.org/10.1080/21690707.2016.1259708.

Linding, R., Russell, R. B., Neduva, V., & Gibson, T. J. (2003). GlobPlot: Exploring protein sequences for globularity and disorder. *Nucleic Acids Research*, *31*(13), 3701–3708. https://doi.org/10.1093/nar/gkg519.

Mabonga, L., & Kappo, A. P. (2019). Protein-protein interaction modulators: Advances, successes and remaining challenges. *Biophysical Reviews*, *11*(4), 559–581. https://doi.org/10.1007/s12551-019-00570-x.

Maginnis, M. S. (2018). Virus-receptor interactions: The key to cellular invasion. *Journal of Molecular Biology*, *430*(17), 2590–2611. https://doi.org/10.1016/j.jmb.2018.06.024.

Marasco, D., & Scognamiglio, P. L. (2015). Identification of inhibitors of biological interactions involving intrinsically disordered proteins. *International Journal of Molecular Sciences*, *16*(4), 7394–7412. https://doi.org/10.3390/ijms16047394.

Mazal, H., & Haran, G. (2019). Single-molecule FRET methods to study the dynamics of proteins at work. *Current Opinion in Biomedical Engineering*, *12*, 8–17. https://doi.org/10.1016/j.cobme.2019.08.007.

Mishra, P. M., Verma, N. C., Rao, C., Uversky, V. N., & Nandi, C. K. (2020). Intrinsically disordered proteins of viruses: Involvement in the mechanism of cell regulation and pathogenesis. *Progress in Molecular Biology and Translational Science*, *174*, 1–78. https://doi.org/10.1016/bs.pmbts.2020.03.001.

Mizianty, M. J., Peng, Z., & Kurgan, L. (2013). MFDp2: Accurate predictor of disorder in proteins by fusion of disorder probabilities, content and profiles. *Intrinsically Disordered Proteins*, *1*(1), e24428. https://doi.org/10.4161/idp.24428.

Necci, M., Piovesan, D., Dosztanyi, Z., & Tosatto, S. C. E. (2017). MobiDB-lite: Fast and highly specific consensus prediction of intrinsic disorder in proteins. *Bioinformatics*, *33*(9), 1402–1404. https://doi.org/10.1093/bioinformatics/btx015.

Ota, M., Koike, R., Amemiya, T., Tenno, T., Romero, P. R., Hiroaki, H., et al. (2013). An assignment of intrinsically disordered regions of proteins based on NMR structures. *Journal of Structural Biology*, *181*(1), 29–36. https://doi.org/10.1016/j.jsb.2012.10.017.

Reimann, J. D., Gardner, B. E., Margottin-Goguet, F., & Jackson, P. K. (2001). Emi1 regulates the anaphase-promoting complex by a different mechanism than Mad2 proteins. *Genes & Development*, *15*(24), 3278–3285. https://doi.org/10.1101/gad.945701.

Rezaei-Ghaleh, N., Blackledge, M., & Zweckstetter, M. (2012). Intrinsically disordered proteins: From sequence and conformational properties toward drug discovery. *Chembiochem*, *13*(7), 930–950. https://doi.org/10.1002/cbic.201200093.

Rotunno, M. S., Lane, M., Zhang, W., Wolf, P., Oliva, P., Viel, C., et al. (2020). Cerebrospinal fluid proteomics implicates the granin family in Parkinson's disease. *Scientific Reports*, *10*(1), 2479. https://doi.org/10.1038/s41598-020-59414-4.

Selvaraj, C., Chandra, I., & Singh, S. K. (2022). Artificial intelligence and machine learning approaches for drug design: Challenges and opportunities for the pharmaceutical industries. *Molecular Diversity*, *26*(3), 1893–1913. https://doi.org/10.1007/s11030-021-10326-z.

Selvaraj, C., Dinesh, D. C., Krafcikova, P., Boura, E., Aarthy, M., Pravin, M. A., et al. (2022). Structural understanding of SARS-CoV-2 drug targets, active site contour map analysis and COVID-19 therapeutics. *Current Molecular Pharmacology*, *15*(2), 418–433. https://doi.org/10.2174/1874467214666210906125959.

Selvaraj, C., Dinesh, D. C., Panwar, U., Abhirami, R., Boura, E., & Singh, S. K. (2021). Structure-based virtual screening and molecular dynamics simulation of SARS-CoV-2 Guanine-N7 methyltransferase (nsp14) for identifying antiviral inhibitors against COVID-19. *Journal of Biomolecular Structure & Dynamics*, *39*(13), 4582–4593. https://doi.org/10.1080/07391102.2020.1778535.

Selvaraj, C., Dinesh, D. C., Panwar, U., Boura, E., & Singh, S. K. (2021). High-throughput screening and quantum mechanics for identifying potent inhibitors against Mac1 domain of SARS-CoV-2 Nsp3. *IEEE/ACM Transactions on Computational Biology and Bioinformatics*, *18*(4), 1262–1270. https://doi.org/10.1109/TCBB.2020.3037136.

Selvaraj, C., Omer, A., Singh, P., & Singh, S. K. (2015). Molecular insights of protein contour recognition with ligand pharmacophoric sites through combinatorial library design and MD simulation in validating HTLV-1 PR inhibitors. *Molecular BioSystems*, *11*(1), 178–189. https://doi.org/10.1039/c4mb00486h.

Selvaraj, C., Panwar, U., Dinesh, D. C., Boura, E., Singh, P., Dubey, V. K., et al. (2020). Microsecond MD simulation and multiple-conformation virtual screening to identify potential anti-COVID-19 inhibitors against SARS-CoV-2 main protease. *Frontiers in Chemistry*, *8*, 595273. https://doi.org/10.3389/fchem.2020.595273.

Smak, P., Chandrabose, S., Tvaroska, I., & Koca, J. (2021). Pan-selectin inhibitors as potential therapeutics for COVID-19 treatment: In silico screening study. *Glycobiology*, *31*(8), 975–987. https://doi.org/10.1093/glycob/cwab021.

Tenchov, R., & Zhou, Q. A. (2022). Intrinsically disordered proteins: Perspective on COVID-19 infection and drug discovery. *ACS Infectious Disease*, *8*(3), 422–432. https://doi.org/10.1021/acsinfecdis.2c00031.

Tokuriki, N., Oldfield, C. J., Uversky, V. N., Berezovsky, I. N., & Tawfik, D. S. (2009). Do viral proteins possess unique biophysical features? *Trends in Biochemical Sciences*, *34*(2), 53–59. https://doi.org/10.1016/j.tibs.2008.10.009.

Uversky, V. N. (2013). A decade and a half of protein intrinsic disorder: Biology still waits for physics. *Protein Science*, *22*(6), 693–724. https://doi.org/10.1002/pro.2261.
Uversky, V. N. (2014). Wrecked regulation of intrinsically disordered proteins in diseases: Pathogenicity of deregulated regulators. *Frontiers in Molecular Biosciences*, *1*, 6. https://doi.org/10.3389/fmolb.2014.00006.
Uversky, V. N. (2020). New technologies to analyse protein function: An intrinsic disorder perspective. *F1000Res*, *9*. https://doi.org/10.12688/f1000research.20867.1.
Uversky, V. N., Oldfield, C. J., & Dunker, A. K. (2008). Intrinsically disordered proteins in human diseases: Introducing the D2 concept. *Annual Review of Biophysics*, *37*, 215–246. https://doi.org/10.1146/annurev.biophys.37.032807.125924.
Uversky, V. N., Oldfield, C. J., Midic, U., Xie, H., Xue, B., Vucetic, S., et al. (2009). Unfoldomics of human diseases: Linking protein intrinsic disorder with diseases. *BMC Genomics*, *10*(Suppl. 1), S7. https://doi.org/10.1186/1471-2164-10-S1-S7.
van der Lee, R., Buljan, M., Lang, B., Weatheritt, R. J., Daughdrill, G. W., Dunker, A. K., et al. (2014). Classification of intrinsically disordered regions and proteins. *Chemical Reviews*, *114*(13), 6589–6631. https://doi.org/10.1021/cr400525m.
Vullo, A., Bortolami, O., Pollastri, G., & Tosatto, S. C. (2006). Spritz: A server for the prediction of intrinsically disordered regions in protein sequences using kernel machines. *Nucleic Acids Research*, *34*(Web server issue), W164–W168. https://doi.org/10.1093/nar/gkl166.
Walsh, I., Martin, A. J., Di Domenico, T., & Tosatto, S. C. (2012). ESpritz: Accurate and fast prediction of protein disorder. *Bioinformatics*, *28*(4), 503–509. https://doi.org/10.1093/bioinformatics/btr682.
Wang, J., Cao, Z., Zhao, L., & Li, S. (2011). Novel strategies for drug discovery based on Intrinsically Disordered Proteins (IDPs). *International Journal of Molecular Sciences*, *12*(5), 3205–3219. https://doi.org/10.3390/ijms12053205.
Wlodawer, A., Minor, W., Dauter, Z., & Jaskolski, M. (2013). Protein crystallography for aspiring crystallographers or how to avoid pitfalls and traps in macromolecular structure determination. *The FEBS Journal*, *280*(22), 5705–5736. https://doi.org/10.1111/febs.12495.
Wright, P. E., & Dyson, H. J. (2015). Intrinsically disordered proteins in cellular signalling and regulation. *Nature Reviews. Molecular Cell Biology*, *16*(1), 18–29. https://doi.org/10.1038/nrm3920.
Xue, B., Dunbrack, R. L., Williams, R. W., Dunker, A. K., & Uversky, V. N. (2010). PONDR-FIT: A meta-predictor of intrinsically disordered amino acids. *Biochimica et Biophysica Acta*, *1804*(4), 996–1010. https://doi.org/10.1016/j.bbapap.2010.01.011.
Yao, X. R., Flaswinkel, H., Reth, M., & Scott, D. W. (1995). Immunoreceptor tyrosine-based activation motif is required to signal pathways of receptor-mediated growth arrest and apoptosis in murine B lymphoma cells. *Journal of Immunology*, *155*(2), 652–661. Retrieved from https://www.ncbi.nlm.nih.gov/pubmed/7608543.
Yoon, M. K., Mitrea, D. M., Ou, L., & Kriwacki, R. W. (2012). Cell cycle regulation by the intrinsically disordered proteins p21 and p27. *Biochemical Society Transactions*, *40*(5), 981–988. https://doi.org/10.1042/BST20120092.
Zhang, J. Z., Mehta, S., & Zhang, J. (2021). Liquid-liquid phase separation: A principal organizer of the cell's biochemical activity architecture. *Trends in Pharmacological Sciences*, *42*(10), 845–856. https://doi.org/10.1016/j.tips.2021.07.003.
Zhang, L. L., Wei, J. Y., Wang, L., Huang, S. L., & Chen, J. L. (2017). Human T-cell lymphotropic virus type 1 and its oncogenesis. *Acta Pharmacologica Sinica*, *38*(8), 1093–1103. https://doi.org/10.1038/aps.2017.17.
Zheng, J., Tan, B. H., Sugrue, R., & Tang, K. (2012). Current approaches on viral infection: Proteomics and functional validations. *Frontiers in Microbiology*, *3*, 393. https://doi.org/10.3389/fmicb.2012.00393.

CHAPTER NINE

Interaction between Sars-CoV-2 structural proteins and host cellular receptors: From basic mechanisms to clinical perspectives

Henry Wade, Qihua Duan, and Qiaozhu Su*
Institute for Global Food Security, School of Biological Sciences, Queen's University Belfast, Belfast, United Kingdom
*Corresponding author: e-mail address: q.su@qub.ac.uk

Contents

Advances in Protein Chemistry and Structural Biology, Volume 132
ISSN 1876-1623
https://doi.org/10.1016/bs.apcsb.2022.05.010

Abstract

Severe acute respiratory syndrome coronavirus 2 (Sars-CoV-2) has caused a global pandemic that has affected the lives of billions of individuals. Sars-CoV-2 primarily infects human cells by binding of the viral spike protein to angiotensin-converting enzyme 2 (ACE2). In addition, novel means of viral entry are currently being investigated, including Neuropillin 1, toll-like receptors (TLRs), cluster of differentiation 147 (CD147), and integrin α5β1. Enriched expression of these proteins across metabolic regulatory organs/tissues, including the circulatory system, liver, pancreas, and intestine contributes to major clinical complications among COVID-19 patients, particularly the development of hypertension, myocardial injury, arrhythmia, acute coronary syndrome and increased coagulation in the circulatory system during and post-infection. Pre-existing metabolic disease, such as cardiovascular disease, obesity, diabetes, and non-alcoholic fatty liver disease, is associated with increased risk of hospitalization, persistent post-infection complications and worse outcomes in patients with COVID-19. This review overviews the biological features of Sars-CoV-2, highlights recent findings that delineate the pathological mechanisms of COVID-19 and the consequent clinical diseases.

Abbreviation

aa amino acid
ACE2 angiotensin converting enzyme 2
Ang angiotensin
Cat cathepsin
CCL chemokine ligand
CD147 cluster of differentiation 147
COVID-19 Coronavirus disease 2019
E protein envelope protein
GI gastrointestinal
HT hypertension
IBD inflammatory bowel disease
IFN interferon
IL interleukin
LDL low-density lipoprotein
M protein membrane protein
Mas mitochondrial assembly
mtDNA mitochondrial DNA
mTOR mammalian target of rapamycin
N protein nucleocapsid protein
NAFLD non-alcoholic fatty liver disease
NASH non-alcoholic steatohepatitis
NF-κb nuclear factor kappa B

NRP	neuropillin
NSP	non-structural protein
ORF	open reading frame
PAH	pulmonary arterial hypertension
PP	polyprotein
RAAS	renin angiotensin aldosterone system
RdRP	RNA-dependent RNA polymerase
RGD	Arg-Gly-Asp
ROS	reactive oxygen species
RTC	replication and transcription complex
S protein	spike protein
Sars-CoV-2	severe acute respiratory syndrome Coronavirus 2
SgRNA	subgenomic RNA
T1D	Type 1 diabetes
T2D	Type 2 diabetes
TLR	Toll-like receptor
TMPRSS2	transmembrane serine protease 2
TNF	tumor necrosis factor
TRAF	tumor necrosis factor receptor-associated factor
TRS	transcription regulatory sequence

1. Genome, proteins and life cycle of Sars-CoV-2

1.1 Genome of Sars-Cov-2

Severe acute respiratory syndrome-coronavirus-2 (Sars-CoV-2) is a positive-sense, single-stranded RNA, enveloped virus belonging to the family Coronaviridae, subfamily Orthocoronavirinae (Masters, 2006) Sars-CoV-2 possesses a remarkably large viral genome of ~30 kb in length. The genome is divided into fourteen open reading frames (ORFs) flanked by cis-acting secondary RNA structures necessary for RNA synthesis at the 5′- and 3′-untranslated regions. The fourteen ORFs of Sars-CoV-2 encode 29 viral proteins, including sixteen non-structural proteins (NSPs), which are encoded within two overlapping polyproteins (pp1a and pp1ab) that span the 5′ two-thirds of the viral genome (Chen, Liu, & Guo, 2020; Kim et al., 2020). The four canonical structural proteins (the spike, envelope, nucleocapsid, and membrane proteins) are encoded within the remaining third of the genome towards the 3′-untranslated region. Interspersed between the structural protein genes are genes encoding accessory proteins which assist in viral infection and replication: ORF3a, ORF3b, ORF6, ORF7a, ORF7b, ORF8b, ORF9b and ORF10.

2. Key proteins of Sars-CoV-2

2.1 Spike (S) protein

The spike (S) protein, a trimeric type I membrane protein, protrudes from the Sars-CoV-2 virion surface and is necessary for recognition of the host cells and initiation of viral fusion into host cells. The heavily glycosylated S proteins of coronaviruses are among the largest recorded, with that of Sars-Cov-2 being 1237 amino acids in length (180–200 kDa) (Huang, Yang, Xu, Xu, & Liu, 2020; Ou et al., 2020). This S protein is divided into two subunits: S1 and S2.

S1 subunit: S1 of the S protein possesses an N-terminal domain and a receptor binding domain (RBD). The RBD of S1 is necessary for host-virus interaction and exists in one of two conformations. In the low energy state of this subunit, termed the "down" state, the RBD is concealed by being angled towards the internal cavity of the trimeric S protein, thus offering protection to this important domain. Upon interaction with the host, the S1 subunit changes conformation via a hinge-like mechanism to reveal the binding site of the RBD and promote S-protein-host-receptor interaction.

S2 subunit: S2 of the S protein composed of four distinct regions: a fusion peptide, two heptad repeats termed HR1 and HR2, and a transmembrane domain. The helical stalk of S2 is formed primarily of HR1 (Tang, Bidon, Jaimes, Whittaker, & Daniel, 2020; Xia, Zhu, Liu, et al., 2020). A unique evolved characteristic of the Sars-CoV-2 virus, compared to that of its most closely related and well-studied counterpart, Sars-CoV, is a polybasic cleavage site (PRRAR) at the S1-S2 boundary that permits efficient cleavage of the S protein by furin protease (Budhraja, Pandey, Kannan, Verma, & Venkatraman, 2021; Peacock et al., 2021; Whittaker, 2021). Proteasomal digestion of the S protein is essential to internalization of the virus and the evolution of this new polybasic cleavage site may be key to the high infectivity, pathogenicity and zoonotic transmission of Sars-CoV-2. Within S2 is a further cleavage site termed S2′, immediately downstream from which is a fusion peptide (Lai, Millet, Daniel, Freed, & Whittaker, 2017). Of note, S2′ cleavage, but not S1 or S2 cleavage is essential to initiate internalization of the virus.

2.2 Envelope (E) protein

The envelope (E) protein is the smallest structural protein of Sars-CoV-2, being only 8–12 kDa. The E protein is composed of three primary domains: a short hydrophilic N-terminus of 7–12 amino acids; a hydrophobic trans-membrane domain ~25 amino acids in length; and a longer hydrophobic C-terminus (Schoeman & Fielding, 2019). The E protein assists viral lysis and subsequent release of the viral genome upon cellular internalization of a virion. In a host cell, the E protein localizes to the endoplasmic reticulum (ER) and Golgi membranes to assist in viral assembly and budding, and to form pentameric ion channels, with little or no selectivity, to enhance viral replication (Corse & Machamer Carolyn, 2000; Nieto-Torres et al., 2014). Notably, E protein mediated pumping of $Ca2+$ ions out of the ER enhances cellular inflammation and may be a significant factor in the unusually large immune response to Sars-CoV-2 infection.

2.3 Nucleocapsid (N) protein

The nucleocapsid (N) protein is the only structural protein within the virion, which protects and localizes the RNA genome towards the membrane by packaging it into a ribonucleoprotein complex (Cubuk et al., 2021). The N protein is composed of two structurally conserved domains: the N-terminal domain and the C-terminal domain; the prior being a monomer, the latter being a dimer (Ye, Lu, & Corbett, 2021). The N-terminal domain is characterized by a positively charged protrusion to facilitate putative RNA binding. This RNA binding characteristic of the N protein allows a degree of protection from intracellular host immune responses as the N protein can suppress RNAi-mediated antiviral responses (Cui et al., 2015).

2.4 Membrane (M) protein

The membrane (M) protein is the most abundant structural protein of Sars-CoV-2, which is the protein to which all other structural proteins bind. The M protein spans the virion lipid membrane and leaves a short NH2-terminal protruding from the virion and a long COOH terminus within the virion (Thomas, 2020). It is interaction between the M and N proteins that promote completion of viral assembly by stabilizing the N protein RNA complex inside the virion (Bianchi et al., 2020).

3. Lifecycle of Sars-CoV-2

3.1 Entry of Sars-CoV-2

Entry of Sars-CoV-2 into the host is initiated by the RBD within S protein of Sars-CoV-2 that interacts with host ACE2 receptor (Yang et al., 2020; Zamorano Cuervo & Grandvaux, 2020). Proteolytic cleavage of the S protein by host-cell-derived proteases is essential to permit fusion of the virus with the host cell membrane. Typically, this is facilitated by transmembrane serine protease 2 (TMPRSS2), with optional catalysis by endosomal cysteine protease cathepsin B (CatB) and/or cathepsin L (CatL) (Padmanabhan, Desikan, & Dixit, 2020; Qiao et al., 2021; Zhao et al., 2021, 2022). Furthermore, the aforementioned novel polybasic furin protease cleavage site between subunits S1 and S2 allows efficient cleavage of the S protein by furin in addition to TMPRSS2 and CatB/L. *In vitro*, it has been demonstrated that TMPRSS2 is preferred over CatB/L for S protein cleavage as inhibition of TMPRSS2 alone is sufficient to block entry of Sars-CoV-2 into primary lung cells (Hoffmann et al., 2020). Means of Sars-CoV-2 invading host cells via relevant protein machinery are utilized in Fig. 1.

***ACE2*:** The ACE2 receptor is expressed across more than 15 organ systems including the lungs, heart, kidneys, gastrointestinal (GI) tract, adipocytes, and pancreas. ACE2 is a component in the renin-angiotensin-aldosterone system (RAAS) (Reynolds et al., 2020): a homeostatic regulator of vascular function. In the RAAS pathway, renal juxtaglomerular cells release renin which cleaves angiotensin to angiotensin-I (AngI), which through the action of angiotensin converting enzyme (ACE) generates angiotensin-II (AngII); a powerful vasoconstrictor and the main effector of the RAAS pathway (Wu et al., 2018). ACE2 functions as a dipeptidyl carboxypeptidase within this system to convert AngII into Ang (1–7), AngI into Ang (1–9), and to hydrolyze peptides (Kuba, Imai, & Penninger, 2013). The primary recognized function of ACE2 is to lower blood pressure via its primary product Ang (1–7), an enzyme that promotes vasodilation by binding the mitochondrial assembly (Mas) receptor (Xu et al., 2017).

***TMPRSS2*:** TMPRSS2 is a 492 amino acid (aa) transmembrane protease of the Hepsin/TMPRSS subfamily of type II transmembrane serine proteases. TMPRSS2 contains an S1 family serine protease domain followed by a scavenger receptor cysteine rich domain. Additionally, there is a low-density lipoprotein receptor class A domain, a predicted transmembrane domain,

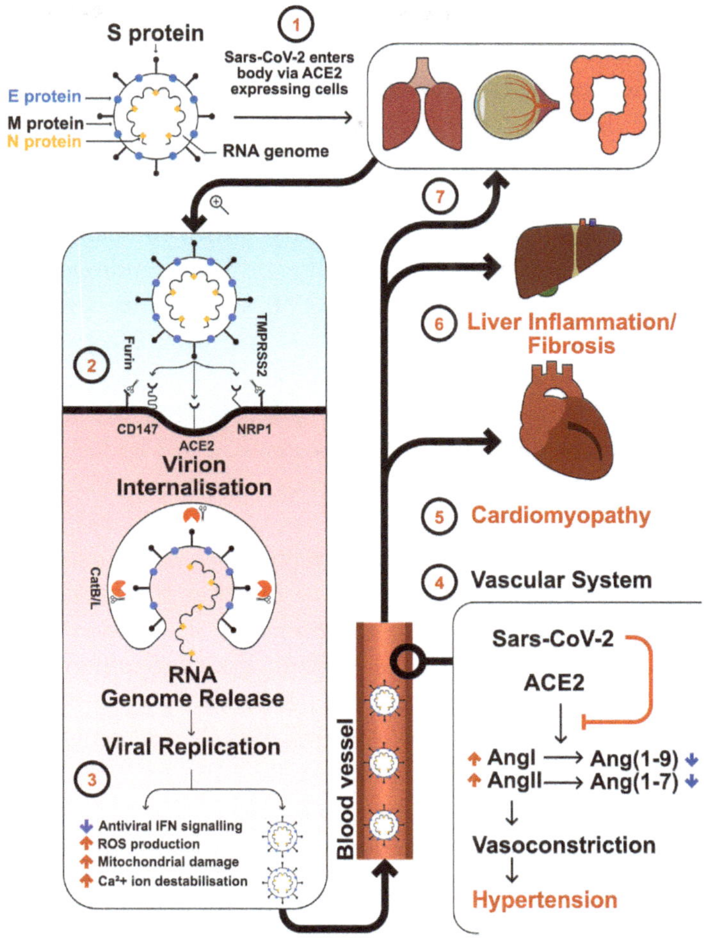

Fig. 1 Sars-CoV-2 induces cardiovascular dysfunction via ACE2 and enhances NAFLD (1) Sars-CoV-2 enters the body via the ACE2 expressing epithelia of the eyes, lungs, and digestive system. (2) Sars-CoV-2 enters cells by binding between the viral S-protein and host-cell surface proteins ACE2, CD147, or NRP1. Digestion of the S-protein by Furin and/or TMPRSS2 is essential for viral internalization. The viral RNA genome is released to facilitate replication and translation of viral proteins that assemble as new viral particles. (3) Translation of viral proteins induces downregulation of antiviral IFN signaling, upregulation of ROS production, mitochondrial damage, and destabilization of calcium ion homeostasis. (4) Newly synthesized virions are released into circulation where the S-protein of Sars-CoV-2 competes with vasoconstrictors AngI and AngII for ACE2 causing accumulation of AngI and AngII and the consequently chronic vasoconstriction, leading to hypertension. (5) Sars-CoV-2 infects the contractile cardiomyocytes of the heart, inducing cardiomyopathy. (6) Sars-CoV-2 infects hepatocytes of the liver, enhancing inflammation and promoting fibrogenesis, particularly in NAFLD/NASH patients. (7) Newly synthesized virions further target the eyes, lungs and the digestive system to amplify the pathological conditions.

and a cytoplasmic domain. TMPRSS2 is expressed across various organs including respiratory system, gastrointestinal system, liver, kidneys, pancreas, prostate, and thymus (Shen, Mao, Wu, Tanaka, & Zhang, 2017). TMPRSS2 primes the Sars-CoV-2 S protein by proteolytic cleavage (Abbasi et al., 2021). Substantial evidence has supported the notion of upregulation of TMPRSS2 expression by androgen may be a mechanism for the observed greater COVID-19 severity in men compared to women, suggesting that agents that downregulate or block androgen mediated TMPRSS2 activity may ameliorate COVID-19 severity (Bonafè et al., 2020; Deng, Rasool, Russell, Natesan, & Asangani, 2021; Qiao et al., 2021; Wambier & Goren, 2020).

***Cathepsin*:** Cathepsins are versatile protease enzymes that function within the low pH environment of lysosomes. CatB and CatL are both cysteine proteases: CatB is 339 aa and in its mature state is composed of a 25–26 kDa heavy chain and 5 kDa light chain. CatL is of similar size and structure being 333 aa in length. In addition to lysosomal activity, Cathepsins are also secreted from cells, allowing CatL to function as a matrix-degrading enzyme associated with chronic inflammation. Expression of CatL is upregulated in COVID-19 and overexpression of CatL in human cell lines increases viral pathogenicity (Zhao et al., 2021).

***Furin*:** Furin is known to cleave some glycoproteins from viral enveloped and increase viral fusion into cells. Furin belongs to the family of Ca^{2+}-dependent protein/prohormone convertases expressed ubiquitously across human tissue and recognizes the R-X-K/R-R peptide motif (Vankadari, 2020). Furin cleavage of the Sars-CoV-2 S protein has been demonstrated to be critical for Sars-CoV-2 replication and pathogenicity as data showed that loss of the furin cleavage site dramatically reduces viral pathogenicity (Johnson et al., 2020).

***Neuropillin-1*:** The Neuropillin 1 (NRP1), a type I cell surface receptor, is a novel means of entry for Sars-CoV-2 (Cantuti-Castelvetri, Ojha, Pedro, et al., 2020; Cantuti-Castelvetri, Ojha, Pedro Liliana, et al., 2020; Daly James et al., 2020). NRP1 is 130 kDa and composed of 850 aa. The structure is highly conserved among vertebrates: it is composed of five structural motifs (a1, a2, b1, b2, and c) that form an ectodomain mosaic. The protein N-terminus contains the tandem CUB domains (a1/a2) adjacent to tandem coagulation factor domains (b1/b2). The remainder of the ectodomain is comprised of the "c" member proximal MAM domain common among metalloproteinases and the N-termini of receptor protein tyrosine phosphatases (Yelland & Djordjevic, 2016). NRP1 is significant in development of the circulatory and nervous systems, influencing angiogenesis by stabilizing vascular endothelial growth factor and its receptor (Lampropoulou & Ruhrberg, 2014; Lee et al., 2002). As such, NRP1 is highly expressed in arteries including the aorta. *In vitro*, it has been demonstrated that, in the absence of ACE2, NRP1 alone is not sufficient to allow entry of Sars-CoV-2, however, presence of NRP1 increases efficiency of Sars-CoV-2 intake (Cantuti-Castelvetri, Ojha, Pedro, et al., 2020; Daly James et al., 2020). Thus, NRP1 expressing tissues, such as those in the circulatory system, are of increased susceptibility to Sars-CoV-2 infection.

CD147: Similarly, the transmembrane glycoprotein, cluster of differentiation (CD147), has been identified as a novel means of host cell entry for Sars-CoV-2 (Wang, Chen, Zhang, et al., 2020). Direct interaction between the Sars-CoV-2 S protein and host CD147 has been observed in vitro and overexpression of CD147 promotes viral infection. Furthermore, expression of CD147 in a cell line that is resistant to Sars-CoV-2 infection due to diminished ACE2 is sufficient to enable Sars-CoV-2 entry into cells (Wang, Chen, Zhou, et al., 2020). However, these claims have been disputed with negative results produced by a second research team (Shilts, Crozier, Greenwood, Lehner, & Wright, 2021). Further investigation is needed to determine the role of CD147 in Sars-CoV-2 infection. Though, of note, CD147 expression is enhanced on cardiomyocytes originating from the hypertrophied left ventricle of transverse-aortic constriction mice (Zhong et al., 2022). This enhancement of CD147 in constricted heart tissue may promote Sars-CoV-2 infection in the cardiovascular system during later stages of COVID-19, and may help explain the association between heart diseases—including hypertension (HT) and arrhythmia—and COVID-19.

Integrin α5β1: Integrins are cell adhesion receptors that comprise 24 αβ heterodimers and are capable of recognizing glycoproteins and cell surface/extracellular matrix ligands. α5β1 is an Arg-Gly-Asp (RGD)-binding integrin expressed by endothelial and epithelial cells (Schaffner, Ray, & Dontenwill, 2013). α5β1 typically acts as a fibronectin receptor and is demonstrated as a facilitator of Sars-CoV-2 cell entry by α5β1-RGD-Sars-CoV-2-RBD binding (Robles et al., 2022).

3.2 Genome transcription and translation

Upon entry of the virus into the host cell, the positive-sense single stranded RNA genome is released to allow translation. Translation of ORF1a and

ORF1b yield two polyproteins, pp1a and pp1ab, from which the sixteen NPSs of the virus are cleaved. Pp1a encodes NSP1-11 whilst pp1ab, which is approximately 1.4 times larger, contains NSP1-10 and NSP12-16 (Chen et al., 2020; V'kovski, Kratzel, Steiner, Stalder, & Thiel, 2021). Release of NSPs from these polyproteins is achieved by proteolytic cleavage by cysteine proteases located within NSP3 and NSP5; release of NSP1 occurs particularly rapidly to enhance viral replication by blocking mRNA export from the host cell nucleus: NSP1 interacts with host mRNA export receptor, the heterodimer NXF1-NXT1 that is necessary for export of nuclear mRNAs. This frees up host ribosomes for translation of viral genes (Zhang et al., 2021).

NSP12-16 compose the viral replication and transcription complex (RTC), containing the core enzymatic functions required for RNA synthesis, RNA proofreading, and RNA modification. It is NSP12, an RNA-dependent RNA polymerase (RdRP) that is essential to further RNA synthesis whilst NSP7 and NSP8 function as cofactors (Naydenova et al., 2021; Peng et al., 2020); NSP14 provides a 3′-5′ exonuclease activity that assists RNA synthesis with a unique proofreading function (Smith, Blanc, Vignuzzi, & Denison, 2013).

The yet to be fully elucidated coronavirus capping machinery is composed of NSP10 (Viswanathan et al., 2020), which functions as a cofactor, NSP13, which provides the RNA 5′-triphosphatase activity (Ivanov et al., 2004), and NSP14 and NSP16, which perform the functions of N7-methyltransferase and 2′-O-methyltransferase respectively (Chen et al., 2009; Vithani et al., 2021). One key enzyme typically involved in the formation of the 5′ cap structure, the guanylyl transferase, has not yet been identified in coronaviruses

3.3 Structural and accessory proteins of Sars-CoV-2

Structural protein genes are encoded in the 3′-third of the RNA genome (Chen et al., 2020). Coronavirus structural proteins assemble and assist in the budding of new virions at the ER-to-Golgi compartments that are suggested to exit the infected cells by exocytosis. Some data suggests β-coronaviruses such as Sars-CoV-2 exit cells via the lysosomal trafficking pathway (Chen et al., 2021).

Between the ORFs of structural protein genes are interspersed accessory protein genes. At least five ORFs encoding accessory genes are encoded in Sars-CoV-2: ORF3, ORF6, ORF7a, ORF7b, ORF8, and ORF9

(Davidson et al., 2020; Finkel et al., 2021). Additionally, ORF10 is postulated to be downstream of the N protein gene (Pancer et al., 2020). However, not all of these are experimentally verified. Accessory genes display high variability among coronavirus groups and usually show no sequence homology with other viral and cellular proteins. Although they are not required for virus replication in cell culture, to some extent, they are conserved within their respective virus species and are suspected to have an important role in virus-host interactions. A common characteristic among Sars-CoV-2 accessory proteins is inhibition of the host cellular antiviral interferon (IFN) response; Sars-CoV-2 accessory proteins are detailed below:

***ORF3*:** ORF3 consists of three family members. ORF3a—The largest accessory protein of Sars-CoV-2 and an integral membrane viroporin that forms ion channels that may promote viral release. It is 275 aa in length and consists of an N-terminal domain, transmembrane domain, and a C-terminal domain folded as an 8-strand barrel (Yu et al., 2004; Yuan et al., 2005). Furthermore, ORF3a presents a TRAF3-binding motif that activates the NLRP3 inflammasome and is a potent stimulator of pro-interleukin (IL)-1β transcription (Siu et al., 2019). ORF3a has been identified as pro-apoptotic in host cells. ORF3b—Primarily localized to the cytoplasm, and is a potent IFN antagonist of 22 aa in length. Inhibition of IFN-1 by ORF3b is associated with disturbing the translocation of IFN1B transcription factor IRF3 to the nucleus (Konno et al., 2020). ORF3c—Function of ORF3c is unclear, however it possesses a conserved transmembrane domain that might suggest lipid bilayer membrane interaction (Jungreis, Sealfon, & Kellis, 2021).ORF3d—The nucleotide sequence of ORF3d overlaps other ORF3 genes from the 5′ of ORF3a to the 3′ or ORF3c. ORF3d is a 57 aa protein of unknown function that interacts with mitochondrial stomatin-like protein 2 (Redondo, Zaldívar-López, Garrido, & Montoya, 2021).

***ORF6*:** ORF6 is composed of 61 aa in length and localizes to membrane-bound organelles including the ER, lysosomes, and vesicles. It acts as an IFN antagonist by interfering with migration of transcription factor STAT to the nucleus, thus blocking IFN activation (Kopecky-Bromberg, Martínez-Sobrido, Frieman, Baric, & Palese, 2007).

***ORF7*:** ORF7 consists of two family members. ORF7a—A type I transmembrane protein formed of 121 aa in a seven-stranded β-sandwich fold. ORF7a is polyubiquitinated at Lys 119, a modification believed to enhance the ability of ORF7a to interfere with the IFN-1 response by inhibition of STAT2 (Li et al., 2020; Su et al., 2006; Xia, Cao, Xie, et al., 2020). Furthermore, ORF7a is observed to bind CD14+ monocytes, reducing their antigen presenting ability, but greatly inducing expression of pro-inflammatory cytokines (Zhou et al., 2021). *ORF7b*—The intended function is yet to be elucidated, but ORF7b is composed of 43 aa and localizes to the Golgi. It is believed not essential for viral replication and has been shown to induce expression of INF-β, IL-6, and tumor necrosis factor (TNF)α in addition to activating type-I IFN signaling via IF3 phosphorylation (Yang, Zhao, Rao, et al., 2021).

ORF8: ORF8 is a 121 aa protein consisting of an N-terminal signal sequence followed by a predicted Ig-like fold (Flower Thomas et al., 2021). Whilst ORF8 expression is not essential to viral replication, it appears to function as an organizer of protein folding in the ER: ORF8 interacts with Torsin-1a (a protein folding quality controller in the ER). Furthermore, ORF8 modulates the unfolded protein response by upregulating ER-resident chaperones binding immunoglobulin protein 78 (Bip) and glucose regulated protein 94 as well as inducing activating transcription factor-6 and inositol requiring enzyme 1, but it does not appear to stimulate the PERK pathway (Rashid, Dzakah, Wang, & Tang, 2021; Valcarcel, Bensussen, Álvarez-Buylla, & Díaz, 2021).

ORF9: ORF9 consists of three family members. ORF9b—There is conflicting nomenclature regarding ORF9b: this ORF has been inconsistently referred to as ORF9a or ORF9b. However, ORF9b has been proposed as the preferred name (Jungreis et al., 2021). Like other Sars-CoV-2 accessory proteins, ORF9b inhibits the IFN response. ORF9b achieves this by localizing to the mitochondrial membrane and reducing cellular tumor necrosis factor receptor-associated factor (TRAF) 3 and TRAF6 levels (Shi et al., 2014). Additionally, ORF9b forms a complex with mitochondrial membrane protein Tom70, a mitochondrial import receptor, consequently impairing cellular anti-viral response (Jiang et al., 2020). *ORF9c*–ORF9c also impairs cell antiviral response by interfering with the IFN response. Moreover, ORF9c interacts with sigma receptors associated with lipid remodeling and ER-stress, as well as nuclear factor (NF)-κB related proteins Nod-like receptor, proteinase activated receptor 2, and Nedd4 family interacting protein 2. The downstream influence of ORF9c is upregulation of MAPK/ERK2 pathway proteins and inhibited immune response caused by impairing antigen presentation and complement signaling (Dominguez Andres et al., 2020).

ORF10: The protein is 38 aa in length. The function of ORF10 is unclear, however, believed to be unessential for viral replication and transmission (Pancer et al., 2020).

3.4 Replication compartments

In the early stages of Coronavirus infection, like many other positive-sense RNA viruses, the infected host cell experiences membrane reorganization with the formation of ER-derived and interconnected perinuclear double-membrane structures (V'kovski et al., 2021). These include double membraned vesicles, open double-membraned spherules, and convoluted membranes. It is the direct interaction between coronavirus NSPs and host cell factors that induced biogenesis of these specialized organelles termed replication compartments or replication organelles. Mechanism for the formation of replication compartments is not fully understood. In Sars-CoV-2 infection, it is proposed that the membrane spanning NPSs, NSP3, NSP4, and NSP6 are responsible for manipulating host membranes into the observed replication compartments (Oudshoorn et al., n.d.; Angelini, Akhlaghpour, Neuman, & Buchmeier, 2013).

Replication compartment formation is a conserved characteristic among coronaviruses that serves to provide a favorable niche rich in necessary macromolecules for RNA synthesis whilst shielding viral replication intermediates from cytosolic immune sensors. It is known that Sars-CoV-2 replicase subunits (NSP3, NSP5, NSP8) anchor themselves to convoluted membranes, however, the exact location of viral RNA replication remains unknown (Ulasli, Verheije, de Haan, & Reggiori, 2010). Upon formation, double-stranded RNA migrates to the interior of double-membrane vesicles via NSP3 formed pores to potentially render newly synthesized viral RNAs available for translation and encapsidation into virions (Wolff et al., 2020). It is unclear when double-stranded viral RNA is separated to form the final single-stranded RNA genome.

3.5 RNA transcription

NSP12-16 comprise the core functions of viral RNA synthesis, proofreading, and modification. The RNA-dependent RNA polymerase (RdRP) residing in NSP12 is the centrepiece of coronavirus RTC and has been suggested as a promising drug target as it is a crucial enzyme in the virus lifecycle both for replication of the viral genome and also for transcription of subgenomic RNAs (sgRNAs) (Naydenova et al., 2021; Peng et al., 2020). Structure of Sars-CoV-2 NSP12 and its cofactors NSP7 and NSP8 show high conservation and similarity with that of Sars-CoV with more than 95% similarity (Biswal et al., 2021). NSP12 is composed of three domains: a Nidovirus RdRP-associated nucleotidyl-transferase domain, a right-handed RdRP domains, and an interface domain (Dwivedy et al., 2021). Viral genomic replication is initiated by synthesis of full-length negative-sense genomic copies which function as templates for the generation of new positive-sense genomic RNA. Newly synthesized genomes are used for translation of more NSPs and RTCs or are packaged into new virions.

Characteristic of coronaviruses, and other members of the Order Nidovirales, is the discontinuous transcription process: in which, a set of nested 3′ to 5′ co-terminal sgRNAs are produced (Kim et al., 2020; Sola, Almazán, Zúñiga, & Enjuanes, 2015). During negative-strand RNA synthesis, the RTC interrupts transcription upon encountering transcription regulatory sequences (TRSs) located upstream to most ORFs in the third of the genome adjacent to the 3′ end. At TRS elements (TRS body), synthesis of negative-strand RNA stops and is re-initiated at the TRS adjacent to a leader sequence (TRS-L) located ~70 nucleotides from the 5′ end of the genome. The discontinuous step of coronavirus RNA synthesis involves interaction between complementary TRSs of the nascent negative strand RNA (negative sense TRS body) and the positive strand genomic RNA (positive sense-TRS-L). Upon re-initiation of RNA synthesis at the TRS-L region, a negative strand copy of the leader sequence is added to the nascent RNA to complete the synthesis of the negative-strand sgRNAs. Creation of negative-strand RNA in this discontinuous manner occurs via production of a set of negative-strand subgenomic RNAs that act as templates for synthesis of positive-sense subgenomic mRNAs; these are translated into structural and accessory proteins. Although the coronavirus subgenomic mRNAs are structurally polycistronic, it is assumed that they are functionally monocistronic and only the first ORF at the 5′ end, absent in the next smaller sgRNA, is translated from each sgRNA (Sola et al., 2015).

4. Molecular mechanisms underlying Sars-CoV-2 induced pathological conditions

Sars-CoV-2 infection increases expression of coagulation markers which favors development of stroke, myocardial infarction, and ischemic diseases, potentially inducing multiple organ dysfunction or failure (Ruiz-Ares et al., 2021). More profound complications extend to thrombosis in large parts of circulation including deep veins, arteries, and cardiac chambers (Long, Brady, Koyfman, & Gottlieb, 2020). This is enhanced by the exceptionally aggressive innate immune response experienced during COVID-19, termed the "cytokine storm", which perpetuates a hypercoagulable state. Individuals with coronary artery disease are at higher risk of complications (e.g., venous thromboembolism and arterial thromboembolism) due to the cytokine storm and exhibit inflamed plaque instability and rupture post Sars-CoV-2 infection (Pan et al., 2021; Song, Li, Xie, Hou, & You, 2020). Additionally, myocarditis is observed in COVID-19 patients, during and after infection, and is associated with impaired cardiac performance that potentially persists indefinitely (Shchendrygina, Nagel, Puntmann, & Valbuena-Lopez, 2021).

4.1 Cytokine storm induced by Sars-CoV-2

Points-of-contact for the Sars-CoV-2 virus are the ACE2 expressing mucus membranes/epithelia of the respiratory system, gastrointestinal system, and eyes. However, the virus has been observed penetrating organs across the body, being identified in biopsies of heart, liver, intestinal, and pancreatic tissues among many others, including immune privileged tissues (e.g., the brain and testes) (Trypsteen, Van Cleemput, Snippenberg, Gerlo, & Vandekerckhove, 2020). Sars-CoV-2 is believed to enter circulation via lesions of the alveoli. Additionally, the virus circulates through the circulatory system partly protected from the immune surveillance of the immune cells within the blood, a trait also observed in Sars-CoV which displays high infectivity of blood cells, macrophages, fibroblasts, and pericytes (Butler et al., 2022; Misiti, 2021). Furthermore, Sars-CoV-2 virion particles have been found in the circulatory system within extracellular vesicles containing the ACE2 receptor.

Upon Sars-CoV-2 tissue infection, there is a prompt infiltration of the pericardium, myocardium, and endocardium by immune cells, including neutrophils, pro-inflammatory monocytes, macrophages, and lymphocytes, converge on sites of viral infection. Cardiomyocytes infected with Sars-CoV-2 have increased expression of chemokine ligand 2 (CCL2), CCL3, and CXC-chemokine ligand 10, thus enhancing monocyte recruitment and secretion of cytokines (Law, Lee, Cheung, Yim, & Lau, 2007; Yang, Nilsson-Payant, Han, et al., 2021). This induces release of pro-inflammatory cytokines, including monocyte chemoattractant protein-1, IL-1β, IL-6, and TNFα. IL-1β is a key regulator of the inflammatory response and simulates the release of other cytokines, such as IL-17, IL-21, and IL-22. IL-1β and TNFα, contribute to COVID-19 associated pain that often manifests as a severe headache and joint pain by inducing cyclooxygenase-2 in the nervous system. Such pain is observed to persist for several weeks after COVID-19 infection (Song et al., 2020).

In the myocardium, the elevated cytokines contribute to myocardial dysfunction and cardiac disease, for example, IL-1β may promote cell proliferation and contribute to cardiomyopathy by thickening layers of heart tissue (Bujak & Frangogiannis, 2009). Activation of the thrombin and complement system in COVID-19 has been found to induce endothelial lesions via excessive accumulation of platelets and white blood cells at sites of Sars-CoV2 infection. This contributes to coagulopathy with elevated D-dimer and fibrin degradation products, leading to micro thrombi (Magro et al., 2020).

The other largely released cytokines, such as TNFα, IFNγ, and IL-4, promote cellular pro-inflammatory pathways via different cytokine receptors. For example, binding of TNFα to the TNFα receptor 1 (TNFR1) recruits several adaptor proteins, out of which TRADD-containing complex activates the IKK complex, thus resulting in IκBα phosphorylation and subsequent ubiquitin-dependent proteasomal degradation and activation of NF-κB (Shi & Sun, 2018). Subsequently, this further activates several well-studied inflammatory and apoptotic signaling pathways, including c-Jun-N-terminal kinase and p38 mitogen activated protein kinases, caspase 8, and the NF-κB pathways that all together damage the cardiomyocytes. In addition, the release of IL-4 is often associated with cardiac fibrosis during HT and heart failure (Peng et al., 2015). Activation of these pathways can further signal through the mammalian target of rapamycin (mTOR) to initiate autophagy in cardiomyocytes.

Virus-induced cell-mediated autoimmune responses can lead to myocarditis and edema of the myocardial interstitial connective tissue. Myocarditis can progress to dilated cardiomyopathy (DCM). DCM patients who harbor cardiotropic viruses and coupled with myocarditis have a much poorer prognosis. Epidemiological studies show myocarditis is present in 20-30% of COVID-19 patients (Akhmerov & Marbán, 2020). Recent studies show COVID-19 patients with cardiovascular disease possessed anti-cardiolipin antibody (Jizzini, Shah, & Zhou, 2020), which against cardiolipin, an inner component of the mitochondrial membrane, and considered an indicator of cellular damage and myocardial injury (Dudek, 2017).

4.2 Perturbation of calcium ion homeostasis by Sars-CoV-2

Calcium is a major secondary messenger in cellular physiology, regulating countless metabolic processes and pathways. Many viruses are capable of co-opting and manipulating host calcium ion systems to support their own replication and survival (Chen, Cao, & Zhong, 2019; Zhou, Frey, & Yang, 2009). The essential role of Ca^{2+} ions in ATP synthesis and neuronal signaling for muscle contraction suggest dysfunction of Ca^{2+} ion homeostasis induced by Sars-CoV-2 infection would contribute to the reduced contractility of cardiomyocytes and general dysfunction of cardiac muscle observed in COVID-19.

In Sars-COV-2 infected endothelial cells, there is reduced expression of voltage-dependent anion channel (VDAC): an ion channel permeable to small solutes such as Ca^{2+} and a regulator of mitochondrial Ca^{2+} flux. Reduction in endothelial cell Ca^{2+} concentrations override release of vasodilator mediators and exacerbates endothelial dysfunction. Furthermore, upregulation of TLR-9 induced by Sars-CoV-2 infection inhibits ER/sarcoplasm reticulum Ca^{2+}-ATPase (SERCA2) activity, leading to reduced uptake of Ca^{2+} into the ER and mitochondria of cardiomyocytes and neurons (Costa et al., 2022; Wang et al., 2015).

4.3 Sars-CoV-2 infection induced mitochondrial oxidative stress

Mitochondria play a central role in host immune responses to viral infection. Mitochondrial dysfunction leads to release of mitochondrial DNA (mtDNA) which is a serious damage associated molecular pattern (DAMP) and the consequently significant induction of inflammation and cytokines. Sars-Cov-2 infection induces mitochondrial dysfunction, including increased production of reactive oxygen species (ROS) and mtDNA, reduced plasma Ca^{2+} concentrations and elevated circulating mtDNA concentrations (Costa et al., 2022).

Changes in mitochondrial membrane potential during mitochondrial dysfunction disrupts synthesis of ATP, induces membrane depolarization, and promotes mitochondrial swelling. *In vitro* infection of endothelial HUVEC cells by Sars-CoV-2 showed elevated gene expression of several components of the electron transport chain and mitochondrial ATP synthesis, including cytochrome B and NADH dehydrogenase. Furthermore, injury-associated intracellular enzyme LDH was also increased upon Sars-CoV-2 infection, suggesting the association between Sars-CoV-2 and mitochondrial damage (Costa et al., 2022).

4.4 Activation of TLR signaling by Sars-CoV-2

COVID-19 patients display elevated plasma mtDNA concentrations which induces immune response via activation of TLR9 (Zhang, Liu, Liu, Ren, & Sun, 2014). TLR9 also responds to unmethylated CpG repeats in RNA, which allow TLR9 to be activated by Sars-CoV-2 RNA that has unmethylated CpG motifs within the E-ORF and ORF10 nucleotide sequences, and mtRNA (Kumar et al., 2022). Similarly, TLR3, TLR7, and TLR8 show high binding affinity to Sars-CoV-2 mRNA (Sartorius, Trovato, Manco, D'Apice, & De Berardinis, 2021). *In silico* studies highlight TLR9 to have no binding affinity to the Sars-CoV-2 S protein in contrast to the high binding affinity identified in TLR4, TLR6, and TLR1 (Choudhury, Das, Patra, & Mukherjee, 2021).

The extent to which TLR9 activation contributes to the hyperinflammation observed in COVID-19 is unclear. However, when considering association between COVID-19 and cardiovascular disease, the exceptional enrichment of myocardial cells with mitochondria may be a major contributing factor to Sars-CoV-2 induced myocardial dysfunction. Further investigation is warranted to delineate the association between Sars-CoV-2 and inflammatory responses in COVID-19 patients.

5. Sars-CoV-2 induced pathological conditions via mediation of ACE2

5.1 Sars-CoV-2 targets the cardiovascular system

COVID-19 and hypertension: The vascular endothelium is a crucial interface between blood and tissue, and plays a critical role in modulating angiogenesis, blood coagulation, and vascular toning. Under viral assault, endothelial cells mediate important innate immune processes, promoting swelling by increasing blood flow, regulating leakage of plasma proteins, and permitting movement of neutrophils, monocytes, and lymphocytes to sites of tissue damage (Rajendran et al., 2013). Sars-CoV-2 induced endothelial dysfunction observed in COVID-19 patients is associated with disrupted coagulation, thrombosis, systemic endotheliitis, increased ROS production, NO deficiency, and elevated D-dimer concentrations (Al-Samkari et al., 2020; Green, 2020; Li et al., 2021).

HT is a major comorbidity among COVID-19 patients, this is also true specifically for pulmonary arterial hypertension (PAH): a progressive condition characterized by abnormally high blood pressure in the pulmonary arteries with a median survival rate of 2.8 years from time of diagnosis if untreated. Development of HT/PAH in COVID-19 patients is believed to be a result of interference/co-opting of the ACE2 receptor. By competing with the ACE2 receptor against AngI and AngII, Sars-CoV-2 blocks conversion of these vasoconstricting angiotensins to their vasodilating metabolites Ang (1–7) and Ang (1–9), consequently promoting elevated blood pressure (Fig. 1.) (Carey, Wang, & Siragy, 2000; Gressens et al., 2021; Kuba et al., 2013). Clinically, this manifests with HT preceding hospitalization being the leading comorbidity in non-surviving COVID-19 patients (Devaux, Rolain, & Raoult, 2020; Tudoran et al., 2021).

Sars-CoV-2 competes with AngI and AngII for ACE2 (Verdecchia, Cavallini, Spanevello, & Angeli, 2020), thus Sars-CoV-2 directly contributes to deterioration of cardiovascular homeostasis by causing an accumulation of vasoconstricting ACE2 ligand, AngII: in the absence of ACE2, AngII will bind angiotensin II type 1 receptor (ET1R) and/or angiotensin II type 2 receptor (AT2R) (Carey et al., 2000; Eguchi, Kawai, Scalia, & Rizzo, 2018). The downstream impact of Sars-CoV-2-induced ACE2 reduction further increases the risk of cardiovascular disease and damage in the form of cardiomyocyte damage/death due to elevated ROS concentrations, apoptosis, hypertrophy and fibrosis, in addition to the classic alveolar damage associated with Sars-CoV viruses (Muñoz et al., 2014; Ye et al., 2007; Zhong et al., 2010). It was postulated that use of RAAS pathway modulators (e.g., ACEi and ARBs) which inhibit production/activity of Angiotensin II, may potentiate the upregulation of ACE2 receptor and increase risk of COVID-19 (Gressens et al., 2021).

COVID-19 and myocarditis: Cardiomyocytes, the primary contractile cells of the heart, are particularly vulnerable to assault by Sars-CoV-2. Cardiomyocytes express all the necessary machinery to enable invasion by Sars-CoV-2 including: ACE2, TMPRSS2, CatB, CatL, and novel means of cell entry including NRP1, CD147, and integrins ($\alpha5\beta1$) (Thum, 2020; Vidak, Javoršek, Vizovišek, & Turk, 2019; Wu, Sun, Trzeciakowski, Meininger, & Muthuchamy, 2006). Infection of the myocardium by Sars-CoV-2 has major implications on host mortality with infected cardiomyocytes displaying diminished contractility due to dysregulation of contractile protein expression and dysregulated Ca^{2+} homeostasis (Danta, 2021).

COVID-19 and Arrythmia: Arrhythmic electrocardiographic abnormalities in the atria and ventricles are also observed in patients infected with Sars-CoV-2, indicating unlocalized myocardial damage (Mountantonakis et al., 2021). This is likely due to myocarditis or direct virus-induced damage to cardiomyocytes. Inflammation of the sinoatrial, atrioventricular nodes or Purkinje may also contribute to arrhythmia. In COVID-19 patients, such arrhythmia is typically recorded as ST-elevation myocardial infarction (STEMI) or NSTEMI, inverted T wave and inverted P wave (Long et al., 2020).

5.2 COVID-19 in obesity and type-2 diabetes

Susceptibility of individuals with metabolic disease to COVID-19 is complicated. For example, obese individuals and those with type-2 diabetes (T2D) often display a weakened immune system and generally increased risk of infection. These groups are at an increased risk of contracting COVID-19 and other infectious illnesses. COVID-19 severity, and consequently rates of hospitalization, are higher in people with type 1 diabetes mellitus (T1D), T2D, or obesity (Barron et al., 2020; Gao et al., 2020; Kwok et al., 2020). A large population-based study including 8885 COVID-19 patients analyzed the association between COVID-19 with various comorbid conditions, such as HT, obesity, diabetes, hyperlipidemia, and non-alcoholic steatohepatitis (NASH) has found strong association between these conditions and incidence of COVID-19, with NASH having the strongest association (Ghoneim, Butt, Hamid, Shah, & Asaad, 2020).

T2D and obesity are metabolic conditions characterized by insulin resistance and chronic low-grade metabolic inflammation (Ferlita et al., 2019; Su et al., 2009; Tam, Morais, & Santosa, 2020). Moreover, the extent of metabolic inflammation is further modified by adipokines, hepatokines, myokines, lipokines, branched-chain amino acids, and microbial metabolites, which may circulate at abnormal levels in some people with T2D or obesity (Stefan & Häring, 2013). Collectively, these acquired abnormalities broadly impair cellular immune function, likely contributing to enhanced activation of the NLR family pyrin domain containing 3 (NLRP3) inflammasome, and a greater susceptibility to infection in vulnerable individuals (Sharma & Kanneganti, 2021). There is considerable interest in whether the expression of viral entry factors essential for SARS-CoV-2 infectivity is enhanced in one or more tissues in patients with T2D or obese people; however, the available data are inconclusive.

The relationship between Sars-CoV-2 and T2D has been described as bidirectional: Sars-CoV-2 may induce de novo T2D in previously non-diabetic patients whilst pre-existing T2D acts as an independent predictor of COVID-19 mortality (Lim, Bae, Kwon, & Nauck, 2021). This is due to the insulin producing pancreatic β-cells expressing ACE2 which is the target of Sars-CoV-2.

There are conflicting reports on the expression of ACE2 in the liver of individuals with T2D: some observing decreased/unchanged hepatic ACE2 expression (Biquard, Valla, & Rautou, 2020), and others observing increased hepatic ACE2 expression (Meijnikman, Bruin, Groen, Nieuwdorp, & Herrema, 2021). Additionally, the effect of T2D on Sars-CoV-2 target proteins ACE2 and TMPRSS2 is observed to vary by sex: men with T2D display unchanged ACE2 and TMPRSS2 expression, whilst in women both proteins decreased compared to non-diabetic individuals. This could partly be explained by the role of androgens in upregulating TMPRSS2 and this is in line with the increased risk of clinical COVID-19 complications observed in men (Deng et al., 2021; Gagliardi, Tieri, Ortona, & Ruggieri, 2020).

Adiponectin (APN) is an adipocytokine that sensitizes insulin receptor signaling, stimulating mitochondrial biogenesis and suppressing inflammation. Pleiotropic effects of APN on multiple tissues are mediated through the membrane-bound adiponectin receptors AdipoR1 and AdipoR2. AdipoR1 sensitizes insulin receptor by inducing calcium influx to activate Ca^{2+}/calmodulin-dependent kinase, AMP kinase (AMPK), and NF-κB with mixed effects on inflammation depending on the circulating APN isoform (Gabler & Spurlock, 2007). Meanwhile, AdipoR2 binding stimulates PPARα signaling to influence fatty acid oxidation.

Obesity is associated with hypoadiponectinemia, which is related to APN gene variants and environmental factors. This phenotype is observed in both insulin-resistant and non-diabetic individuals as well as various insulin-resistant animal models (Prakash, Mittal, Awasthi, Agarwal, & Srivastava, 2013). Conceivably, as insulin resistance progresses to chronic T2D and APN resistance, the subsequence of compensatory hyperadiponectinemia might emerge, which could involve dysfunctional AdipoR2 signaling, a phenotype observed in an insulin-resistant transgenic mouse model (Bauche et al., 2006). Accordingly, it has been proposed that Sars-CoV-2 induced pattern recognition triggers complex inflammatory signaling, transforming hypoadiponectinemia to pathological hyperadiponectinemia (Ho et al., 2021).

5.3 COVID-19 in non-alcoholic fatty liver disease

Non-alcoholic fatty liver disease (NAFLD) is a broad term encompassing a variety of liver pathological conditions, including hepatic steatosis, NASH, liver fibrosis and liver cirrhosis (Kumar, Duan, Wu, Harris, & Su, 2021). Generally, NAFLD is associated with abnormal accumulation of lipids in the liver that results from increased de novo lipogenesis, high adipose tissue lipolysis, and insulin resistance. As such, NAFLD is intrinsically associated with T2D and obesity.

Frequent mild liver injury is observed in patients with COVID-19, while patients with NAFLD had a higher chance of developing severe COVID-19 (Sarin et al., 2020). A retrospective study conducted on 202 COVID-19 patients admitted to hospital has reported liver injury in 50% of patients during admission, which increased to 75.2% during hospitalization (Huang, Zhu, Xue, et al., 2020). The injury mainly occurred on hepatocytes. Huang et al. further report that, in 280 COVID-19 patients, 30.7% of the patients had NAFLD whilst 35.7% had abnormal liver function at the time of hospital admission. However, the patients with or without NAFLD developed comparable complications and other clinical outcomes and no liver failure occurred during hospitalization. A retrospective case-control study has also reported similar observation which identified NAFLD among 31% of COVID-19 patients (Huang, Zhu, Wang, et al., 2020).

In a subgroup of NAFLD patients, fibrosis was reported as an additional major risk factor (Targher et al., 2020). ACE2 exhibits anti-inflammatory and anti-fibrotic effects, so under transition from relatively mild non-fibrotic NAFLD such as hepatic steatosis to the more severe fibrotic NAFLD such as NASH, ACE2 expression may change. This is consistent with observations in which ACE2 and TMPRSS2 expression has been shown to correlate with NAFLD disease score (Fondevila et al., 2021).

Sars-CoV-2 entry factors, including ACE2 and TMPRSS2, are differently affected by T2D compared to NAFLD. In obese women, those with T2D have significantly lower expression of ACE2 and TMPRSS2 compared to obese women without T2D. In contrast, obese individuals with NASH have higher expression of both ACE2 and TMPRSS2 compared to obese individuals without NASH. This demonstrates that different forms of metabolic syndrome will have different effects on the susceptibility of people to COVID-19, and that the advanced stages of NAFLD may specifically predispose patients to COVID-19 (Fondevila et al., 2021).

5.4 COVID-19 in inflammatory bowel disease

In the gastrointestinal system, ACE2 is expressed in the oral mucosa, the esophagus, stomach, pancreas, duodenum, ileum, colon, and rectum (Lehmann et al., 2021). In the intestine specifically, ACE2 is expressed at the epithelial brush border, muscularis mucosa, muscularis propria, microvascular endothelium, and vascular smooth muscle. In human intestinal biopsies, catalytic activity of ACE2 is identified as higher in the terminal ileum compared to the colon; ACE2 controls bicarbonate secretion and absorption of dietary ions and glucose (Ferreira-Duarte, Estevinho, Duarte-Araújo, Magro, & Morato, 2020). Also, in the colon, AngII regulates motility to promote contraction of the colon. ACE2 product Ang (1–7) has anti-inflammatory and antifibrotic effects in the gastrointestinal system (Zizzo et al., 2017).

These factors make the GI system a prime site for infection by Sars-CoV-2. Individuals with inflammatory bowel disease (IBD) are at increased risk of COVID-19 complications with evidence of IBD patients suffering COVID-19 being significantly higher risk of developing venous thromboembolism. There is a 2–3-fold increased risk of developing venous thromboembolism in IBD patients compared to the healthy population in a hospitalized setting (Mahmud et al., 2021).

6. Conclusions

COVID-19 is most commonly associated with the characteristic development of severe acute respiratory symptoms. However, the extensive expression of ACE2 throughout tissues of the circulatory system and metabolic regulatory organs has led to growing recognition of secondary COVID-19 complications and comorbidities including cardiovascular disease, T2D, NAFLD, and potentially IBD. Particular attention should be paid to the long-term and permanent impact of COVID-19 on the cardiovascular system, in which, Sars-CoV-2 can induce exceptional damage by disrupting the RAAS system that regulates blood pressure. Additionally, novel interactions between the Sars-CoV-2 S protein and cellular receptors and proteins have highlighted several potential means of cellular entry for Sars-CoV-2, including NRP1, TLRs, and integrins.

7. Future perspective

Development of COVID-19 therapies was initially limited due to poor understanding of the virus (Cao et al., 2020; López-Medina et al., 2021; Mitjà et al., 2021). However, several effective agents have since been made available including Molnupiravir (Jayk Bernal et al., 2022), Paxlovid (Mahase, 2021), and Remedisvir (Gottlieb et al., 2022), which have been endorsed by the FDA and EUA. Additionally, the corticosteroid dexamethasone is indicated as an effective agent for reducing the hyperinflammation induced by the cytokine storm (Horby et al., 2021); similarly, immune treatment including anti-IL-6 monoclonal antibodies, IL-1 antagonists, and Janus kinase inhibitors are being utilized to ease COVID-19 associated inflammation (Conti et al., 2020; Huet et al., 2020; Michot et al., 2020; Richardson et al., 2020). Currently vaccines are being utilized as an effective preventative measure against the development of severe acute respiratory symptoms requiring hospitalization. However, there is controversy surrounding the ability of some vaccines to predispose individuals, e.g., those with IBD, to development of venous thromboembolism (Mahmud et al., 2021). Since obesity is one of the major risk factors uniting COVID-19 with metabolic disease and cardiovascular disease (Gao et al., 2020), it seems poignant to investigate the effectiveness of diet and lifestyle as a preventative measure against COVID-19.

Acknowledgments

Q. Su is supported by the British Heart Foundation (UK) (PG/19/86/34788) and Northern Ireland Chest Heart & Stroke (UK) (2019_08). H. Wade is a recipient of PhD research scholarships from the Department for the Economy in Northern Ireland (UK).

References

Abbasi, A. Z., Kiyani, D. A., Hamid, S. M., Saalim, M., Fahim, A., & Jalal, N. (2021). Spiking dependence of SARS-CoV-2 pathogenicity on TMPRSS2. *Journal of Medical Virology*, *93*(7), 4205–4218. https://doi.org/10.1002/jmv.26911.

Akhmerov, A., & Marbán, E. (2020). COVID-19 and the Heart. *Circulation Research*, *126*(10), 1443–1455. https://doi.org/10.1161/CIRCRESAHA.120.317055.

Al-Samkari, H., Karp Leaf, R. S., Dzik, W. H., Carlson, J. C. T., Fogerty, A. E., Waheed, A., et al. (2020). COVID-19 and coagulation: bleeding and thrombotic manifestations of SARS-CoV-2 infection. *Blood*, *136*(4), 489–500. https://doi.org/10.1182/blood.2020006520.

Angelini, M. M., Akhlaghpour, M., Neuman, B. W., & Buchmeier, M. J. (2013). Severe acute respiratory syndrome coronavirus nonstructural proteins 3, 4, and 6 induce double-membrane vesicles. *MBio*, *4*(4). https://doi.org/10.1128/mBio.00524-13.

Barron, E., Bakhai, C., Kar, P., Weaver, A., Bradley, D., Ismail, H., et al. (2020). Associations of type 1 and type 2 diabetes with COVID-19-related mortality in England: A whole-population study. *The Lancet Diabetes & Endocrinology*, *8*(10), 813–822. https://doi.org/10.1016/S2213-8587(20)30272-2.

Bauche, I. B., Ait El Mkadem, S., Rezsohazy, R., Funahashi, T., Maeda, N., Miranda, L. M., et al. (2006). Adiponectin downregulates its own production and the expression of its AdipoR2 receptor in transgenic mice. *Biochemical and Biophysical Research Communications*, *345*(4), 1414–1424. https://doi.org/10.1016/j.bbrc.2006.05.033.

Bianchi, M., Benvenuto, D., Giovanetti, M., Angeletti, S., Ciccozzi, M., & Pascarella, S. (2020). Sars-CoV-2 envelope and membrane proteins: Structural differences linked to virus characteristics? *BioMed Research International, 2020*, 4389089. https://doi.org/10.1155/2020/4389089.

Biquard, L., Valla, D., & Rautou, P.-E. (2020). No evidence for an increased liver uptake of SARS-CoV-2 in metabolic-associated fatty liver disease. *Journal of Hepatology, 73*(3), 717–718. https://doi.org/10.1016/j.jhep.2020.04.035.

Biswal, M., Diggs, S., Xu, D., Khudaverdyan, N., Lu, J., Fang, J., et al. (2021). Two conserved oligomer interfaces of NSP7 and NSP8 underpin the dynamic assembly of SARS-CoV-2 RdRP. *Nucleic Acids Research, 49*(10), 5956–5966. https://doi.org/10.1093/nar/gkab370.

Bonafè, M., Prattichizzo, F., Giuliani, A., Storci, G., Sabbatinelli, J., & Olivieri, F. (2020). Inflamm-aging: Why older men are the most susceptible to SARS-CoV-2 complicated outcomes. *Cytokine & Growth Factor Reviews, 53*, 33–37. https://doi.org/10.1016/j.cytogfr.2020.04.005.

Budhraja, A., Pandey, S., Kannan, S., Verma, C. S., & Venkatraman, P. (2021). The polybasic insert, the RBD of the SARS-CoV-2 spike protein, and the feline coronavirus—Evolved or yet to evolve. *Biochemistry and Biophysics Reports, 25*, 100907. https://doi.org/10.1016/j.bbrep.2021.100907.

Bujak, M., & Frangogiannis, N. G. (2009). The role of IL-1 in the pathogenesis of heart disease. *Archivum Immunologiae et Therapiae Experimentalis, 57*(3), 165–176. https://doi.org/10.1007/s00005-009-0024-y.

Butler, D., Coyne, D., Pomeroy, L., Williams, P., Holder, P., Carterson, A., et al. (2022). Confirmed circulation of SARS-CoV-2 in Irish blood donors prior to first national notification of infection. *Journal of Clinical Virology, 146*, 105045. https://doi.org/10.1016/j.jcv.2021.105045.

Cantuti-Castelvetri, L., Ojha, R., Pedro, L. D., Djannatian, M., Franz, J., Kuivanen, S., et al. (2020). Neuropilin-1 facilitates SARS-CoV-2 cell entry and provides a possible pathway into the central nervous system. *bioRxiv*. https://doi.org/10.1101/2020.06.07.137802. 2020.2006.2007.137802.

Cantuti-Castelvetri, L., Ojha, R., Pedro Liliana, D., Djannatian, M., Franz, J., Kuivanen, S., et al. (2020). Neuropilin-1 facilitates SARS-CoV-2 cell entry and infectivity. *Science, 370*(6518), 856–860. https://doi.org/10.1126/science.abd2985.

Cao, B., Wang, Y., Wen, D., Liu, W., Wang, J., Fan, G., et al. (2020). A trial of lopinavir-ritonavir in adults hospitalized with severe Covid-19. *The New England Journal of Medicine, 382*(19), 1787–1799. https://doi.org/10.1056/NEJMoa2001282.

Carey, R. M., Wang, Z.-Q., & Siragy, H. M. (2000). Role of the angiotensin type 2 receptor in the regulation of blood pressure and renal function. *Hypertension, 35*(1), 155–163. https://doi.org/10.1161/01.HYP.35.1.155.

Chen, Y., Cai, H., Pan, J. A., Xiang, N., Tien, P., Ahola, T., et al. (2009). Functional screen reveals SARS coronavirus nonstructural protein nsp14 as a novel cap N7 methyltransferase. *Proceedings of the National Academy of Sciences, 106*(9), 3484–3489. https://doi.org/10.1073/pnas.0808790106.

Chen, X., Cao, R., & Zhong, W. (2019). Host calcium channels and pumps in viral infections. *Cell, 9*(1), 94. https://doi.org/10.3390/cells9010094.

Chen, Y., Liu, Q., & Guo, D. (2020). Emerging coronaviruses: Genome structure, replication, and pathogenesis. *Journal of Medical Virology, 92*(4), 418–423. https://doi.org/10.1002/jmv.25681.

Chen, D., Zheng, Q., Sun, L., Ji, M., Li, Y., Deng, H., et al. (2021). ORF3a of SARS-CoV-2 promotes lysosomal exocytosis-mediated viral egress. *Developmental Cell, 56*(23), 3250–3263. e3255. https://doi.org/10.1016/j.devcel.2021.10.006.

Choudhury, A., Das, N. C., Patra, R., & Mukherjee, S. (2021). In silico analyses on the comparative sensing of SARS-CoV-2 mRNA by the intracellular TLRs of humans. *Journal of Medical Virology, 93*(4), 2476–2486. https://doi.org/10.1002/jmv.26776.

Conti, P., Ronconi, G., Caraffa, A., Gallenga, C. E., Ross, R., Frydas, I., et al. (2020). Induction of pro-inflammatory cytokines (IL-1 and IL-6) and lung inflammation by Coronavirus-19 (COVI-19 or SARS-CoV-2): Anti-inflammatory strategies. *Journal of Biological Regulators and Homeostatic Agents, 34*(2), 327–331. https://doi.org/10.23812/conti-e.

Corse, E., & Machamer Carolyn, E. (2000). Infectious bronchitis virus e protein is targeted to the golgi complex and directs release of virus-like particles. *Journal of Virology, 74*(9), 4319–4326. https://doi.org/10.1128/JVI.74.9.4319-4326.2000.

Costa, T. J., Potje, S. R., Fraga-Silva, T. F. C., da Silva-Neto, J. A., Barros, P. R., Rodrigues, D., et al. (2022). Mitochondrial DNA and TLR9 activation contribute to SARS-CoV-2-induced endothelial cell damage. *Vascular Pharmacology, 142*, 106946. https://doi.org/10.1016/j.vph.2021.106946.

Cubuk, J., Alston, J. J., Incicco, J. J., Singh, S., Stuchell-Brereton, M. D., Ward, M. D., et al. (2021). The SARS-CoV-2 nucleocapsid protein is dynamic, disordered, and phase separates with RNA. *Nature Communications, 12*(1), 1936. https://doi.org/10.1038/s41467-021-21953-3.

Cui, L., Wang, H., Ji, Y., Yang, J., Xu, S., Huang, X., et al. (2015). The nucleocapsid protein of coronaviruses acts as a viral suppressor of RNA silencing in mammalian cells. *Journal of Virology, 89*(17), 9029–9043. https://doi.org/10.1128/jvi.01331-15.

Daly James, L., Simonetti, B., Klein, K., Chen, K.-E., Williamson Maia, K., Antón-Plágaro, C., et al. (2020). Neuropilin-1 is a host factor for SARS-CoV-2 infection. *Science, 370*(6518), 861–865. https://doi.org/10.1126/science.abd3072.

Danta, C. C. (2021). SARS-CoV-2, hypoxia, and calcium signaling: The consequences and therapeutic options. *ACS Pharmacology & Translational Science, 4*(1), 400–402. https://doi.org/10.1021/acsptsci.0c00219.

Davidson, A. D., Williamson, M. K., Lewis, S., Shoemark, D., Carroll, M. W., Heesom, K. J., et al. (2020). Characterisation of the transcriptome and proteome of SARS-CoV-2 reveals a cell passage induced in-frame deletion of the furin-like cleavage site from the spike glycoprotein. *Genome Medicine, 12*(1), 68. https://doi.org/10.1186/s13073-020-00763-0.

Deng, Q., Rasool, R. U., Russell, R. M., Natesan, R., & Asangani, I. A. (2021). Targeting androgen regulation of TMPRSS2 and ACE2 as a therapeutic strategy to combat COVID-19. *iScience, 24*(3), 102254. https://doi.org/10.1016/j.isci.2021.102254.

Devaux, C. A., Rolain, J.-M., & Raoult, D. (2020). ACE2 receptor polymorphism: Susceptibility to SARS-CoV-2, hypertension, multi-organ failure, and COVID-19 disease outcome. *Journal of Microbiology, Immunology and Infection, 53*(3), 425–435. https://doi.org/10.1016/j.jmii.2020.04.015.

Dominguez Andres, A., Feng, Y., Campos, A. R., Yin, J., Yang, C.-C., James, B., et al. (2020). SARS-CoV-2 ORF9c is a membrane-associated protein that suppresses antiviral responses in cells. *bioRxiv*. https://doi.org/10.1101/2020.08.18.256776. 2020.2008.2018.256776.

Dudek, J. (2017). Role of cardiolipin in mitochondrial signaling pathways [review]. *Frontiers in Cell and Development Biology, 5*. https://doi.org/10.3389/fcell.2017.00090.

Dwivedy, A., Mariadasse, R., Ahmad, M., Chakraborty, S., Kar, D., Tiwari, S., et al. (2021). Characterization of the NiRAN domain from RNA-dependent RNA polymerase provides insights into a potential therapeutic target against SARS-CoV-2. *PLoS Computational Biology, 17*(9), e1009384. https://doi.org/10.1371/journal.pcbi.1009384.

Eguchi, S., Kawai, T., Scalia, R., & Rizzo, V. (2018). Understanding angiotensin II type 1 receptor signaling in vascular pathophysiology. *Hypertension, 71*(5), 804–810. https://doi.org/10.1161/HYPERTENSIONAHA.118.10266.

Ferlita, S., Yegiazaryan, A., Noori, N., Lal, G., Nguyen, T., To, K., et al. (2019). Type 2 diabetes mellitus and altered immune system leading to susceptibility to pathogens, especially mycobacterium tuberculosis. *Journal of Clinical Medicine, 8*(12). https://doi.org/10.3390/jcm8122219.

Ferreira-Duarte, M., Estevinho, M. M., Duarte-Araújo, M., Magro, F., & Morato, M. (2020). Unraveling the role of ACE2, the binding receptor for SARS-CoV-2 in inflammatory bowel disease. *Inflammatory Bowel Diseases, 26*(12), 1787–1795. https://doi.org/10.1093/ibd/izaa249.

Finkel, Y., Mizrahi, O., Nachshon, A., Weingarten-Gabbay, S., Morgenstern, D., Yahalom-Ronen, Y., et al. (2021). The coding capacity of SARS-CoV-2. *Nature, 589*(7840), 125–130. https://doi.org/10.1038/s41586-020-2739-1.

Flower Thomas, G., Buffalo Cosmo, Z., Hooy Richard, M., Allaire, M., Ren, X., & Hurley James, H. (2021). Structure of SARS-CoV-2 ORF8, a rapidly evolving immune evasion protein. *Proceedings of the National Academy of Sciences, 118*(2), e2021785118. https://doi.org/10.1073/pnas.2021785118.

Fondevila, M. F., Mercado-Gómez, M., Rodríguez, A., Gonzalez-Rellan, M. J., Iruzubieta, P., Valentí, V., et al. (2021). Obese patients with NASH have increased hepatic expression of SARS-CoV-2 critical entry points. *Journal of Hepatology, 74*(2), 469–471. https://doi.org/10.1016/j.jhep.2020.09.027.

Gabler, N. K., & Spurlock, M. E. (2007). Adiponectin attenuates reactive oxygen species accumulation and NFkB translocation in human cardiac myocytes during inflammation. *The FASEB Journal, 21*(5), A583. https://doi.org/10.1096/fasebj.21.5.A583-c.

Gagliardi, M. C., Tieri, P., Ortona, E., & Ruggieri, A. (2020). ACE2 expression and sex disparity in COVID-19. *Cell Death Discovery, 6*(1), 37. https://doi.org/10.1038/s41420-020-0276-1.

Gao, F., Zheng, K. I., Wang, X.-B., Sun, Q.-F., Pan, K.-H., Wang, T.-Y., et al. (2020). Obesity is a risk factor for greater COVID-19 severity. *Diabetes Care, 43*(7), e72–e74. https://doi.org/10.2337/dc20-0682.

Ghoneim, S., Butt, M. U., Hamid, O., Shah, A., & Asaad, I. (2020). The incidence of COVID-19 in patients with metabolic syndrome and non-alcoholic steatohepatitis: A population-based study. *Metabolism Open, 8*, 100057. https://doi.org/10.1016/j.metop.2020.100057.

Gottlieb, R. L., Vaca, C. E., Paredes, R., Mera, J., Webb, B. J., Perez, G., et al. (2022). Early remdesivir to prevent progression to severe Covid-19 in outpatients. *The New England Journal of Medicine, 386*(4), 305–315. https://doi.org/10.1056/NEJMoa2116846.

Green, S. J. (2020). Covid-19 accelerates endothelial dysfunction and nitric oxide deficiency. *Microbes and Infection, 22*(4-5), 149–150. https://doi.org/10.1016/j.micinf.2020.05.006.

Gressens, S. B., Leftheriotis, G., Dussaule, J.-C., Flamant, M., Levy, B. I., & Vidal-Petiot, E. (2021). Controversial roles of the renin angiotensin system and its modulators during the COVID-19 pandemic [review]. *Frontiers in Physiology, 12*. https://doi.org/10.3389/fphys.2021.624052.

Ho, G., Ali, A., Takamatsu, Y., Wada, R., Masliah, E., & Hashimoto, M. (2021). Diabetes, inflammation, and the adiponectin paradox: Therapeutic targets in SARS-CoV-2. *Drug Discovery Today, 26*(8), 2036–2044. https://doi.org/10.1016/j.drudis.2021.03.013.

Hoffmann, M., Kleine-Weber, H., Schroeder, S., Krüger, N., Herrler, T., Erichsen, S., et al. (2020). SARS-CoV-2 cell entry depends on ACE2 and TMPRSS2 and is blocked by a clinically proven protease inhibitor. *Cell, 181*(2), 271–280. e278. https://doi.org/10.1016/j.cell.2020.02.052.

Horby, P., Lim, W. S., Emberson, J. R., Mafham, M., Bell, J. L., Linsell, L., et al. (2021). Dexamethasone in hospitalized patients with Covid-19. *The New England Journal of Medicine, 384*(8), 693–704. https://doi.org/10.1056/NEJMoa2021436.

Huang, Y., Yang, C., Xu, X.-F., Xu, W., & Liu, S.-W. (2020). Structural and functional properties of SARS-CoV-2 spike protein: Potential antivirus drug development for COVID-19. *Acta Pharmacologica Sinica*, *41*(9), 1141–1149. https://doi.org/10.1038/s41401-020-0485-4.

Huang, R., Zhu, L., Wang, J., Xue, L., Liu, L., Yan, X., et al. (2020). Clinical features of patients with COVID-19 with nonalcoholic fatty liver disease. *Hepatology Communications*, *4*(12), 1758–1768. https://doi.org/10.1002/hep4.1592.

Huang, R., Zhu, L., Xue, L., Liu, L., Yan, X., Wang, J., et al. (2020). Clinical findings of patients with coronavirus disease 2019 in Jiangsu province, China: A retrospective, multi-center study. *PLoS Neglected Tropical Diseases*, *14*(5), e0008280. https://doi.org/10.1371/journal.pntd.0008280.

Huet, T., Beaussier, H., Voisin, O., Jouveshomme, S., Dauriat, G., Lazareth, I., et al. (2020). Anakinra for severe forms of COVID-19: A cohort study. *Lancet Rheumatology*, *2*(7), e393–e400. https://doi.org/10.1016/s2665-9913(20)30164-8.

Ivanov, K. A., Thiel, V., Dobbe, J. C., van der Meer, Y., Snijder, E. J., & Ziebuhr, J. (2004). Multiple enzymatic activities associated with severe acute respiratory syndrome coronavirus helicase. *Journal of Virology*, *78*(11), 5619–5632. https://doi.org/10.1128/jvi.78.11.5619-5632.2004.

Jayk Bernal, A., Gomes da Silva, M. M., Musungaie, D. B., Kovalchuk, E., Gonzalez, A., Delos Reyes, V., et al., & Group, M. O.-O. S. (2022). Molnupiravir for oral treatment of Covid-19 in nonhospitalized patients. *The New England Journal of Medicine*, *386*(6), 509–520. https://doi.org/10.1056/NEJMoa2116044.

Jiang, H.-W., Zhang, H.-N., Meng, Q.-F., Xie, J., Li, Y., Chen, H., et al. (2020). SARS-CoV-2 Orf9b suppresses type I interferon responses by targeting TOM70. *Cellular & Molecular Immunology*, *17*(9), 998–1000. https://doi.org/10.1038/s41423-020-0514-8.

Jizzini, M., Shah, M., & Zhou, K. (2020). SARS-CoV-2 and anti-cardiolipin antibodies. *Clinical Medicine Insights: Case Reports*, *13*. https://doi.org/10.1177/1179547620980381.1179547620980381.

Johnson, B. A., Xie, X., Kalveram, B., Lokugamage, K. G., Muruato, A., Zou, J., et al. (2020). Furin cleavage site is key to SARS-CoV-2 pathogenesis. *bioRxiv: The Preprint Server for Biology*. https://doi.org/10.1101/2020.08.26.268854. 2020.2008.2026.268854.

Jungreis, I., Nelson, C. W., Ardern, Z., Finkel, Y., Krogan, N. J., Sato, K., et al. (2021). Conflicting and ambiguous names of overlapping ORFs in the SARS-CoV-2 genome: A homology-based resolution. *Virology*, *558*, 145–151. https://doi.org/10.1016/j.virol.2021.02.013.

Jungreis, I., Sealfon, R., & Kellis, M. (2021). SARS-CoV-2 gene content and COVID-19 mutation impact by comparing 44 Sarbecovirus genomes. *Nature Communications*, *12*(1), 2642. https://doi.org/10.1038/s41467-021-22905-7.

Kim, D., Lee, J.-Y., Yang, J.-S., Kim, J. W., Kim, V. N., & Chang, H. (2020). The architecture of SARS-CoV-2 transcriptome. *Cell*, *181*(4), 914–921. e910. https://doi.org/10.1016/j.cell.2020.04.011.

Konno, Y., Kimura, I., Uriu, K., Fukushi, M., Irie, T., Koyanagi, Y., et al. (2020). SARS-CoV-2 ORF3b is a potent interferon antagonist whose activity is increased by a naturally occurring elongation variant. *Cell Reports*, *32*(12), 108185. https://doi.org/10.1016/j.celrep.2020.108185.

Kopecky-Bromberg, S. A., Martínez-Sobrido, L., Frieman, M., Baric, R. A., & Palese, P. (2007). Severe acute respiratory syndrome coronavirus open reading frame (ORF) 3b, ORF 6, and nucleocapsid proteins function as interferon antagonists. *Journal of Virology*, *81*(2), 548–557. https://doi.org/10.1128/jvi.01782-06.

Kuba, K., Imai, Y., & Penninger, J. M. (2013). Multiple functions of angiotensin-converting enzyme 2 and its relevance in cardiovascular diseases. *Circulation Journal*, *77*(2), 301–308. https://doi.org/10.1253/circj.cj-12-1544.

Kumar, S., Duan, Q., Wu, R., Harris, E. N., & Su, Q. (2021). Pathophysiological communication between hepatocytes and non-parenchymal cells in liver injury from NAFLD to liver fibrosis. *Advanced Drug Delivery Reviews*, *176*, 113869. https://doi.org/10.1016/j.addr.2021.113869.

Kumar, A., Goyal, N., Saranathan, N., Dhamija, S., Saraswat, S., Menon, M. B., et al. (2022). The slowing rate of CpG depletion in SARS-CoV-2 genomes is consistent with adaptations to the human host. *Molecular Biology and Evolution*, *39*(3), msac029. https://doi.org/10.1093/molbev/msac029.

Kwok, S., Adam, S., Ho, J. H., Iqbal, Z., Turkington, P., Razvi, S., et al. (2020). Obesity: A critical risk factor in the COVID-19 pandemic. *Clinical Obesity*, *10*(6), e12403. https://doi.org/10.1111/cob.12403.

Lai, A. L., Millet, J. K., Daniel, S., Freed, J. H., & Whittaker, G. R. (2017). The SARS-CoV fusion peptide forms an extended bipartite fusion platform that perturbs membrane order in a calcium-dependent manner. *Journal of Molecular Biology*, *429*(24), 3875–3892. https://doi.org/10.1016/j.jmb.2017.10.017.

Lampropoulou, A., & Ruhrberg, C. (2014). Neuropilin regulation of angiogenesis. *Biochemical Society Transactions*, *42*(6), 1623–1628. https://doi.org/10.1042/bst20140244.

Law, A. H. Y., Lee, D. C. W., Cheung, B. K. W., Yim, H. C. H., & Lau, A. S. Y. (2007). Role for nonstructural protein 1 of severe acute respiratory syndrome coronavirus in chemokine dysregulation. *Journal of Virology*, *81*(1), 416–422. https://doi.org/10.1128/JVI.02336-05.

Lee, P., Goishi, K., Davidson Alan, J., Mannix, R., Zon, L., & Klagsbrun, M. (2002). Neuropilin-1 is required for vascular development and is a mediator of VEGF-dependent angiogenesis in zebrafish. *Proceedings of the National Academy of Sciences*, *99*(16), 10470–10475. https://doi.org/10.1073/pnas.162366299.

Lehmann, M., Allers, K., Heldt, C., Meinhardt, J., Schmidt, F., Rodriguez-Sillke, Y., et al. (2021). Human small intestinal infection by SARS-CoV-2 is characterized by a mucosal infiltration with activated CD8 + T cells. *Mucosal Immunology*, *14*(6), 1381–1392. https://doi.org/10.1038/s41385-021-00437-z.

Li, F., Li, J., Wang, P.-H., Yang, N., Huang, J., Ou, J., et al. (2021). SARS-CoV-2 spike promotes inflammation and apoptosis through autophagy by ROS-suppressed PI3K/AKT/mTOR signaling. *Biochimica et Biophysica Acta (BBA) - Molecular Basis of Disease*, *1867*(12), 166260. https://doi.org/10.1016/j.bbadis.2021.166260.

Li, J. Y., Liao, C. H., Wang, Q., Tan, Y. J., Luo, R., Qiu, Y., et al. (2020). The ORF6, ORF8 and nucleocapsid proteins of SARS-CoV-2 inhibit type I interferon signaling pathway. *Virus Research*, *286*, 198074. https://doi.org/10.1016/j.virusres.2020.198074.

Lim, S., Bae, J. H., Kwon, H.-S., & Nauck, M. A. (2021). COVID-19 and diabetes mellitus: From pathophysiology to clinical management. *Nature Reviews Endocrinology*, *17*(1), 11–30. https://doi.org/10.1038/s41574-020-00435-4.

Long, B., Brady, W. J., Koyfman, A., & Gottlieb, M. (2020). Cardiovascular complications in COVID-19. *The American Journal of Emergency Medicine*, *38*(7), 1504–1507. https://doi.org/10.1016/j.ajem.2020.04.048.

López-Medina, E., López, P., Hurtado, I. C., Dávalos, D. M., Ramirez, O., Martínez, E., et al. (2021). Effect of ivermectin on time to resolution of symptoms among adults with mild COVID-19: A randomized clinical trial. *JAMA*, *325*(14), 1426–1435. https://doi.org/10.1001/jama.2021.3071.

Magro, C., Mulvey, J. J., Berlin, D., Nuovo, G., Salvatore, S., Harp, J., et al. (2020). Complement associated microvascular injury and thrombosis in the pathogenesis of severe COVID-19 infection: A report of five cases. *Translational Research: The Journal of Laboratory and Clinical Medicine*, *220*, 1–13. https://doi.org/10.1016/j.trsl.2020.04.007.

Mahase, E. (2021). Covid-19: Pfizer's paxlovid is 89% effective in patients at risk of serious illness, company reports. *BMJ*, *375*, n2713. https://doi.org/10.1136/bmj.n2713.

Mahmud, N., Weiss, A., Trivedi, C., Yang, Y.-X., Lewis, J., & Khan, N. (2021). Risk of venous thromboembolism among patients with inflammatory bowel disease who contract severe acute respiratory syndrome Coronavirus 2. *Gastroenterology*, *161*(5), 1709–1711.e1701. https://doi.org/10.1053/j.gastro.2021.06.012.

Masters, P. S. (2006). The molecular biology of coronaviruses. *Advances in Virus Research*, *66*, 193–292. https://doi.org/10.1016/s0065-3527(06)66005-3.

Meijnikman, A. S., Bruin, S., Groen, A. K., Nieuwdorp, M., & Herrema, H. (2021). Increased expression of key SARS-CoV-2 entry points in multiple tissues in individuals with NAFLD. *Journal of Hepatology*, *74*(3), 748–749. https://doi.org/10.1016/j.jhep.2020.12.007.

Michot, J. M., Albiges, L., Chaput, N., Saada, V., Pommeret, F., Griscelli, F., et al. (2020). Tocilizumab, an anti-IL-6 receptor antibody, to treat COVID-19-related respiratory failure: A case report. *Annals of Oncology*, *31*(7), 961–964. https://doi.org/10.1016/j.annonc.2020.03.300.

Misiti, F. (2021). SARS-CoV-2 infection and red blood cells: Implications for long term symptoms during exercise. *Sports Medicine and Health Science*, *3*(3), 181–182. https://doi.org/10.1016/j.smhs.2021.07.002.

Mitjà, O., Corbacho-Monné, M., Ubals, M., Alemany, A., Suñer, C., Tebé, C., et al. (2021). A cluster-randomized trial of hydroxychloroquine for prevention of Covid-19. *The New England Journal of Medicine*, *384*(5), 417–427. https://doi.org/10.1056/NEJMoa2021801.

Mountantonakis, S. E., Saleh, M., Fishbein, J., Gandomi, A., Lesser, M., Chelico, J., et al. (2021). Atrial fibrillation is an independent predictor for in-hospital mortality in patients admitted with SARS-CoV-2 infection. *Heart Rhythm*, *18*(4), 501–507. https://doi.org/10.1016/j.hrthm.2021.01.018.

Muñoz, M. C., Burghi, V., Miquet, J. G., Giani, J. F., Banegas, R. D., Toblli, J. E., et al. (2014). Downregulation of the ACE2/Ang-(1–7)/Mas axis in transgenic mice overexpressing GH. *Journal of Endocrinology*, *221*(2), 215–227. https://doi.org/10.1530/JOE-13-0497.

Naydenova, K., Muir Kyle, W., Wu, L.-F., Zhang, Z., Coscia, F., Peet Mathew, J., et al. (2021). Structure of the SARS-CoV-2 RNA-dependent RNA polymerase in the presence of favipiravir-RTP. *Proceedings of the National Academy of Sciences*, *118*(7), e2021946118. https://doi.org/10.1073/pnas.2021946118.

Nieto-Torres, J. L., DeDiego, M. L., Verdiá-Báguena, C., Jimenez-Guardeño, J. M., Regla-Nava, J. A., Fernandez-Delgado, R., et al. (2014). Severe acute respiratory syndrome coronavirus envelope protein ion channel activity promotes virus fitness and pathogenesis. *PLoS Pathogens*, *10*(5), e1004077. https://doi.org/10.1371/journal.ppat.1004077.

Ou, X., Liu, Y., Lei, X., Li, P., Mi, D., Ren, L., et al. (2020). Characterization of spike glycoprotein of SARS-CoV-2 on virus entry and its immune cross-reactivity with SARS-CoV. *Nature Communications*, *11*(1), 1620. https://doi.org/10.1038/s41467-020-15562-9.

Oudshoorn, D., Rijs, K., Limpens Ronald, W. A. L., Groen, K., Koster Abraham, J., Snijder Eric, J., Kikkert, M., Bárcena, M., & Denison Mark, R. Expression and cleavage of Middle East Respiratory Syndrome Coronavirus nsp3-4 polyprotein induce the formation of double-membrane vesicles that mimic those associated with coronaviral RNA replication. MBio, 8(6), e01658-e01617. doi:https://doi.org/10.1128/mBio.01658-17.

Padmanabhan, P., Desikan, R., & Dixit, N. M. (2020). Targeting TMPRSS2 and Cathepsin B/L together may be synergistic against SARS-CoV-2 infection. *PLoS Computational Biology*, *16*(12), e1008461. https://doi.org/10.1371/journal.pcbi.1008461.

Pan, P., Shen, M., Yu, Z., Ge, W., Chen, K., Tian, M., et al. (2021). SARS-CoV-2 N protein promotes NLRP3 inflammasome activation to induce hyperinflammation. *Nature Communications*, *12*(1), 4664. https://doi.org/10.1038/s41467-021-25015-6.

Pancer, K., Milewska, A., Owczarek, K., Dabrowska, A., Kowalski, M., Łabaj, P. P., et al. (2020). The SARS-CoV-2 ORF10 is not essential in vitro or in vivo in humans. *PLoS Pathogens*, *16*(12), e1008959. https://doi.org/10.1371/journal.ppat.1008959.

Peacock, T. P., Goldhill, D. H., Zhou, J., Baillon, L., Frise, R., Swann, O. C., et al. (2021). The furin cleavage site in the SARS-CoV-2 spike protein is required for transmission in ferrets. *Nature Microbiology*, *6*(7), 899–909. https://doi.org/10.1038/s41564-021-00908-w.

Peng, Q., Peng, R., Yuan, B., Zhao, J., Wang, M., Wang, X., et al. (2020). Structural and biochemical characterization of the nsp12-nsp7-nsp8 core polymerase complex from SARS-CoV-2. *Cell Reports*, *31*(11), 107774. https://doi.org/10.1016/j.celrep.2020.107774.

Peng, H., Sarwar, Z., Yang, X. P., Peterson, E. L., Xu, J., Janic, B., et al. (2015). Profibrotic role for interleukin-4 in cardiac remodeling and dysfunction. *Hypertension*, *66*(3), 582–589. https://doi.org/10.1161/hypertensionaha.115.05627.

Prakash, J., Mittal, B., Awasthi, S., Agarwal, C. G., & Srivastava, N. (2013). Hypoadiponectinemia in obesity: Association with insulin resistance. *Indian Journal of Clinical Biochemistry: IJCB*, *28*(2), 158–163. https://doi.org/10.1007/s12291-012-0246-3.

Qiao, Y., Wang, X.-M., Mannan, R., Pitchiaya, S., Zhang, Y., Wotring Jesse, W., et al. (2021). Targeting transcriptional regulation of SARS-CoV-2 entry factors ACE2 and TMPRSS2. *Proceedings of the National Academy of Sciences*, *118*(1), e2021450118. https://doi.org/10.1073/pnas.2021450118.

Rajendran, P., Rengarajan, T., Thangavel, J., Nishigaki, Y., Sakthisekaran, D., Sethi, G., et al. (2013). The vascular endothelium and human diseases. *International Journal of Biological Sciences*, *9*(10), 1057–1069. https://doi.org/10.7150/ijbs.7502.

Rashid, F., Dzakah, E. E., Wang, H., & Tang, S. (2021). The ORF8 protein of SARS-CoV-2 induced endoplasmic reticulum stress and mediated immune evasion by antagonizing production of interferon beta. *Virus Research*, *296*, 198350. https://doi.org/10.1016/j.virusres.2021.198350.

Redondo, N., Zaldívar-López, S., Garrido, J. J., & Montoya, M. (2021). SARS-CoV-2 accessory proteins in viral pathogenesis: Knowns and unknowns [mini review]. *Frontiers in Immunology*, *12*. https://doi.org/10.3389/fimmu.2021.708264.

Reynolds, H. R., Adhikari, S., Pulgarin, C., Troxel, A. B., Iturrate, E., Johnson, S. B., et al. (2020). Renin–angiotensin–aldosterone system inhibitors and risk of Covid-19. *New England Journal of Medicine*, *382*(25), 2441–2448. https://doi.org/10.1056/NEJMoa2008975.

Richardson, P., Griffin, I., Tucker, C., Smith, D., Oechsle, O., Phelan, A., et al. (2020). Baricitinib as potential treatment for 2019-nCoV acute respiratory disease. *Lancet (London, England)*, *395*(10223), e30–e31. https://doi.org/10.1016/S0140-6736(20)30304-4.

Robles, J. P., Zamora, M., Adan-Castro, E., Siqueiros-Marquez, L., Martinez de la Escalera, G., & Clapp, C. (2022). The spike protein of SARS-CoV-2 induces endothelial inflammation through integrin α5β1 and NF-κB signaling. *The Journal of Biological Chemistry*, *101695*. https://doi.org/10.1016/j.jbc.2022.101695.

Ruiz-Ares, G., Jimenez-Valero, S., Fernández-Prieto, A., Arbas-Redondo, E., Díaz-Pollán, B., Navia, P., et al. (2021). Concurrent stroke and myocardial infarction after mild COVID-19 infection. *The Neurologist*, *26*(3), 86–89. https://doi.org/10.1097/NRL.0000000000000311.

Sarin, S. K., Choudhury, A., Lau, G. K., Zheng, M.-H., Ji, D., Abd-Elsalam, S., et al. (2020). Pre-existing liver disease is associated with poor outcome in patients with SARS CoV2 infection: The APCOLIS Study (APASL COVID-19 Liver Injury Spectrum Study). *Hepatology International, 14*(5), 690–700. https://doi.org/10.1007/s12072-020-10072-8.

Sartorius, R., Trovato, M., Manco, R., D'Apice, L., & De Berardinis, P. (2021). Exploiting viral sensing mediated by Toll-like receptors to design innovative vaccines. *NPJ Vaccines, 6*(1), 127. https://doi.org/10.1038/s41541-021-00391-8.

Schaffner, F., Ray, A. M., & Dontenwill, M. (2013). Integrin α5β1, the fibronectin receptor, as a pertinent therapeutic target in solid tumors. *Cancers, 5*(1), 27–47. https://doi.org/10.3390/cancers5010027.

Schoeman, D., & Fielding, B. C. (2019). Coronavirus envelope protein: Current knowledge. *Virology Journal, 16*(1), 69. https://doi.org/10.1186/s12985-019-1182-0.

Sharma, B. R., & Kanneganti, T.-D. (2021). NLRP3 inflammasome in cancer and metabolic diseases. *Nature Immunology, 22*(5), 550–559. https://doi.org/10.1038/s41590-021-00886-5.

Shchendrygina, A., Nagel, E., Puntmann, V. O., & Valbuena-Lopez, S. (2021). COVID-19 myocarditis and prospective heart failure burden. *Expert Review of Cardiovascular Therapy, 19*(1), 5–14. https://doi.org/10.1080/14779072.2021.1844005.

Shen, L. W., Mao, H. J., Wu, Y. L., Tanaka, Y., & Zhang, W. (2017). TMPRSS2: A potential target for treatment of influenza virus and coronavirus infections. *Biochimie, 142*, 1–10. https://doi.org/10.1016/j.biochi.2017.07.016.

Shi, C.-S., Qi, H.-Y., Boularan, C., Huang, N.-N., Abu-Asab, M., Shelhamer, J. H., et al. (2014). SARS-coronavirus open reading frame-9b suppresses innate immunity by targeting mitochondria and the MAVS/TRAF3/TRAF6 signalosome. *Journal of Immunology (Baltimore, Md.: 1950), 193*(6), 3080–3089. https://doi.org/10.4049/jimmunol.1303196.

Shi, J.-H., & Sun, S.-C. (2018). Tumor necrosis factor receptor-associated factor regulation of nuclear factor κB and mitogen-activated protein kinase pathways [review]. *Frontiers in Immunology, 9*. https://doi.org/10.3389/fimmu.2018.01849.

Shilts, J., Crozier, T. W. M., Greenwood, E. J. D., Lehner, P. J., & Wright, G. J. (2021). No evidence for basigin/CD147 as a direct SARS-CoV-2 spike binding receptor. *Scientific Reports, 11*(1), 413. https://doi.org/10.1038/s41598-020-80464-1.

Siu, K.-L., Yuen, K.-S., Castano-Rodriguez, C., Ye, Z.-W., Yeung, M.-L., Fung, S.-Y., et al. (2019). Severe acute respiratory syndrome Coronavirus ORF3a protein activates the NLRP3 inflammasome by promoting TRAF3-dependent ubiquitination of ASC. *The FASEB Journal, 33*(8), 8865–8877. https://doi.org/10.1096/fj.201802418R.

Smith, E. C., Blanc, H., Vignuzzi, M., & Denison, M. R. (2013). Coronaviruses lacking exoribonuclease activity are susceptible to lethal mutagenesis: Evidence for proofreading and potential therapeutics. *PLoS Pathogens, 9*(8), e1003565. https://doi.org/10.1371/journal.ppat.1003565.

Sola, I., Almazán, F., Zúñiga, S., & Enjuanes, L. (2015). Continuous and discontinuous RNA synthesis in coronaviruses. *Annual Review of Virology, 2*(1), 265–288. https://doi.org/10.1146/annurev-virology-100114-055218.

Song, P., Li, W., Xie, J., Hou, Y., & You, C. (2020). Cytokine storm induced by SARS-CoV-2. *Clinica Chimica Acta, 509*, 280–287. https://doi.org/10.1016/j.cca.2020.06.017.

Stefan, N., & Häring, H.-U. (2013). The role of hepatokines in metabolism. *Nature Reviews Endocrinology, 9*(3), 144–152. https://doi.org/10.1038/nrendo.2012.258.

Su, Q., Tsai, J., Xu, E., Qiu, W., Bereczki, E., Santha, M., et al. (2009). Apolipoprotein B100 acts as a molecular link between lipid-induced endoplasmic reticulum stress and hepatic insulin resistance. *Hepatology, 50*(1), 77–84. https://doi.org/10.1002/hep.22960.

Su, Q., Wang, S., Baltzis, D., Qu, L.-K., Wong Andrew, H.-T., & Koromilas Antonis, E. (2006). Tyrosine phosphorylation acts as a molecular switch to full-scale activation of the eIF2α RNA-dependent protein kinase. *Proceedings of the National Academy of Sciences*, *103*(1), 63–68. https://doi.org/10.1073/pnas.0508207103.

Tam, B. T., Morais, J. A., & Santosa, S. (2020). Obesity and ageing: Two sides of the same coin. *Obesity Reviews*, *21*(4), e12991. https://doi.org/10.1111/obr.12991.

Tang, T., Bidon, M., Jaimes, J. A., Whittaker, G. R., & Daniel, S. (2020). Coronavirus membrane fusion mechanism offers a potential target for antiviral development. *Antiviral Research*, *178*, 104792. https://doi.org/10.1016/j.antiviral.2020.104792.

Targher, G., Mantovani, A., Byrne, C. D., Wang, X.-B., Yan, H.-D., Sun, Q.-F., et al. (2020). Risk of severe illness from COVID-19 in patients with metabolic dysfunction-associated fatty liver disease and increased fibrosis scores. *Gut*, *69*(8), 1545. https://doi.org/10.1136/gutjnl-2020-321611.

Thomas, S. (2020). The structure of the membrane protein of SARS-CoV-2 resembles the sugar transporter SemiSWEET. *Pathogens & Immunity*, *5*(1), 342–363. https://doi.org/10.20411/pai.v5i1.377.

Thum, T. (2020). SARS-CoV-2 receptor ACE2 expression in the human heart: Cause of a post-pandemic wave of heart failure? *European Heart Journal*, *41*(19), 1807–1809. https://doi.org/10.1093/eurheartj/ehaa410.

Trypsteen, W., Van Cleemput, J., Snippenberg, W. V., Gerlo, S., & Vandekerckhove, L. (2020). On the whereabouts of SARS-CoV-2 in the human body: A systematic review. *PLoS Pathogens*, *16*(10), e1009037. https://doi.org/10.1371/journal.ppat.1009037.

Tudoran, C., Tudoran, M., Lazureanu, V. E., Marinescu, A. R., Pop, G. N., Pescariu, A. S., et al. (2021). Evidence of pulmonary hypertension after SARS-CoV-2 infection in subjects without previous significant cardiovascular pathology. *Journal of Clinical Medicine*, *10*(2). https://doi.org/10.3390/jcm10020199.

Ulasli, M., Verheije, M. H., de Haan, C. A., & Reggiori, F. (2010). Qualitative and quantitative ultrastructural analysis of the membrane rearrangements induced by coronavirus. *Cellular Microbiology*, *12*(6), 844–861. https://doi.org/10.1111/j.1462-5822.2010.01437.x.

V'kovski, P., Kratzel, A., Steiner, S., Stalder, H., & Thiel, V. (2021). Coronavirus biology and replication: Implications for SARS-CoV-2. *Nature Reviews Microbiology*, *19*(3), 155–170. https://doi.org/10.1038/s41579-020-00468-6.

Valcarcel, A., Bensussen, A., Álvarez-Buylla, E. R., & Díaz, J. (2021). Structural analysis of SARS-CoV-2 ORF8 protein: Pathogenic and therapeutic implications [mini review]. *Frontiers in Genetics*, *12*. https://doi.org/10.3389/fgene.2021.693227.

Vankadari, N. (2020). Structure of furin protease binding to SARS-CoV-2 spike glycoprotein and implications for potential targets and virulence. *The Journal of Physical Chemistry Letters*, *11*(16), 6655–6663. https://doi.org/10.1021/acs.jpclett.0c01698.

Verdecchia, P., Cavallini, C., Spanevello, A., & Angeli, F. (2020). The pivotal link between ACE2 deficiency and SARS-CoV-2 infection. *European Journal of Internal Medicine*, *76*, 14–20. https://doi.org/10.1016/j.ejim.2020.04.037.

Vidak, E., Javoršek, U., Vizovišek, M., & Turk, B. (2019). Cysteine cathepsins and their extracellular roles: Shaping the microenvironment. *Cell*, *8*(3), 264. https://doi.org/10.3390/cells8030264.

Viswanathan, T., Arya, S., Chan, S.-H., Qi, S., Dai, N., Misra, A., et al. (2020). Structural basis of RNA cap modification by SARS-CoV-2. *Nature Communications*, *11*(1), 3718. https://doi.org/10.1038/s41467-020-17496-8.

Vithani, N., Ward, M. D., Zimmerman, M. I., Novak, B., Borowsky, J. H., Singh, S., et al. (2021). SARS-CoV-2 Nsp 16 activation mechanism and a cryptic pocket with pan-coronavirus antiviral potential. *Biophysical Journal*, *120*(14), 2880–2889. https://doi.org/10.1016/j.bpj.2021.03.024.

Wambier, C. G., & Goren, A. (2020). Severe acute respiratory syndrome coronavirus 2 (SARS-CoV-2) infection is likely to be androgen mediated. *Journal of the American Academy of Dermatology*, *83*(1), 308–309. https://doi.org/10.1016/j.jaad.2020.04.032.

Wang, K., Chen, W., Zhang, Z., Deng, Y., Lian, J.-Q., Du, P., et al. (2020). CD147-spike protein is a novel route for SARS-CoV-2 infection to host cells. *Signal Transduction and Targeted Therapy*, *5*(1), 283. https://doi.org/10.1038/s41392-020-00426-x.

Wang, K., Chen, W., Zhou, Y.-S., Lian, J.-Q., Zhang, Z., Du, P., et al. (2020). SARS-CoV-2 invades host cells via a novel route: CD147-spike protein. *bioRxiv*. https://doi.org/10.1101/2020.03.14.988345. 2020.2003.2014.988345.

Wang, H., Quirion, R., Little, P. J., Cheng, Y., Feng, Z. P., Sun, H. S., et al. (2015). Forkhead box O transcription factors as possible mediators in the development of major depression. *Neuropharmacology*, *99*, 527–537. https://doi.org/10.1016/j.neuropharm.2015.08.020.

Whittaker, G. R. (2021). SARS-CoV-2 spike and its adaptable furin cleavage site. *The Lancet Microbe*, *2*(10), e488–e489. https://doi.org/10.1016/S2666-5247(21)00174-9.

Wolff, G., Limpens, R., Zevenhoven-Dobbe, J. C., Laugks, U., Zheng, S., de Jong, A. W. M., et al. (2020). A molecular pore spans the double membrane of the coronavirus replication organelle. *Science*, *369*(6509), 1395–1398. https://doi.org/10.1126/science.abd3629.

Wu, C.-H., Mohammadmoradi, S., Chen, J. Z., Sawada, H., Daugherty, A., & Lu, H. S. (2018). Renin-angiotensin system and cardiovascular functions. *Arteriosclerosis, Thrombosis, and Vascular Biology*, *38*(7), e108–e116. https://doi.org/10.1161/ATVBAHA.118.311282.

Wu, X., Sun, Z., Trzeciakowski, J. P., Meininger, G. A., & Muthuchamy, M. (2006). Abstract 536: Mechanical properties of the interaction between fibronectin and α5β1-integrin on cardiomyocytes studied by atomic force microscopy. *Circulation*, *114*(Suppl. 18), II_84. https://doi.org/10.1161/circ.114.suppl_18.II_84.

Xia, H., Cao, Z., Xie, X., Zhang, X., Chen, J. Y., Wang, H., et al. (2020). Evasion of Type I Interferon by SARS-CoV-2. *Cell Reports*, *33*(1), 108234. https://doi.org/10.1016/j.celrep.2020.108234.

Xia, S., Zhu, Y., Liu, M., Lan, Q., Xu, W., Wu, Y., et al. (2020). Fusion mechanism of 2019-nCoV and fusion inhibitors targeting HR1 domain in spike protein. *Cellular & Molecular Immunology*, *17*(7), 765–767. https://doi.org/10.1038/s41423-020-0374-2.

Xu, J., Fan, J., Wu, F., Huang, Q., Guo, M., Lv, Z., et al. (2017). The ACE2/angiotensin-(1–7)/Mas receptor axis: Pleiotropic roles in cancer [review]. *Frontiers in Physiology*, *8*. https://doi.org/10.3389/fphys.2017.00276.

Yang, L., Nilsson-Payant, B. E., Han, Y., Jaffré, F., Zhu, J., Wang, P., et al. (2021). Cardiomyocytes recruit monocytes upon SARS-CoV-2 infection by secreting CCL2. *Stem Cell Reports*, *16*(9), 2274–2288. https://doi.org/10.1016/j.stemcr.2021.07.012.

Yang, J., Petitjean, S. J. L., Koehler, M., Zhang, Q., Dumitru, A. C., Chen, W., et al. (2020). Molecular interaction and inhibition of SARS-CoV-2 binding to the ACE2 receptor. *Nature Communications*, *11*(1), 4541. https://doi.org/10.1038/s41467-020-18319-6.

Yang, R., Zhao, Q., Rao, J., Zeng, F., Yuan, S., Ji, M., et al. (2021). SARS-CoV-2 accessory protein ORF7b mediates tumor necrosis factor-α-induced apoptosis in cells [original research]. *Frontiers in Microbiology*, *12*. https://doi.org/10.3389/fmicb.2021.654709.

Ye, Q., Lu, S., & Corbett, K. D. (2021). Structural basis for SARS-CoV-2 nucleocapsid protein recognition by single-domain antibodies [original research]. *Frontiers in Immunology*, *12*. https://doi.org/10.3389/fimmu.2021.719037.

Ye, J., Zhang, B., Xu, J., Chang, Q., McNutt, M. A., Korteweg, C., et al. (2007). Molecular pathology in the lungs of severe acute respiratory syndrome patients. *The American Journal of Pathology*, *170*(2), 538–545. https://doi.org/10.2353/ajpath.2007.060469.

Yelland, T., & Djordjevic, S. (2016). Crystal structure of the neuropilin-1 MAM domain: Completing the neuropilin-1 ectodomain picture. *Structure (London, England: 1993), 24*(11), 2008–2015. https://doi.org/10.1016/j.str.2016.08.017.

Yu, C. J., Chen, Y. C., Hsiao, C. H., Kuo, T. C., Chang, S. C., Lu, C. Y., et al. (2004). Identification of a novel protein 3a from severe acute respiratory syndrome coronavirus. *FEBS Letters, 565*(1-3), 111–116. https://doi.org/10.1016/j.febslet.2004.03.086.

Yuan, X., Li, J., Shan, Y., Yang, Z., Zhao, Z., Chen, B., et al. (2005). Subcellular localization and membrane association of SARS-CoV 3a protein. *Virus Research, 109*(2), 191–202. https://doi.org/10.1016/j.virusres.2005.01.001.

Zamorano Cuervo, N., & Grandvaux, N. (2020). ACE2: Evidence of role as entry receptor for SARS-CoV-2 and implications in comorbidities. *eLife, 9*, e61390. https://doi.org/10.7554/eLife.61390.

Zhang, J. Z., Liu, Z., Liu, J., Ren, J. X., & Sun, T. S. (2014). Mitochondrial DNA induces inflammation and increases TLR9/NF-κB expression in lung tissue. *International Journal of Molecular Medicine, 33*(4), 817–824. https://doi.org/10.3892/ijmm.2014.1650.

Zhang, K., Miorin, L., Makio, T., Dehghan, I., Gao, S., Xie, Y., et al. (2021). Nsp1 protein of SARS-CoV-2 disrupts the mRNA export machinery to inhibit host gene expression. *Science Advances*, 7(6), eabe7386. https://doi.org/10.1126/sciadv.abe7386.

Zhao, H., Lu, L., Peng, Z., Chen, L.-L., Meng, X., Zhang, C., et al. (2022). SARS-CoV-2 Omicron variant shows less efficient replication and fusion activity when compared with Delta variant in TMPRSS2-expressed cells. *Emerging Microbes & Infections, 11*(1), 277–283. https://doi.org/10.1080/22221751.2021.2023329.

Zhao, M.-M., Yang, W.-L., Yang, F.-Y., Zhang, L., Huang, W.-J., Hou, W., et al. (2021). Cathepsin L plays a key role in SARS-CoV-2 infection in humans and humanized mice and is a promising target for new drug development. *Signal Transduction and Targeted Therapy, 6*(1), 134. https://doi.org/10.1038/s41392-021-00558-8.

Zhong, J., Basu, R., Guo, D., Chow, F. L., Byrns, S., Schuster, M., et al. (2010). Angiotensin-converting enzyme 2 suppresses pathological hypertrophy, myocardial fibrosis, and cardiac dysfunction. *Circulation, 122*(7), 717–728. https://doi.org/10.1161/CIRCULATIONAHA.110.955369.

Zhong, F.-Y., Zhao, Y.-C., Zhao, C.-X., Gu, Z.-C., Lu, X.-Y., Jiang, W.-L., et al. (2022). The role of CD147 in pathological cardiac hypertrophy is regulated by glycosylation. *Oxidative Medicine and Cellular Longevity, 2022*, 6603296. https://doi.org/10.1155/2022/6603296.

Zhou, Y., Frey, T. K., & Yang, J. J. (2009). Viral calciomics: Interplays between Ca2+ and virus. *Cell Calcium, 46*(1), 1–17. https://doi.org/10.1016/j.ceca.2009.05.005.

Zhou, Z., Huang, C., Zhou, Z., Huang, Z., Su, L., Kang, S., et al. (2021). Structural insight reveals SARS-CoV-2 ORF7a as an immunomodulating factor for human CD14(+) monocytes. *iScience, 24*(3), 102187. https://doi.org/10.1016/j.isci.2021.102187.

Zizzo, M. G., Auteri, M., Amato, A., Caldara, G., Nuzzo, D., Di Carlo, M., et al. (2017). Angiotensin II type II receptors and colonic dysmotility in 2,4-dinitrofluorobenzenesulfonic acid-induced colitis in rats. *Neurogastroenterology and Motility, 29*(6). https://doi.org/10.1111/nmo.13019.

FSC
www.fsc.org
MIX
Papier | Fördert gute Waldnutzung
FSC® C083411

Kolibri 360 GmbH
Ferdinand-Jühlke-Straße 7
99095 Erfurt, Deutschland
produktsicherheit@kolibri360.de